Biomechanics III

Medicine and Sport

Vol. 8

Editor: E. JOKL, Lexington, Ky.
Assistant Editor: M. HEBBELINCK, Brussels

University Park Press · Baltimore · London · Tokyo

Biomechanics III

Editors: S. CERQUIGLINI, A. VENERANDO, Rome, and J. WARTENWEILER, Zürich

With 207 figures and 32 tables

University Park Press · Baltimore · London · Tokyo · 1973

Originally published by S. Karger AG, Basel, Switzerland
Distributed exclusively in the United States of America and Canada by University
Park Press, Baltimore, Maryland (ISBN 0-8391-0531-2)

Library of Congress Catalog Card Number 73-9257

Medicine and Sport

University Park Press distributes all volumes in USA and Canada

Vol. 2 Biomechanics I
Technique of Drawings of Movement and Movement Analysis.
First International Seminar of Biomechanics, Zurich 1967.
Edited by J. WARTENWEILER, Zurich; E. JOKL, Lexington, Ky. and
M. HEBBELINCK, Brussels
XXIV + 352 p., 216 fig., 17 tab., 1968.

Vol. 6 Biomechanics II
2nd International Seminar on Biomechanics, Eindhoven
(The Netherlands) 1969.
Editors: J. VREDENBREGT, Eindhoven; J. WARTENWEILER, Zürich
XI + 332 p., 177 fig., 21 tab., 1971

S. Karger · Basel · München · Paris · London · New York · Sydney
Arnold-Böcklin-Strasse 25, CH-4011 Basel (Switzerland)

Printed in Switzerland by Graphische Betriebe Coop Schweiz, Basel
Blocks: Steiner & Co., Basel
ISBN 3-8055-1406-9

Contents

Contents

I. Methods

1. Data Processing

2. Motion Systems and Models

II. Fundamental Research

III. EMG

IV. Applied Biomechanics in Sport

V. Applied Biomechanics in Rehabilitation

Opening Address

S. Cerquiglini

Ladies and Gentlemen,

It is a very great honour for me to welcome to the city of Rome everyone who has gathered here in our country from the farthest parts of the world to take part in the 3rd International Seminar on Biomechanics.

Such an honour is derived from the privilege that the authoritative members of the Research Committee of the International Council for Sports and Physical Education of UNESCO have kindly granted me and the other members of the organizing committee, which elected me as its president, entrusting us with the responsibility of starting the 3rd International Meeting among students of this science, after the meetings held in Zurich and Eindhoven, which were so successful and advantageous.

It is definitely right to state, and a reason for being proud to show why, Italy, among other pleasant places, has a claim for being chosen as a meeting place by present experts of biomechanics. In saying this, I do not mean to refer to the fascination that the natural beauties and artistic treasures of this country have for those who come from abroad, even less do I want to refer to the opinion, which is untenable, that studies of biomechanics flourish more vigorously in Italy than elsewhere. The reason that makes this a worthy choice is considerably more important and significant for the very nature of the subject from which our biennial meetings draw inspiration. It is simply and nothing more than this: Italy is the native place of biomechanics.

In an age such as the present one, in which scientific problems have become so urgent as to recruit an ever-increasing number of ultraspecialized scholars and, at the same time, to impose the *ex novo* creation of new branches of doctrine and practice across cultural syntheses which involve the interaction of disciplines until now widely separated, it often happens that we forget or, even worse, that we are unacquainted with the ancient roots which certain branches of our science have, but which we like to think are very new.

It is only fair to prevent experts of biomechanics from making such a mistake. And which opportunity, I ask myself, could be more fitting to this aim which is anything but useless, other than that of an international meeting, and which place is more suited than the present one?

Whatever the significance is, that is understood or claimed to be given to the subject, biomechanics has a single and very noble paternity: the genius of LEONARDO DA VINCI.

We know that the significance of biomechanics can be so wide that it is identified with the whole chapter of biophysics, which includes mechanical phenomena produced at all levels in living structures from molecules to macro-organisms. It can be so conventionally specific as to make it coincide both with ergonomy or human engineering, bionics and, finally, with the science of motility by which man expresses his life or relationships, as was decided by the promoters of these seminars when they met for the first time in Zurich in 1967.

In fact, anyone who only glances over the collection of the writings and notes of LEONARDO, in which the verbal expression is interlaced almost inextricably with the graphic expression so as to form the material of a very original speech in which science and art blend in an incomparable humanism, is immediately aware that all these aspects of renewed biomechanics were not only realized and generally stated by that great man, but also to a great extent rigorously treated within the limits of the mathematical knowledge of the time.

But the Italian priority, in this field, does not exhaust itself with the prodigious conception of LEONARDO. Near this very famous hill, in St. Pantaleo's church, close to the college of the religious founders of the *Scholae Piae*, rest, for ever, the mortal remains of a man who devoted his life as a doctor and mathematician to studying the motion of living beings.

Aided by the modest financial support of the learned Queen CHRISTINE of Sweden, who had appointed him as her personal physician and member of the academy which she founded, BORELLI spent the last years of his life in that college and was able to write his most famous masterpiece *De motu animalium*. This book, which the author had mediated over for at least a quarter of a century and which showed his teaching of mathematics in the universities of Messina, Pisa and Florence, as everyone knows, is monumental in the history of biological literature. And it is in this work, in fact, that for the first time a mechanical solution of physiological problems has been systematically sought.

The importance that the work of this follower of GALILEO GALILEI and teacher of MARCELLO MALPIGHI has had for the progress of natural sciences, and particularly for physiology and medicine, is historically proved by the fact that the author is considered to be the founder of the famous school of iatrophysics.

As is well-known, not only the science we now call biophysics, but also the important fact that the very mechanistic or positivistic concept of the whole living phenomenology started from this school.

Ladies and gentlemen, forgive me if I have taken advantage of the privilege of being one of the first to speak at this meeting, to remind you of certain glories of Italic science of nearly half a millenium ago. I felt this my duty, not only as organizer of this seminar and as an Italian but also, and above all, as a keen scholar of biomechanical research. If, as observers, researchers or merely onlookers of nature, we scrutinize any phenomenon open to our knowledge in its deepest essence, we find that everything is reduced to motion and to a pattern of motion. Apart from this there is no sensitive reality.

Now, from what we are given to know, any vital manifestation is nothing but motion, a very complicated, orderly and guided motion. At the peak of organic perfection is man, whose bodily machine is only capable of mechanical actions. But even if it is true, as SHERRINGTON reminds us 'all man can do is to move things' it is equally true that by this sole instrument he has created the whole of his universe. And the means is anything but simple. Nothing in this world of material processes is more complicated, more refined and potentially perfectible than human motility. Allied to the mind, that only through the body can show itself, it has already subjected Earth and, as dramatically shown by the numerous footmarks left on the surface of the moon, human motion does not cease to manipulate the cosmos. Biomechanical knowledge identifies itself, as LEONARDO intended, with the knowledge of the entire nature of man.

Medicine and Sport, vol. 8: Biomechanics III, pp. 1–64 (Karger, Basel 1973)

Keynote Address:
The Physical Structure of Mind
A Psychological Study of the Body Image on an Anatomical Basis[1]

E. JOKL, President
Research Committee, ICSPE-UNESCO

'Two Ways of One Mind'

In an essay 'Two Ways of One Mind', published in 1940, SHER-RINGTON discussed 'the sensual basis of awareness of the limb'. There is, he pointed out, a two-fold design of mind's communication: one which establishes contact with the outside world; another with its own body. The first is mediated through the *'five senses'* vision, hearing, touch, smell and taste; the second through the *proprioceptive* system[2].

1   This is the third of a series of papers on the neurological basis of human performances presented as opening addresses of the International Symposia on Biomechanics. The first, 'Acquisition of Skill' appeared in volume 1, the second, 'Motor Functions of the Human Brain', in volume 2 of 'Biomechanics' (Karger, Basel 1967 and 1971).
2   In speaking of *'The Five Senses'* as well as of *'Proprioception'* I use quasi-symbolical expressions. Contemporary physiology classifies 'sensations' more specifically as follows:
I. Special senses served by the cranial nerves
    1. Vision
    2. Audition
    3. Taste
    4. Olfaction
    5. Vestibular
II. Superficial or cutaneous sensations served by the cutaneous branches of spinal and certain cranial nerves
    1. Touch-pressure
    2. Warmth
    3. Cold
    4. Pain *(continuation of footnote on p. 2)*

The 'five senses' were known already to ARISTOTLE; the existence of a 'sixth sense' or 'muscle sense' was first postulated in 1826 by CHARLES BELL, co-discoverer with FRANCOIS MAGENDIE of the fact that the nerves which connect muscles with the spinal cord contain afferent as well as efferent fibers. During the ensuing years sensory receptors were discovered first in muscles, subsequently also in tendons, joints, and connective tissue. Thus, the term 'sense of muscle' was substituted by the term 'sense of movement'.

BASTIAN, the neurologist, introduced the term 'kinesthesis' which has remained popular among psychologists ever since. 'Kinesthesis' today includes more than consideration of afferent impulses from muscles, tendons, joints and subcutaneous tissues. BASTIAN himself realized that the 'sense of movement' is or may be engendered by a variety of receptors, including cutaneous and visual. BASTIAN'S view of kinesthesis was comprehensive. Among the bodily components which mediate the 'sense of movement' he included those whose disturbance through cerebral disease interferes with the comprehension of speech.

The Concept of Force

Both kinesthesis and proprioception relate to the experience of force: man's sense of movement depends upon his own activity. 'If I experience the resistance of a heavy table', H. HENSEL writes, 'it is my own muscular effort which engenders the experience.' Force is the only concept of sensory physiology that is used likewise in physics. The idea of 'force' is derived from phenomenology. There is no other way to define 'force', even though its range and scale are not synonymous in phenomenology and in physics. HENSEL speaks of 'sinnesphysiologische Eigenmetrik', 'a metric scale whose validity applies specifically to sensory physiology'. The physicist is not bound to this 'Eigenmetrik'. For example, the atomic scientist assumes the existence of particles too small for man's sense of movement to perceive.

(continuation of footnote 2 on p.1)
III. Deep sensations served by muscular branches of spinal nerves and certain cranial nerves
    1. Muscle, tendon, and joint sensibility, or position sense
    2. (Deep pain)
    3. (Deep pressure)
IV. Visceral sensations served by fibers traversing the autonomic nervous system
    1. Organic sensation, e.g., hunger, thirst
    2. Visceral pain

But the concept of even such infinitesimal entities is derived from fraction-alization of force-units which the human individual is able to 'feel'. Basi-cally, sensory and physical forces are thus commensurable. It amounts to the same if I say 'willed effort causes a muscle pull', or if I say 'muscle pull is the physical stimulus for the experience of movement'.

Studies of the 'muscle sense', 'sense of movement', 'kinesthesis', and the 'proprioceptive system' raise the issue of 'the sensual basis of awareness of the limb' and beyond that the question of 'the experience of the body'. Con-temporary neurophysiology is unable to provide answers because of its virtually exclusive pre-occupation with histology, biochemistry, electro-physiology, and electron-microscopy. It was thus that philosophy once again assumed its role as participant in the discussion on neuropsychiatry so as it had participated in it until the end of the nineteenth century. Existential philosophy rather than neurophysiology pointed out that we experience ourselves in a macroscopic world. My body, ERWIN STRAUS wrote, is not 'known' to me so as I 'know' others. I cannot see my eyes, my face, my ears, my back, and the soles of my feet – unless I look into a mirror or assume postures of an unusual kind. Nevertheless, most of us take it for granted that the knowledge each of us possesses of his body as well as that of others is derived in the same manner. The latter is obtained through the 'five senses', chief among them vision; while in the acquisition and main-tenance of the knowledge I have of my own body the proprioceptive system plays a major part.

'Sensation of Innervation' and 'Experience of Effort'

In 1664, DESCARTES wrote that the mind perceives distance through the angle formed by the axes of the eyes in looking at an object, 'so as a blind man perceives distance through the angles formed by two sticks held in his hands'. Since then, there have been those who believe that the visual perception of distance to which DESCARTES referred is mediated by a 'sense of innervation' conveyed by the extraocular muscles; and others who held that perception of distance depends on 'experience of effort'. The first concept is based on the supposition that subjective experiences accompany the innervation of the eye muscles, and that such experiences are associated with the 'efferent' phase of motor activation and that they thus are 'confined to the brain'. The other theory finds itself supported by physiological data obtained in analytical studies of sensory receptors in muscle tissue, tendons, etc. It assumes that perception of distance or of posture and movement in general presupposes feedback from spindles and Golgi tendon organs.

The theory of sensations of innervation or Bewegungsideen rests on the assumption that the will is an experienced entity which accompanies innervation and can thus be noted before the movement occurs. The concept was first presented in 1811 by STEINBUCH. It was retained after evidence had been produced to the effect that the muscle is an executive as well as a sense organ, a discovery which established a physical basis for the feedback theory. Among those who agreed with STEINBUCH'S thesis was JOHANNES MÜLLER who in 1838 wrote:

'The idea of weight and pressure in raising bodies or in resisting forces may in part originate from the consciousness of the amount of nervous energy transmitted directly from the brain rather than indirectly from a sensation in the muscles themselves.'

However, MÜLLER held the view that the 'sense of movement' is mediated in a two-fold manner, through 'Bewegungsideen' as well as through afferent impulses originating in muscle. In 1846, WILLIAM HAMILTON referred to sensations of innervation as 'locomotive faculties' which, he explained, provide the 'mental energy' that initiates movements as well as mental experiences. In 1851, E. F. W. WEBER argued in favor of a two-way basis of the sense of movement.

As far as the eye muscles are concerned the validity of the first theory to the exclusion of that of Bewegungsideen seemed supported until a few years ago by statements to the effect that there are no spindles in extraocular muscles. In 1894, SHERRINGTON reported he was unable to discover spindles in extraocular muscles of chimpanzees. More recently, however, spindles and Golgi receptors have been demonstrated in human extraocular muscles. The case thus called for re-evaluation.

That spatial awareness is possible through *vision* can of course not be doubted. The two retinal projections are interpreted stereoscopically. At the same time, the general capacity of proprioceptors to assess spatially distributed components of the body and its environment can likewise not be doubted. Thus, the interdependence between proprioception and vision in perceiving distances required consideration. In his *Physiologic Optics*, HELMHOLTZ in 1867 described experiments in which the visual image was displaced through prisms. He reported that the disorientation thus caused was progressively reduced as the subjects repeated their efforts; and that after removal of the prisms all visual dimensions were misinterpreted in that they appeared to be situated in directions opposite to those in which the prisms had distorted them.

Vision and Proprioception

The way in which sensory experiences and voluntary movements are interconnected was studied more recently by H. L. TEUBER and R. HELD. Subjects wearing prism-goggles that caused deformation of their visual fields adapted themselves to the situation only if they were allowed to move about; while subjects who were passively transported (fig. 13) were unable to do likewise. The linkage between visual and motor processes in the central nervous system was evident (see also p. 14).

In 1964, P. A. MERTON summarized the present knowledge on position sense as conveyed through the human eyes. Awareness of our extremities in darkness, he wrote, depends on messages from the limbs themselves. In the special case of the eyes, however, afferent information supplied by spindles and Golgi receptors of the extraocular muscles competes with information supplied through retinal vision. HELMHOLTZ in his above-quoted book wrote already that information about the position of the eyes as derived from sense endings in the eye muscles may be blotted out by visual impression. He drew this conclusion from the following observations: Suppose one is looking at stationary objects. As voluntary eye movements transfer the fixation points from one object to another, the objects appear to remain stationary, although their retinal images have, of course, moved. Since sense receptors in the extraocular muscles must have been activated in the process, it is evident that in interpreting changing retinal impressions, the mind makes allowance for the movements of the eyes. Otherwise the shift of the retinal image as well as the action of the eye muscles would give a sensation of movement of the seen objects.

It is of interest in this context to refer to reports given by astronauts on return to earth after prolonged deprivation of gravity that they suffered from disturbances of orientation as well as of precision of movements though their visual apparatus was un-impaired.

Bell, Adrian, Plessner and Penfield

The visual situation is of a special kind also because of the dependence of eye movements on vestibular controls. But the latter are capable of influencing the *entire* motor system of man, not only the ocular muscles. When in 1826 CHARLES BELL proposed the existence of a 'sixth sense' or 'muscle sense' he did so because he assumed – quite correctly as we now know – that the muscle is a motor executive as well as a sense organ. During the second half of the 17th century histologists described sensory receptors not only in muscles but also in tendons, joints and connective tissue, even in periostal membranes. The term 'muscle sense' was replaced by the term 'sense of movement'. Still later it was found that vestibular mechanisms form an integral part of the proprioceptive system. In 1914, ROBERT BARANY received a Nobel Prize for the clarification of the inter-relation under reference. In 1967, the Ciba Foundation published a monograph entitled 'Myotatic, Kinesthetic and Vestibular Mechanisms' which reviews the subject in great detail.

The peripheral receptors that convey the 'Sense of Movement' are well known. They comprise primarily those which specifically belong to the 'proprioceptive system', namely muscle spindles and Golgi tendon organs whose functional analysis we owe largely to the Cambridge School of Neurophysiology under the leadership of Lord ADRIAN who received a Nobel Prize in 1932; as well as Vater-Pacini bodies and free nerve endings; fibers originating from the latter as well as those originating from cutaneous receptors jointly enter the spinal cord through the dorsal roots (fig. 9).

Information-aggregates of the kind under reference are flexible in that they may utilize for the purpose of cognition a great variety of sensory receptor systems such as proprioceptive, cutaneous, labyrinthine, vestibular and visual. The general principle of perception thus mentioned was formulated by H. PLESSNER in 1923 in an essay entitled 'Einheit der Sinne' ('Indivisibility of Sensory Experiences'). Cognition, PLESSNER argued, is derived from diverse components. At one time or another all sensory modalities participate in cognition. In spite of this diversity, he wrote, we experience outside objects as well as our own body as entities. Of the specific and multiple elements from which sensory experiences are synthesized we are not aware.

The scientific dispute between adherents of the theory of 'sensations of innervation' and the theory of 'muscle sense' which raged throughout the 19th century, was accompanied by philosophical discussions of the problem of 'will and action'. The will, psychologists held, is an experienced entity of which 'sensations of innervation' are but one among many examples. In 1690 already LOCKE expounded his views on the relationship between experience and ideas. HUME in 1739 and JAMES MILL in 1829 stressed the primacy of sensations, saying that ideas were copies of them. There was general agreement that willing is conscious: we know that we move, or that we are about to move. If consciousness is sensory, it was argued, there had to be 'willing sensations'.

An unexpected datum proving STEINBUCH'S hypothesis correct became available when in 1952 WILDER PENFIELD, the neurosurgeon reported that in response to stimulation of precentral and post-central cortical gyri, patients experienced 'a sense of movement of a part of the body although no movement could be seen'. A more impressive proof of the validity of the theory of 'Sensation of Innervation' or 'Bewegungsideen' cannot be imagined. That the feedback concept is correct has been conclusively proved by recordings of impulses originating in extero- and proprioceptors (fig. 14), of their passage in nerve fibers (fig. 15), and of their modifiability through variation of stimuli as well as of their 'arrival' in the somatosensory cortex (fig. 16).

Phantom Limb, the Agnosias and Anosognosia

In 1906, WERNICKE introduced the term 'somatopsyche' in reference to 'consciousness of one's own body'. Five years later, HEAD and HOLMES referred to the 'postural scheme' of the body. In 1922, PICK used the word 'Körperschema' which in 1923 SCHILDER of Vienna chose as title of a monograph. Twelve years later, the same author published a revised edition of his book under the title 'The Image and Appearance of the Human Body'. French psychiatry was acquainted with the issue only in 1939 when LHERMITTE presented a monograph on 'L'image de notre corps'.

The analysis of three neurological entities deepened the understanding of the body image: the *phantom limb*, the *agnosias* and the *anosognosias*.

The term 'phantom limb' was introduced in 1871 by WEIR MITCHELL, an American physician. It means 'perception of a missing limb following amputation'. The word 'agnosia' was coined by SIGMUND FREUD in 1891. It denotes abnormalities of perceptual recognition due to cerebral lesions; 'anosognosia' was described by BABINSKI in 1914, a condition seen in patients with destructive afflictions of the right cerebral hemisphere who were ignorant of the hemiplegia of the left side of their body.

All gnostic phenomena presuppose awareness. The latter is mediated through the *ascending reticular arousal system* (ARAS). Among the sensory modalities through which orientation of the body and its parts in space is effected, *proprioception* occupies a more important position than *vision*.

Phantom Limb

Perception of a missing limb was noted in 1537 by AMBROISE PARÉ. A hundred years later, RENÉ DESCARTES wrote in his 'Discourse on Method', as follows:

'I have learned from some persons whose arms or legs have been cut off that they sometimes seemed to feel pain in the part which had been amputated, which made me think I could not be quite confident that it was a certain member which pained me even although I felt pain in it... In the same way when I feel pain in my foot, my knowledge of physics teaches me that this sensation is communicated by means of nerves dispersed through the foot which being extended like cords from there to the brain when they are affected in the foot, at the same time affect the inmost portion of the brain which is their extremity and place of origin, and there excite a sensation of pain represented as existing in the foot... If there is any cause which excites, not in the foot but in some part of the nerves which are extended between the foot and the brain or even in the brain itself, the same action which usually is produced when the foot is detrimentally affected, pain will be experienced as though it were in the foot.'

The phantom limb appears immediately after amputation. Often it is experienced for the rest of the amputee's life though this is not the rule. Between 98 and 99 % of all amputees have phantom experiences. After the operation patients are known to look under the blanket to find out whether the amputation has indeed been done. The phantom may appear intermittently. When it does its 'presence' may be so realistic that it causes the amputee to stumble on getting up because he 'forgets'. Usually the phantom limb is experienced as 'normal' in respect of size, length, weight, position and movement. Vision plays no part in the phantom experience. In their dreams, amputees perceive themselves as normal.

In 1892, CHARCOT described a patient who felt his engagement ring on a finger of an amputated hand. In 1913, OPPENHEIM saw a patient with tabes dorsalis who had lightning pains in a phantom limb. In 1943, HENRY COHEN published a paper entitled *Reference of Cardiac Pain to a Phantom Left Arm* containing the story of a patient with angina pectoris causing severe radiating pain in the left arm. The attacks continued after the arm had been amputated. The radiating pain was felt in the absent extremity. A like observation was made in 1950 by P. CHRISTIAN: A 58-year-old cardiac patient whose left arm had been amputated continued to have angina pectoris attacks accompanied by phantom pain in the left hand. In 1949, BORNSTEIN gave an account of patients with sciatica in phantom limbs. In 1969, H. PLÜGGE reported on a man whose right leg had been amputated because of arteriosclerotic gangrene. Following the surgery, the patient suffered from claudication in the left leg as well as in the right phantom leg. MALONEY saw a patient with urethritis and phantom leg pain following amputation; after treatment of the urethritis the phantom disappeared. In another patient whose leg had been amputated for sarcoma, phantom limb sensations occurred consistently during voiding, also during cystoscopy and rectal palpation [J. amer. med. Ass. *216:*684, 1971; *218:* 1052, 1971; J. Phys. med. Rehab. 7:360, 1966].

Painful phantom limb is a rare occurrence, observed in but 1 or 2% of all amputees. It is thus interesting that the first medical report in 1649 on the persistent body image by AMBROISE PARÉ mentioned it.

In 1862, GUENIOT described the phenomenon of 'telescoping': the phantom limb shortens over the months, with the result that the hand is felt getting closer to the shoulder.

The phantom limb occupies a position of importance for the theory of the body image. It is literally a transcendental phenomenon in that the experience of the non-existing extremity extends beyond its demonstrable portion. Attempts to explain the phantom limb have been made in terms of 'peripheral', 'central', and 'psychological' hypotheses, none of them satisfying.

The Agnosias

The term agnosia was proposed by FREUD (1891) to describe a special kind of discorder of recognition. The syndrome has been extensively discussed in the literature. Agnostic states are distinguished by cognitive failure due to inability to integrate sensory information. The following definition of the term was given by FREDERIKS in 1969:

'Agnosia is the impaired recognition of an object which is sensorially presented while at the same time the impairment cannot be reduced to sensory defects, mental deterioration, disorders of consciousness and attention, or to a non-familiarity with the object.'

Agnostic states have been described in respect of each of 'the five senses' as well as in respect of the 'body image'. However, it is doubtful whether agnosia ever occurs as a strictly exclusive affliction of a single sensory category. Furthermore, the agnosias are probably always associated with at least slight disturbances of the body image and apraxia. According to JOHANNES LANGE, 'a sharp distinction between agnosia and apraxia is

impossible'. An impressive illustration of the latter has been presented by
MacDonald Critchley noted in a woman of 57 years with biparietal
atrophy (fig. 41).

The agnosias and apraxias due to cerebral lesions manifest themselves
as disturbances of complex cognitive and executive functions, invariably of
an *acquired* kind. Clinically and pathologically they can be distinguished
from syndromes due to similar cerebral lesions that cause release of *reflex
mechanisms* of a kind normally masked by cortical controls. Jokl and
Schepers have presented a detailed account of a case belonging to the
latter category. A patient, 37 years old, suffered a severe head injury which
damaged the portion of the left cerebral hemisphere delineated in figure 22.
A *support reaction* was found on the homolateral arms as well as marked
impairment of the associated *arm-swinging movement* during walking on
the contralateral side[3]. There was no evidence of agnosia and apraxia.
Comparison of the localization of the injury of the brain of the patient with
that of those whose destruction caused agnostic and apraxic states in Liepmann's patients shows that in the case of Jokl and Schepers the greater
part of the lesion was situated anterior to the central fissure, sparing almost
entirely the areas marked 1-2-3 in Liepmann's diagram (fig. 20).

Anosognosia

Shortly before World War I, Babinski reported on a syndrome
observed following cerebral hemiplegia, consisting of 'patients ignoring or
seeming to ignore the existence of the paralysis'. Babinski coined for it
the word 'anosognosia' applying the generic concept of 'nosos' (disease)

---

3   The kinetic as well as the static components of the positive and negative components of the support reaction were demonstrated. Hyperextension of the hand
elicited forcible extension of the entire arm which thus was transformed into a rigid
pillar. Only if finger and wrist joints were passively flexed could the static fixation
be relaxed and flexion of the arm effected. So triggered, a mass flexor movement
occurred which included elbow and shoulder joints. From this newly established static
position, extension of the arm was possible only if the fingers and wrist joints were
extended. Marked impairment of associated arm-swinging movements during walking
was in evidence on the right side of the body, i.e. contralaterally to the cerebral
lesion. At the commencement of walking the right arm which at rest was held in
spastic fixation close to the body performed abortive swinging movements of a clumsy
incoordinated character. The left (homolateral) arm was held normally at rest and
moved normally during walking.

exclusively to cases of hemiplegia. The term anosognosia has subsequently been used also in reference to patients who ignore blindness, deafness and other disabilities. However, for purposes of the discussion of the body image it will be restricted to the meaning BABINSKI gave it; that is, to anosognosia for hemiplegia, almost invariably of the left side of the body, associated with hemianesthesia following lesions of the non-dominant cerebral hemisphere, always including the parietal lobe.

BABINSKI emphasized that the patients' peculiar defect cannot be explained by the assumption of clouded consciousness, organic dementia or memory disturbances. ERWIN STRAUS pointed out that organized behavior and interest of patients with anosognosia in their surroundings contrasts with their failure to perceive part of their body.

The Nature of Experience

The foregoing discussions concerning the history of the concept of the body image, of the sustained dialogue between physiologists and psychologists on sensation of innervation and experience of effort, as well as of the clinical observations around which today the discussion of the body image centers challenge physician and philosopher alike in their efforts to deduce theories on the mind. Strictly speaking all the available information ultimately relates to structure. In its entirety it renders possible a distinction between sensory awareness and cognition in respect of all sensory modalities. The latter are linked to movement. To exemplify, movements of the hands are prerequisites for 'Braille reading'. Awareness of being touched is something else. Movements of the eyes are part of the process of 'searching, looking and finding'. Mere differentation between light and dark is something else.

Other issues present themselves for consideration: Cognition presupposes integration of different sensory systems, a concept of which PLESSNER first wrote in 1923; and awareness is a basic state upon which specific sensory experiences are superimposed, as it were. Until not too long ago, awareness was discussed from various angles, for instance in considering the unconscious by psychoanalysis, in neurophysiological studies of sleep, and in attempts on the part of clinical neurologists to understand mechanisms underlying loss of memory and of consciousness.

It was an event of major importance when in 1949 MORUZZI and MAGOUN described the ascending reticular arousal system (ARAS) as primary mediator of awareness. Before 1949, awareness was thought to be a function of the cerebral cortex. Now the brain stem was shown to be the chief regulator of wakefulness and sleep. ARAS – described later in this essay – subserves not only the cortex, but also regulates muscle tone and autonomic processes such as blood pressure whose adaptation is essential for the assumption of the upright posture. The differential design of the integrative control of awareness, muscle tone and blood pressure offers itself for analysis in clinical syndromes such as cataplectic loss of tone unaccompanied by unconsciousness, the psycho-motor epilepsies, and the orthostatic collapses.

Thirdly, maintenance and awareness of posture presuppose a fixed system of spatial orientation mediated through vision and proprioception. This system is not the same as the Euclidian with which physics operates. The difference between the two shows the two-fold nature of experience and of the difference between the 'theater of the event' of existence of which JEAN FERNEL wrote 400 years ago, as against 'the event itself'.

Three issues relate to the problem under discussion, all of them referred to before but once again summarized: 'Einheit der Sinne', ARAS, and posture in as far as it determines the body image.

Einheit der Sinne

The scope of the concept of the 'Einheit der Sinne' of which PLESSNER wrote in 1923 has recently been extended through analyses of outstanding performances in sport by athletes with sensory impairments. The study of creative motor acts of the kind encountered in sport necessitates entry into a territory which the massive literature on a-praxia, a-gnosia, a-lexia, a-nosognosia has left untouched. Among model situations arising not infrequently in sport are record feats by athletes who according to clinical criteria have to be classified as 'crippled' or 'handicapped'. Existing industrial labor legislation makes them eligible for receipt of disability payments. One such case was that of the Hungarian champion in target pistol shooting, Karoly Takacs, who at the height of his competitive career lost the right arm through an accident. He continued training with the left and won two Olympic Gold Medals.

In his book, 'Physical Background of Perception', ADRIAN presented figure 26 showing the crossing of sensory and motor pathways to and from the cerebral hemispheres in conformity with the laterality of the target. It illustrates the integration of vision and proprioception in the execution of the motor act in the case under description.

We have continued to study the respective significance of the two sensory sectors in the execution of differentiated acquired motor acts. Figure 27 depicts *blind champion gymnasts* in highly differentiated actions which evidently can be performed without participation of the visual system. Proprioceptive control mechanisms play, of course, a decisive part also in gymnastic performances by persons with *unimpaired vision*. Nevertheless, the fact that vision is indispensable in the performance by *normal* gymnasts of feats such as those shown in figure 27 cannot be doubted. In fact, Olympic contestants insist upon bright illumination of the halls in which their competitions are held. Skillful gymnasts find themselves badly handicapped if they are asked to perform with their eyes covered. The impairment persists even with practice. I do not know of any gymnast with normal vision who without use of his eyes can perform the exercises executed by the blind champions.

Other observations which have deepened our knowledge of the Einheit of vision and proprioception in the execution of motor acts of great complexity is that of outstanding athletes with major uncorrected errors of refraction. I have described the case of a well-known cricket player with sph. −10.5 6/6 on both eyes; of a tournament tennis player with sph. −9.0 left and sph. −8.5 right with cycl. 1.0 85°; and of a tennis player with keratoconus. All of them performed with great distinction without glasses. Their superior motor skill enabled them to make optimal use of retinal images so poor that physically less skillfull persons likewise afflicted are rendered by them helpless even when supplied with corrective lenses. Evidently proprioception can compensate for impaired vision within a range much greater than has so far been assumed.

The Ascending Reticular Activating System (ARAS)

'Experience' presupposes wakefulness. This statement referred to before is of broad relevance in neurophysiology and clinical neurology. For example, if a patient's consciousness is clouded, it is impossible to test for phantom limb, agnosia or anosognosia.

To reiterate, in 1949, MORUZZI and MAGOUN described a mechanisms of arousal mediated through activation of the brain stem. Electroencephalography revealed characteristic alterations during change from wakefulness to sleep. MORUZZI and MAGOUN showed that electrical stimulation of the portion of the brain stem marked in figures 28-31 simulated like alterations. It was thus that they came to identify the 'ascending reticular activating (or arousal) system' (ARAS). Prior to 1949, wakefulness was considered a function of the cerebral cortex. Now it was evident that the cortex plays but a secondary role in that it first must be 'turned on' by ARAS, even though once 'turned on' it can send out impulses to the reticular formation. ARAS is capable of responding to impulses reaching it from the periphery of the body as well as from the cortex. ARAS also excites or inhibits the tone of the musculature, especially that which counteracts gravity. Its failure results in collapses such as those seen during cataplectic seizures. Under physiological conditions increase of tone accompanies awakening while inhibition of tone accompanies sleep.

The functional integrity of ARAS is an indispensable prerequisite of all gnostic experiences. Even contents of dreams and elaborations of 'the unconscious' have invariably been described by scientists who were awake.

Posture and Body Image

The existence of a body image presupposes frames of reference through which the position of head, trunks and limbs can be related to the outside world. These frames of reference are not necessarily established by vision. HELMHOLTZ demonstrated that correct visual perception of space does not depend upon permanently fixed retinal images. In fact, the latter change with all movements of the body. Nevertheless, our surroundings serve us as reliable background for orientation, whether or not we move head or eyes. HELMHOLTZ conducted experiments in which he displaced visual images by prisms, showing that subjects adapt themselves to the new situation provided they are allowed to move. Enforced inactivation rendered adaptation impossible.

The problem was subsequently studied in great detail by IVO KÖHLER, RICHARD HELD and LUCAS TEUBER whose experimental results corroborated the validity of HELMHOLTZ'S statements. It is the pull of gravity involving myotatic, kinesthetic and vestibular mechanisms and not vision which is the *primary* prerequisite for the maintenance of posture. We are in the true sense of the word terrestrial creatures. Significant as the complex visual apparatus is for orientation, its function as link between the 'I' and the 'outside world' ranks second in importance to the action of earth's gravity upon the proprioceptive system. The latter is of special significance for homo sapiens because of his upright posture.

The Seat of The Mind

The question of 'the seat of the mind' has been discussed since antiquity. However, until the turn of the 18th century answers were invariably derived from superstition and phantasy. Factual observations were not presented. For more than a millenium preceding the advent of scientists of the caliber of HARVEY (1578-1657) and GALILEO (1564-1642), nature was thought to be ruled by the stars which it was said also guided man and his mind. The human soul like the heavenly spheres, so it was believed were animated by 'divine inspirations'. The world in which MARTIN LUTHER (1483-1546) grew up was altogether directed by the planets as RICHARD FRIEDENTHAL has recently pointed out. The philosopher PICO DE MIRANDOLA, the 15th-century philosopher, observed: 'if astrologers believe in anything, it is the planets, not God.' A hundred years later, KEPLER (1571-1630) was still convinced that man's mind was a replica of the heavenly spheres, a view reiterated by ROBERT FLUDD (1574-1637) whose picture 'A Universe Within the Mind' (fig. 33) showed God, earth and man and their alleged counterparts, 'intellect, imagination and sensation'. According to SCHILLER, Wallenstein's decisions hinged on forecasts by his astrologer, Seni, whose views on man and his nature are reflected in the following:

'...Five is the soul of man. So as man is a mixture of good and bad, five is the first number containing a divisible and an indivisible part.'

When in 1940 SHERRINGTON said that 'astrology was slain by its child, astronomy', he addressed his learned colleagues in the Royal Society. To the populace, astrology remained and remains ultimate evidence to the effect that the stars rule our lives and our minds. Many even among those who do not believe in astrology present views on the subject that are similarly fantastic. In 1815, Dr. GALL and Dr. SPURZHEIM[4] published a book

4   GALL and SPURZHEIM brought the concept of cerebral localization into such disrepute that it delayed the acceptance of the classical discoveries of cortical organization, first BROCA'S description of the *'motor'* center of speech in the left *frontal* lobe in 1863, followed by WERNICKE'S account of its *'sensory'* complement in the left *temporal* lobe in 1874.

entitled 'The Physiognomical System, founded on an anatomical and physiological examination of the nervous system' which gained adherents all over the world. It is an utterly nonsensical treatise (fig. 34). As late as 1938, a diagram was published pretending to be a clue for the understanding of the human mind. It was prepared by no less a person than SIGMUND FREUD who prior to the turn of the 19th century had proved himself to be a competent neurologist (fig. 35). Of the fact that the diagram, like his system of 'psychoanalysis', is wholly divorced from anatomical features of the brain his disciples took no cognisance. In a letter to the Editor of the Journal of the American Medical Association of January 17, 1972, J. CHUSID commented on the current popularity in the United States of Yoga. He quoted the Encyclopaedia Britannica to the effect that 'the aim of Yoga should be to raise the feminine kundalini ('the serpent force') at the base of the spine from one of 'the seven psychic centers or chakra' to another until it unites with the 'masculine sahasrara' atop the skull when full salvation is achieved (fig. 36).

This kind of absurdity enjoys a receptive press today. According to CHUSID, the *New York Times* of July 2, 1971, reported that religious leaders of different denominations discovered that Yoga enhances faith; the *New York Post* of November 1, 1971, stated that Harvard physicians found 'transcendental meditation' effective for drug abuse; and the *New York News* of September 26, 1971, devoted a two-page centerfold spread to the 'Yoga Day' activities in Central Park which were co-sponsored by the City of New York on September 18, 1971.

Evidently the discussion on man's mind proceeds on two levels: One on which superstitious irrationality is allowed to move without restraint; the other which looks upon facts and interprets them in the tradition of ARISTOTLE and SPINOZA, respecting the laws of logical thought. An example of the second is the effort to come to grips with the result of studies of the highest cerebral functions of man whose disturbances manifest themselves in various forms of aphasia, apraxia and agnosia. In as far as clinical neurology is concerned in its quest after diagnostic accuracy, the effort has been singularly fruitful. But in as far as it encompasses the search for attachment of mind to the body, science tells us nothing. In his last recorded statement, made in the 95th year of his blessed life, SHERRINGTON considered this fact important enough to reiterate. At the conclusion of this essay I shall comment on the issue which represents the great vacuum of our knowledge.

All information we possess on the working of the brain is derived from studies of persons afflicted with *diseases*. Neuroanatomy and neurophysiology's efforts to analyze *normal* structures and functions have given us information that throws light solely on what JEAN FERNEL would have called 'the theater of the event', not on 'the event itself'.

Following his description of 'sensory aphasia' in 1874, WERNICKE designed the two schemata of the mechanism of speech shown in figures 37 a and b in which he distinguished a concept center (C) from a sensory center (s) and a motor center (m). The afferent impulses would reach an acoustical projection field, an area of 'primary identification' where memory images of articulated sounds, the sound images of words of HELMHOLTZ, were stored. From this area, impulses would be conducted over transcortical pathways (sC) to the concept center, the locus of 'secondary identification', i.e. of the meaning of words. In speaking, efferent impulses initiated in the concept center would reach the Broca area over the transcortical pathway (cm).

The diagram permitted WERNICKE to localize the various forms of cortical, subcortical, transcortical, and conduction aphasia observed or postulated by him and the other 'diagram makers', to use HEAD'S disparaging term. While the idea of a cortical center suggests precise localization, WERNICKE emphasized that the hypothetical concept center can extend over wide areas of the hemispheres. He divided 'the concept center' into two different zones: one of sensory images (A) (Anschauungsvorstellungen) and another one of goal images (Z) (Zielvorstellungen) (fig. 37 b). In this, the section sAZm represents the psychological reflex arc. The part Zm is the psychomotor pathway. Under normal conditions, WERNICKE assumed Zm is under the control of Z. Under pathological conditions, through a process of 'sejunction', the transcortical pathway Zm is broken loose from Z. It is blocked in cases of apraxia.

WERNICKE'S schemata have retained their usefulness until today. Recently, GESCHWIND has emphasized the fact that the clinical and pathological studies by BROCA and WERNICKE once and for all established the distinction between aphasias due to lesions of grey matter and those due to lesions of white matter. It was thus that the concepts of 'cortical' and 'conduction' syndromes as separate clinical and pathological entities originated. The history of *aphasia* is of course the great model of the history of *gnostic and praxic phenomena* in general. Notwithstanding significant advancements made during the past decades in the interpretation of the different forms of the two groups of clinical syndromes, WERNICKE'S diagrams are still valid as schemata.

In an article entitled 'Current Concepts of Aphasia', published in 1971 [New Engl. J. Med., March 15], GESCHWIND introduced the terms 'fluent' and 'non-fluent' for disturbances of language-output due to lesions behind the central sulcus and Broca's area, respectively, dividing them into 4 types: 'Wernicke', 'Conduction', 'Anomic', and 'Isolation'. These are essentially the same as those of which WERNICKE knew. It is only in respect of the *clinical classification* of the aphasias – and also of the agnosias and apraxias –, not of their neuro-anatomical design that changes have been made since the days of WERNICKE and LIEPMANN.

It is due to the efforts of JOHANNES LANGE and NORMAN GESCHWIND that LIEPMANN'S work on the relationship of cognition and movement, on their mediation through the brain, and on the indivisibility of gnostic and praxic functions has been fully appreciated. LANGE'S monograph 'Agnosien und Apraxien' appeared in 1936. It was devoted almost in its entirety to a review of LIEPMANN'S writings. GESCHWIND'S papers on 'Disconnexion Syndromes', and his historical account of CARL WERNICKE'S contributions were published in the past decade. They acquainted English speaking neurologists for the first time with the achievements of the Breslau school around the turn of our century.

'Eine scharfe Trennung von Agnosie und Apraxie', LANGE stated, 'ist nicht möglich' ('A precise delineation between agnosia and apraxia is not possible.'). This fact was first realized on account of LIEPMANN'S descriptions of the interdependence of gnostic and praxic involvement in reading and writing; of the fact that disturbances of motor execution cause disturbances of cognition and vice versa, and of the significance of this interdependence for the understanding of the alexias and agraphias. LIEPMANN made a reference to the perfection through training of acquired motor skills. During sustained practice, he pointed out, progressively less voluntary effort is needed to initiate progressively more automatized motor acts.

LIEPMANN emphasized that agnosia and apraxia can be studied only in patients whose intellectual faculties are unimpaired and that in normal subjects cognition and action cannot be disentangled.

LIEPMANN confirmed that the left cerebral hemisphere is the dominant half of the brain, agreeing with BASTIAN'S assertion that it is 'the brain of the brain'. The left hemisphere also mediates the 'praxic capacities of both cerebral halves'. In a patient with isolated tactile agnosia and apraxia of the left side of the body, LIEPMANN diagnosed a lesion in the corpus callosum which was subsequently confirmed at autopsy. The left cerebral hemisphere's connexion with the right had thus been interrupted with the result that gnostic and praxic functions of the left side of the body were lost.

Interestingly, non-recognition of objects in the presence of unimpaired sensory reception was first noted in animals: In 1877, MUNK reported that dogs after removal of part of the optical cortex failed to respond to situations to which they had responded normally prior to the surgery: The latter had left their vision unimpaired. But deprived of the optical cortex their master had become a stranger to them; gestured commands and threats no longer conveyed their meaning. The animals were visually agnostic.

CHARCOT likened the functions whose disturbance causes agnosia and apraxia, to KANT's 'facultas signatrix'. In 1879, he described a patient with loss of 'mental imagery' due to a lesion of the occipital cortex. LISSAUER, one of WERNICKE's co-workers in Breslau, was the first to distinguish between 'receptive' and 'associative' blindness, a condition to which KLEIST – also trained as a member of WERNICKE's team – referred to as 'Seelenblindheit', inability to link perception with pre-existing visual knowledge and loss of the capacity to interpret it.

Transcendental Images

What about experience? It cannot be documented. Many a student of neurophysiology deals with sensory data as electric impulses. As regards their conversion into experience, biophysics fails him. In looking for mind as energy he is unable to detect how energy is translated into mind. He does not know how radiant energy via nerves engenders seeing, or how the experience of heat or pain or movement originates. There is complete ignorance concerning the transmutation of physical processes into mental. Of energy that is mind we know nothing. It is one of the noteworthy achievements of the natural sciences to have delineated the scope of insight obtainable with the energy-concept of physics, as contrasted with the unattainability of a 'scientific' definition of mind. The energy concept has proved to be a weapon for man's conquest of the earth and of the space beyond it. It has brought us to the moon. But is has made clear that the far-reaching notion 'energy' is, as it stands, powerless to describe mind.

No knowledge is available pertaining to neuro-anatomical or neurophysiological equivalents of human performances. In my paper *'Acquisition of Skill'*, I have alluded to their scope:

'Not only can human beings carry out movements in the image of their thinking, but they can also improve upon their execution. The scope of this improvement ranges from a child's drawings to Michelangelo's paintings in the Sistine chapel; from a beginner's renderings on the piano to a concert presentation by Arthur Rubinstein; from a school boy's efforts at the carpenter's bench to the building of wooden churches in Northern Finland; and from a young track and field contestant's performances to the feats accomplished by Olympic decathlon finalists.'

The vast body of information on the nervous system of man now available, e. g., in the German *Handbuch der Normalen und Pathologischen Physiologie,* in the *Handbook of Physiology* of the American Physiological Society, and in the new *Handbook of Neurology* gives no clue to the understanding of contributions of the class of BACH and SHAKESPEARE. Though SHERRINGTON'S dictum 'Mind is invariably embodied mind' applies to them too, the resources of the natural sciences do not explain them. They are transcendental images rendered amenable to cognition:

And as imagination bodies forth
The form of things unknown
The poet's pen turns them to shapes
And gives to airy nothing
A local habitation and a name.

(Shakespeare, A Midsummer Night's Dream, IV, 1.)

Biographies [5]

1. AMBROISE PARÉ (1510-1590). The earliest observation relating to the problem of the body image was made in the 16th century by AMBROISE PARÉ who in 1537 gave a description of what is now called the phantom limb. Born in Bourg-Hersent in France, PARÉ studied surgery in a barber shop – the medical profession at the time considered it beneath its dignity to carry out bloody operations. PARÉ obtained an appointment in the Hôtel-Dieu in Paris where his skill came to the attention of Marshal MONTEJAN. The later invited him to join his army when the third war broke

---

5   Biographical sketches of the following investigators whose work is relevant to the theme of this paper have been presented in my paper *'Motor Functions of the Human Brain'* [1971]: RENÉ DESCARTES (1596-1650), THOMAS WILLIS (1621-1675), FRANZ JOSEF GALL (1758-1828), PIERRE FLOURENS (1794)-1867), PIERRE PAUL BROCA (1824-1880), CARL WERNICKE (1848-1904), JOHN H. JACKSON (1835-1911), EDUARD HITZIG (1838-1907), CHARLES SHERRINGTON (1857-1952), RUDOLF MAGNUS (1873-1927), K. BRODMANN (1868-1918), HANS BERGER (1873-1941), HARVEY CUSHING (1864-1939), OTFRID FOERSTER (1873-1941), ERWIN STRAUS (born 1891), WALTER HESS (born 1881), PETROVITCH PAVLOV (1849-1936), SIGMUND FREUD (1856-1939), F. J. J. BUYTENDIJK (born 1887), E. DOUGLAS ADRIAN (born 1889).

out between Francis I und Charles V in 1536. During the ensuing campaign PARÉ observed soldiers experiencing their limbs after amputation. It is of interest to note that PARÉ described first the *painful* phantom which as we know today is an uncommon, modality of the syndrome.

HENRY E. SIGERIST considered PARÉ one of the founders of modern surgery, saying that he was intelligent and knowledgeable, also motivated by a sense of obligation and piety. 'Je le pensai, Dieu le guarist.'

2. ABRAHAM VATER (1684-1751) was born in Wittenberg, Germany, where he taught anatomy since 1719. His name is known to students of anatomy because of an account he gave of sensory receptors that had been discovered by ANTONIO PACCHIONI (1665-1726). His studies were further elaborated by F. PACINI (1812-1833).

3-4. FRANCOIS MAGENDIE (1783-1855) and CHARLES BELL (1774-1842) demonstrated that the anatomical arrangement of the spinal nerve roots reflects a functional principle: the anterior roots carry efferent, the posterior roots afferent impulses. Since the nerves that connect muscles with the spinal cord contain fibers can be traced to both anterior and posterior roots, the question was posed whether the muscle is an organ of execution as well as a source of sensory information. It was thus that the traditional view of the five senses of man was extended. In 1826, CHARLES BELL postulated a sixth sense, 'muscle sense'! Precision of movements, he explained. depends on motor innervation originating in the brain as well as on muscular sensations originating in the muscle.

'Between the brain and the muscles there is a circle of nerves; one nerve conveys the influence from the brain to the muscle, another gives the sense of the condition of the muscle to the brain.'

5. RUDOLF ALBERT VON KÖLLIKER (1817-1905). Early in his career VON KÖLLIKER reported that Pacinian corpuscles occur in proximity to striated muscles. In 1862, he described receptors in voluntary muscles which were subsequently named 'spindles' by KUEHNE [Arch. Path. Anat., Physiol., Klin. Med., vol. 27 and 28, 1863]. KÖLLIKER called the histological entities under references 'nerve-buds' which he thought at first were 'centers of growth'. His above-mentioned discoveries were made in the course of a sustained and comprehensive study of the histological design of the nervous system which enabled him to prove the validity of the assertion presented by PURKYNJE, REMAK and HANNOVER that nerve fibers arise from nerve cells. In 1899, KÖLLIKER introduced CAJAL to the German Anatomical Society, an event which led to the latter's lifelong friendship with CHARLES SHERRINGTON. To the cooperation of KÖLLIKER, CAJAL and SHERRINGTON we owe the concept of the neuron whose identification played the same role in the development of neurology as the identification of the molecule played in the development of chemistry.

KÖLLIKER was born in Zürich in 1817. He studied in Zürich, Bonn, Heidelberg, Berlin where he worked under JOHANNES MÜLLER who recommended him for the chair of Physiology and Microscopic and Comparative Anatomy in Würzburg. This position he occupied for almost half a century. He retired in 1902 at the age of 85.

KÖLLIKER wrote two textbooks of histology: *'Mikroskopische Anatomie'* (Leipzig 1850-1854, 3 vol.) and *'Handbuch der Gewebelehre des Menschen'* (Leipzig 1852).

6. WILLY KUEHNE (1837-1900) was one of the great neuro-histologists of the 19th century. He described the muscle spindle as well as the muscle end plate, contributions which represent building stones of the edifice of the present concept of the nervous system.

KUEHNE was born in Hamburg. He studied under WÖHLER, R. WAGNER, WEBER, HENLE, VIRCHOW, CLAUDE BERNARD, BRÜCKE and DU BOIS-REYMOND. In 1871, he was appointed successor of HELMHOLTZ as professor of physiology in Heidelberg, in preference to WUNDT who had expected to receive the post. Among KUEHNE'S friends were JAKOB VON UEXKULL and OTTO COHNHEIM. One of his most distinguished pupils was RUDOLF MAGNUS.

KUEHNE'S account of the spindle appeared in 1863. He recognized that the spindle is a receptor for muscle sense of which CHARLES BELL had spoken much earlier. In 1892, RUFFINI presented proof that the spindles are neural tissue and that nerve fibers lead directly from them to the spinal cord. Details of the sensory function of spindles were elucidated in 1894 by SHERRINGTON in experiments in which posterior spinal roots were sectioned.

7. HERMANN VON HELMHOLTZ (1821-1894) was one of the most versatile scientists of the 19th century. Among his many contributions to the issue under discussion is that he lent support to STEINBUCH'S theory of innervation of efferent tracts in voluntary action establishing sensations. Such sensations, HELMHOLTZ assumed, 'arise wholly in the brain'. A number of important theories relating to 'physiological optics' were formulated by HELMHOLTZ including those pertaining to the part played by the retina and the eye muscles in distance vision. HELMHOLTZ exerted a lasting influence upon psychology which he felt should not develop into an independent science. Psychology must be studied, he postulated, as part of physiology and interpreted in terms of philosophical concepts.

In 1843, HELMHOLTZ qualified as a physician. During the ensuing five years he worked as an army surgeon. In 1848, ALEXANDER VON HUMBOLDT and JOHANNES MUELLER were instrumental in obtaining permission of the military authorities to allow HELMHOLTZ to relinquish his post as army surgeon and accept the position as lecturer at the Berlin Academy of Art, in succession to his friend BRUECKE. In 1849, at the age of 28, he was called to Königsberg as professor of physiology and general pathology; in 1856 to the chair of physiology in Bonn; and 1858 to the same chair in Heidelberg. In 1887, he was made director of the Physikalisch-Technische Reichsanstalt in Berlin. Among his assistants there were HEINRICH HERTZ and EMIL WARBURG whose son OTTO was destined to elevate biology to a strict science.

Professor EDWIN G. BORING of Harvard University, who was a lifelong admirer of HELMHOLTZ, wrote of his influence on psychology and sensory physiology as follow: 'HELMHOLTZ was not a systematic psychologist, but his work in vision brought him to the problem of visual perception and thus of perception in general. In many ways, perception has been the central problem of systematic psychology, and it thus comes about that HELMHOLTZ occupies an important position in the history of systematic psychological thought. That HELMHOLTZ was primarily an experimentalist simply goes to show that systematization and experimentation cannot be separated in the history of psychology, or for that matter, in the history of science.

HELMHOLTZ stood for psychological empiricism. He belongs thus systematically more with British thought than with German, in the tradition of JOHN LOCKE down to J. S. MILL, rather than in the tradition of LEIBNITZ, KANT, and FICHTE. German philosophical psychology had stressed intuitionism – that is to say, the doctrine of innate ideas, of a priori judgments, of native categories of the understanding. British psychology was built around empiricism, the doctrine of the genesis of the mind through individual experience. HELMHOLTZ took his stand with the latter group against the reigning German philosophy of KANT and FICHTE. Later within psychology the opposition became that between geneticism and nativism in perception, with LOTZE, HELMHOLTZ and WUNDT representing geneticism, with MÜLLER, HERING and STUMPF representing nativism.'

8. CAMILLO GOLGI (1843-1926), the great Italian neurohistologist, discovered the musculo-tendinous end organ named after him. Born in Cortena, he studied medicine in Pavia where he worked for some time under the direction of CESARE LOMBROSO. His interest in neuroanatomy was aroused through reading VIRCHOW'S 'Cellularpathologie'. In 1873, he was appointed resident physician in Abbiategrasso, a small town away from all academic activity. There he conducted research at night by candlelight in the kitchen of his home using a simple microscope and a few self-made instruments. During the winter months he worked in a thick overcoat.

In 1875, he returned to the University of Pavia where he remained for more than 40 years.

In 1906, GOLGI received a Nobel Prize in Medicine, jointly with CAJAL who at that time described his relationship with GOLGI as that of 'two Siamese brothers attached at the back', thus bringing to a happy end a furious polemic which had taken place between them during the preceding years. He died at an age of 83.

9. WEIR MITCHELL (1829-1914) of Philadelphia coined the term *phantom limb* for the phenomenon of experience of non-existing limbs after amputation. He did so in an article which appeared in Lippincott's Magazine in 1871 [8: 563 ff.]. MITCHELL who was well acquained with European medicine – he spent time in Paris where he met CLAUDE BERNARD and CHARCOT – served as a surgeon in the Civil War. On his military experiences he published several reports, among them one in 1864 on 'gunshot wounds and other injuries of nerves', and one in 1872 on 'reflex paralysis'. In the latter he described 'sudden motor loss resulting from wounds of the forebrain'.

10. CHARLES SHERRINGTON (1857-1952) at age 36 when his reputation as one of England's leading experimental neurophysiologists was already established, together with GRÜNBAUM, had completed his study on segmental stratification of dermatomes and was engaged with the analysis of the functional design of the motor cortex of chimpanzees.

11. JOSEPH FRANCOIS BABINSKI (1857-1932). In a meeting of the Paris Neurological Society held shortly before World War I, BABINSKI reported on two hemiplegic patients – both with paralysis of the left side – who 'ignored' their condition. Together with the motor paralysis there was complete hemianesthesia. BABINSKI emphasized that the patients' peculiar behavior was not to be explained as a result of

clouded consciousness, organic dementia, or memory disturbance. In fact, the interest they took in their surroundings and the appropriateness of their expressions stood in sharp contrast to their failure to notice or perceive their paralyses. The 'agnosias' appeared altogether artless and unaffected.

BABINSKI gave the syndrome the name *anosognosia*. Its description was but one of many original and painstaking neurological discoveries made by him, the most well-known among them, of course, the account he gave in 1896 of the 'cutaneous plantar reflex'. Recognition of anosognosia as a neuropsychological entity of its own likewise represented a contribution of major importance.

Early in his medical career BABINSKI was 'Chef de Clinique' under CHARCOT. With his exceptional intelligence and critical faculty, BABINSKI was aware of the questionable validity of the demonstrations of 'hysterical' patients and of the fallacious interpretations given to them by CHARCOT.

TH. ALAJOUANINE, one of BABINSKI'S erstwhile pupils wrote in 1950 that the memory of the experiences in the Salpetrière were foremost in BABINSKI'S mind throughout his life, 'like Banco's ghost in Macbeth'. Thus, the precision which distinguished BABINSKI'S accounts of neurological syndromes and their subjective accompaniments.

12. ROBERT BARANY (1876-1936) pioneered in the clarification of the role played by the vestibular apparatus of the inner ear in the control of movements. Early in his career, BARANY worked with KRAEPELIN, the neuropsychiatrist, before joining ADAM POLITZER in Vienna, the otolaryngologist. It was there that he undertook his studies of rhythmic nystagmus following thermic stimulation of the ears, and of vertigo caused by a variety of mechanisms, among those subserved by labyrinth and cerebellum. BARANY held that the cerebellum controls somatomotor localization. The vermis, he thought, coordinates movements of the trunk, the cerebral hemispheres those of the extremities, the flocular cortex those of the eyes. The vestibular and labyrinthine apparatus, he wrote, is an integral part of the body's proprioceptive system.

While serving with the Austrian army during World War I, BARANY was taken prisoner by the Russians who confined him to the fortress of Przemysl. There, notification reached him of the award to him of a Nobel Prize in medicine. At the request of Prince Carl of Sweden, the Czar allowed BARANY to travel to Stockholm to attend the ceremonies.

In 1917, BARANY accepted an appointment as professor at the University of Upsala which he held until his death in 1936.

13. HENRY HEAD (1861-1940) was one of the most versatile neurologists of modern times. As a student he visited the German University of Prague and the University of Halle where he attended lectures by HERING. His M.D. thesis submitted in 1892 'On Disturbance of Sensation, with special reference to the *pain of visceral disease*' contained the first of several *'schemata'* he proposed over the years in an effort to elucidate structure and function of the nervous system. Following the publication in 1896 by CHARLES SHERRINGTON of the results of experiments with monkeys in whom posterior nerve roots were severed to delineate corresponding *dermatomes,* HEAD took up the study of the same question in human subjects, choosing for the purpose patients with herpes zoster in whom he *mapped areas of skin eruption* caused by inflammation of single nerve roots. For two years HEAD took up residence in a mental hospital in order to collect the required information on herpes that was then common in paretics. When in 1925 FOERSTER compared dermatomes obtained by him with the method of section of nerve roots, he was impressed with the remarkable accuracy of HEAD'S observations. Still another attempt at 'schematization' was made by HEAD in a description of the *segmental topography of afferent pathways within the spinal cord.* DENNY-BROWN wrote that 'HEAD'S investigations were devoted almost wholly to the sensory system, where he brought order out of chaos at every level by his vivid thought and refined clinical method'. In 1911, HEAD, together with GORDON HOLMES, published a lengthy paper on *sensory disturbances from cerebral lesions.* From these studies he developed his concept of a *central body schema* which depended, he wrote, at least in part upon two separate sensory systems – protopathic and epicritic.

In the second volume of his collective writings which appeared in 1920 under the title 'Studies in Neurology', HEAD defined his concept of the body schema as follows:

'Every recognisable change in posture enters consciousness already charged with its relation to something which has gone before, and the final product is directly perceived as a measured postural change. For this combined standard, against which all subsequent changes in posture are estimated, before they enter consciousness, we have proposed the word 'schema'. Man perpetually builds up a model of himself, which constantly changes. Every new posture or movement is recorded on this plastic schema and the activity of the cortex brings every fresh group of sensations evoked by altered posture into relation with it. Immediate postural recognition tends to destroy such schemata and so disturbs the certainty of spatial recognition.'

14. PAUL SCHILDER (1881-1940) was born in Vienna and died in New York. He came to the United States in the wake of the European upheavals that brought to the New World an unprecedented wealth of scientific knowledge and tradition. Like his teacher SIGMUND FREUD, SCHILDER was associated over several years with WAGNER VON JAUREGG. At the time he entered neuropsychiatry, SCHILDER was a philosopher well grounded in physiology and thoroughly acquainted with German as well as English neurological literature.

SCHILDER borrowed the term 'body scheme' ('Körperschema') from HENRY HEAD using it as title of a monograph published in 1923. In it he described clinical observations of distortion of awareness of the body, arguing that if a seemingly ubiquitous accompaniment of human existence such as the knowledge we normaly have of our body can undergo major modifications, it is necessary to ask how man ordinarily experiences his body.

In 1930, SCHILDER accepted the position of Professor of Psychiatry at Bellevue Hospital in New York. In 1935, he published a second monograph *The Image and Appearance of the Human Body* in which his earlier concepts were elaborated. In it he pointed out that the writings on the subject by KANT, BERGSON, FREUD, WILLIAM JAMES and JOHN DEWEY were of a purely speculative nature and thus of little relevance for the interpretation of the neurological phenomena under study by him. Likewise psychoanalytical theories of the body image which he had discussed at length with FREUD he considered irrelevant.

'I do not think', he wrote, 'that FREUD'S view to the effect that our desires try to lead us back to a previous state is a true description of inner and outer experiences.'

He expressed his indebteness to VON FREY, VON WEIZSACKER and HENRY HEAD. The following is a passage from SCHILDER:

'The image of the human body means the picture of our own body which we form in our mind, that is to say the way in which the body appears to ourselves. There are sensations which come from the muscles and their sheaths, indicating the deformation of the muscle; sensations coming from the innervation of the muscles (energy sense, VON FREY); and sensations coming from the viscera. Beyond that there is the immediate experience that there is a unity of the body. This unity is perceived, yet it is more than a perception. We call it a schema of our body or bodily schema, or, following HENRY HEAD who emphasizes the importance of the knowledge of the position of the body, 'postural model of the body'. The body schema is the tridimensional image everybody has about himself. We may call it 'body-image'. The term indicates that we are not dealing with a mere sensation or imagination. There is a self-appearance of the body. It indicates also that, although it has come through the senses, it is not a mere perception. There are mental pictures and representations involved in it, but it is not mere representation.'

15. JOHANNES LANGE (1891-1937). For several years, LANGE worked with KRAE-PELIN with whom he wrote a well-known textbook of psychiatry. In 1930, LANGE became Professor of Neuropsychiatry in Breslau where he established close cooperation with OTFRID FOERSTER. With the latter he discussed the design of his monograph '*Agnosien und Apraxien*' (153 pp.) which appeared in 1936 in volume 6 of the *Handbuch der Neurologie,* under FOERSTER'S editorship. The monograph ranks as the authoritative statement on the subject in the German literature. In it LANGE set out to present LIEPMANN'S work in its entirety, adding to it clinical and pathological evidence that had become available since LIEPMANN'S death in 1925. Prior to the appearance of the monograph published by GESCHWIND in 1965, LANGE'S treatise was the most complete and authoritative statement on the subject. It still is so in the German literature.

While LANGE'S contributions to neurology and psychiatry cover an exceptionally great variety of subjects, the presentation of the above-mentioned monograph alone allocates him a place among the founders of neurology.

16. HUGO KARL LIEPMANN (1863-1925) studied philosophy as well as medicine. He worked one year with KARL WEIGERT, the neuro-anatomist before accepting an appointment at WERNICKE'S clinic in Breslau in 1895. There his interest became focussed upon cognitive and executive disturbances in patients with lesions of the parietal, temporal and occipital lobes (fig. 20). His descriptions of the apraxias and agnosias, their interdependence, their causation by afflictions of the dominant hemisphere, and of the role of the corpus callosum in the development of connexion syndromes have remained unsurpassed. Figure 21, drawn by LIEPMANN, illustrates how lesions of the corpus callosum can cause isolated apraxia and tactile agnosia of the left side of the body, resulting from severance of the right cerebral hemisphere from the dominant half of the brain. In 1899, LIEPMANN left Breslau to accept a post as psychiatrist in the mental institution of Dalldorf outside Berlin where he worked until his death in 1925.

The following is a list of LIEPMANN'S main publications:
Das Krankheitsbild der Apraxie ('motorischen Asymbolie'). Mschr. Psychiat. *15:* 102, 182 (1900). – Diskussionsbemerkung zu OPPENHEIM. Neu. Zbl. *21:* 617 (1902); *23:* 490, 664 (1904). – Über Störungen des Handelns bei Gehirnkranken (Karger, Berlin 1905). – Mschr. Psychiat. *17:* 289 (1905); *19:* 217 (1906). – Neurol. Zbl. *26:* 473, 937 (1907). – Med. Klin. *1:* 725, 765 (1907). – Drei Aufsätze aus dem Apraxiegebiet (Karger, Berlin 1908). – Dtsch. med. Wschr. *2:* 1470 (1908). – Apraktische Störungen; in CURSCHMANN Lehrbuch der Nervenkrankheiten, p. 487 (Julius Springer, Berlin 1909). – Neurol. Zbl. *30:* 345 (1911). – Dtsch. med. Wschr. *2:* 1249, 1308 (1911). – Mschr. Psychiat. *34:* 485 (1913); *35:* 490 (1914). – Apraxie; in BRUGSCH Ergebnisse der gesamten Medizin (Urban & Schwarzenberg, Berlin/Wien 1920). – Mschr. Psychiat. *71:* 169 (1929).

17. HELMUT PLESSNER'S (1892) contribution to the understanding of sensory perception corroborated HELMHOLTZ'S view that psychology ought to be studied with the methods of physiology and interpreted in the framework of philosophical concepts. Starting out his career as a student of the exact natural sciences he eventually devoted his efforts to philosophy. In 1923, he published a book entitled 'Die Einheit der Sinne, Grundlinien einer Ästhesiologie des Geistes' (Bonn 1923) (The Indivisibility of the Senses. Mind as an aesthesiological phenomenon) in which he formulated a comprehensive theory on the nature of sensory phenomena in their dependence on structure and function of the nervous system. In several books and essays written during his long and distinguished career as a university professor in Holland and Germany, PLESSNER demonstrated that HUSSERL'S dream of phenomenology being made 'an objective science of human existence' could be realized. PLESSNER'S lifelong interest in neurology had been established early in his life by his father, a recognized authority in the clinical speciality under reference. With E. HUSSERL, MAX WEBER, GEORGE SIMMEL, MAX SCHELER, VIKTOR VON WEIZSACKER, E. VON GEBSATTEL, and ERWIN STRAUS, PLESSNER developed phenomenology into an academic discipline of its own.

18. WILDER PENFIELD (1891) made major physiological and pathophysiological contributions to neurology during his long and distinguished career as a surgeon. His investigations of the motor and somatosensory cortex of man represent unsurpassed achievements. Accounts of them are contained in several books, chief among them 'Epilepsy and Cerebral Localization' (1941), 'The Cerebral Cortex of Man' (1950), 'Epilepsy and the Functional Anatomy of the Human Brain' (1954), and 'The Excitable Cortex in Conscious Man' (1958).

Applying the method of evoked potentials introduced by ADRIAN in 1925, PENFIELD and FOERSTER studied sensory representation in the parietal and frontal lobes of the human cortex. The areas of vocalization and handiness, which are exclusively found in man, were thus identified. PENFIELD mapped the entire sensory and motor sequences and noted that postcentral and precentral 'fields' overlap. For purposes of the discussion under reference it is relevant to quote PENFIELD'S statement:

'that the major cortical representation of somatic sensation (proprioceptive and discriminatory) is in the postcentral gyrus. But there is also a minor representation in the precentral gyrus. Ablation of the precentral gyrus does not interfere with sensory perception. Ablation of the postcentral gyrus abolishes it, at least initially.'

Incidentally, as it were, PENFIELD settled the argument whether the 'sense of movement' is engendered by proprioceptive feedback or 'directly in the brain'. In some of his patients, PENFIELD reported, electrical stimulation caused 'a sense of movement of parts of the body although no movement could be seen'. Since no feedback could have been responsible for the sensation, the validity of the theory first formulated by STEINBUCH in 1811 was established.

To illustrate the representation of sensory and motor sequences on the cortical level, PENFIELD drew the now famous homunculus shown in figure 17.

Like CHARLES SHERRINGTON, HENRY DALE and EDGAR ADRIAN, PENFIELD received the Order of Merit, England's highest honor. Among other distinctions awarded to him are the Lister Medal of the Royal College of Surgeons, the Otfrid Foerster Medal of the German Neurosurgical Society, the Gold Medal of the Royal Society of Medicine, and the Erb Medal of the 'Deutsche Gesellschaft für Neurologie'.

19. MacDONALD CRITCHLEY, President since 1965 of the World Federation of Neurology, is one of the most distinguished clinical neurologists of the 20th century. A worthy representative of the great tradition of the National Hospital at Queen Square in London, CRITCHLEY'S contributions cover a wide field. Several among them are of specific relevance to the theme under review: In 1950, the Lancet published his paper *'The Body Image in Neurology'*. In 1953 appeared his monograph of 480 pages entitled 'The Parietal Lobes', unsurpassed in scope, knowledge, style and didactic excellence. The fact that it has become possible to comprehend neurological entities such as those associated with the designations of the four lobes of the human cortex is due to clinicians of the caliber of CRITCHLEY. Three major works on reading disturbances, 'Developmental Dyslexia', 'Aphasiology and other aspects of Language', and 'The Dyslexic Child', have been written by CRITCHLEY during the past decade. CRITCHLEY has also been one of the architects of the *Handbook of Neurology*.

20. NORMAN GESCHWIND (born 1926) ranks among the leading neurologists in the United States today. In recognition of his clinical acumen as well as of a number of remarkable scientific contributions, he was appointed Professor of Neurology at the Harvard Medical School at the early age of 42. In 1965, Brain published two papers by him entitled 'Disconnexion Syndromes in Animals and Man' which contain the most up-to-date account on the agnosias, apraxias and related disturbances, comprehensively dealing with their anatomical background, with results of animal experiments, with clinical observations and their philosophical implications. In a number of historical essays, GESCHWIND had described the contributions of CARL WERNICKE and the Breslau school of neurology whose last surviving representative, KURT GOLDSTEIN collaborated with him the collection of source material, including the work of HUGO LIEPMANN which had previously not been fully known in the English literature.

6377

1 Ambroise Paré (1510-1590)

2 Abraham Vater (1684-1751)

3 Francois Magendie (1783-1855)

4 Charles Bell (1774-1842)

5 RUDOLF ALBERT VON KÖLLIKER (1817-1905)

6 WILLY KUEHNE (1837-1900)

7 HERMANN VON HELMHOLTZ (1821-1894)

8 CAMILLO GOLGI (1843-1926)

9 Weir Mitchell (1829-1914)

10 Charles Sherrington (1857-1952)

11 Joseph Francois Babinski (1857-1932)

12 Robert Barany (1876-1936)

13 HENRY HEAD (1861-1940)

14 PAUL SCHILDER (1881-1940)

15 JOHANNES LANGE (1891-1937)

16 HUGO KARL LIEPMANN (1863-1925)

17 HELMUT PLESSNER (1892)   *18* WILDER PENFIELD (1891)

19 MACDONALD CRITCHLEY   *20* NORMAN GESCHWIND (born 1926)

Fig.1. Four title pages.

Historical Illustrations

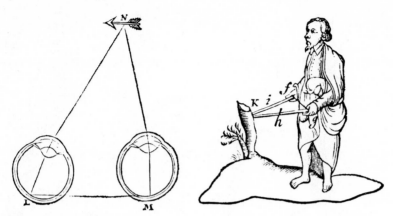

Fig. 2. DESCARTES on distance vision. In 1637, DESCARTES wrote that we are aware of distance by perceiving the directions of our eyes just as a blind man assesses distance through staves held in his hands. In 1709, BERKELEY expressed the same view saying that perception of strain of the muscles of the eyes renders awareness of distance possible. When in 1826 CHARLES BELL introduced the term 'muscle sense', he presented a sound physiological argument in support of DESCARTES' and BERKELEY'S ideas: nerves connecting muscles with the spinal cord contain afferent as well as efferent fibers [from DESCARTES' L'homme, 1649].

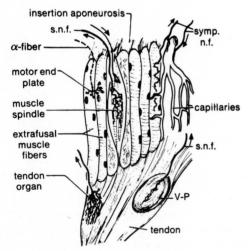

Fig. 3. Three proprioceptors *in situ: muscle spindles, tendon organ,* and *Vater-Pacinian corpuscle* (V-P). The spindle is arranged 'in parallel' with the extrafusal muscle fibers; the tendon organ functions with the muscle 'in series'. Vater-Pacinian corpuscles respond to pressure [from BRODAL].

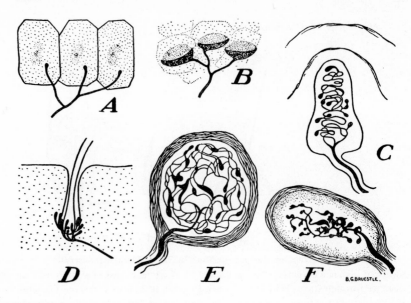

Fig. 4. Diagram of the principal cutaneous receptors. The structure of all end organs is variable and individual drawings are schematic. *A:* Free nerve endings from the cornea of the eye. Nerves terminate within the cell. Similar endings are found in the skin. Others terminate as networks (WOOLLARD). *B:* MERKEL'S tactile disc (from the pig's snout). *C:* MEISSNER'S tactile corpuscle. *D:* Basket-ending at the root of a hair follicle. *E:* End bulb of KRAUSE from human conjunctiva. *F:* GOLGI-MAZZONI corpuscle from human skin (pressure) [from JOHN FULTON].

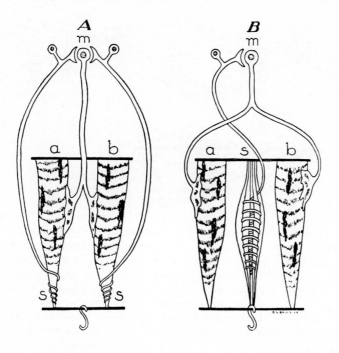

Fig. 5. Golgi tendon organ and muscle spindle. The Golgi organ (A) and the spindle (B) are the two chief proprioceptors conveying 'muscle sense'. The tendon organ is disposed 'in series', the spindle 'in parallel' with the tension-supporting element. The functional design of the two systems is detailed in the figure: a and b = muscle fibers; m = motor horn cell; s = sensory organ [from JOHN FULTON and J. PI-SUNER].

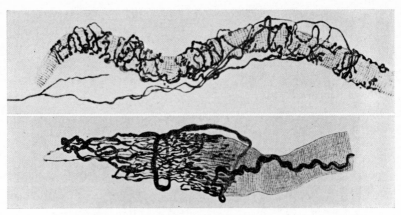

Fig. 6. A: Spindle. *B:* Golgi tendon organ in human extraocular muscle [Dogiel, A. S.: Arch. mikroskop. Anat. 68, 1906]. The demonstration by Dogiel in 1906 of spindles and Golgi organs in human extraocular muscles represented a contribution of importance since in 1896 Sherrington had stated he had been unable to discover receptors in the eye muscle of chimpanzees. Dogiel's findings reopened the discussion of the part played in estimating distance by vision and proprioception. The following is a list of publications by Dogiel on peripheral receptors: a) Methylenblautinktion der motorischen Nervenendigungen in den Muskeln der *Amphibien* und *Reptilien.* Arch. mikroskop. Anat. *35:* 305 (1890). – b) Die Nervenendigungen in Tastkörperchen. Arch. Anat. Physiol. (anat. Abt.), p. 182 (1891). – c) Die sensiblen Nervenendigungen im Herzen und in den Blutgefässen der *Säugetiere.* Arch. mikroskop. Anat. *52:* 44 (1892). – d) Die Nervenendigungen in Meissnerschen Tastkörperchen. Int. Mschr. Anat. Physiol. *9: 1* (1892). – e) Die Nervenendigungen in der Haut der äusseren Genital-organe des Menschen. Arch. mikroskop. Anat. *41:* 585 (1892). – f) Zur Frage über den Bau der Herbstschen Körperchen und die Methylenblaufixierung nach Bethe. Z. wiss. Zoll. *66:* 358 (1899). – g) Die Nervenendigungen im Bauchfell, in den Sehnen, den Muskelspindeln und dem Centr. tend. des Diaphragmas beim Menschen und den *Säugetieren.* Arch. mikroskop. Anat. *59:* 1 (1901). – h) Über die Nervenendapparate in der Haut des Menschen. Z. wiss. Zool. *75:* 46 (1903). – i) Über die Nervenendigungen in den Grandryschen und Herbstschen Körperchen im Zusammenhang mit der Neu-ronentheorie. Anat. Anz. *25:* 558 (1904). – k) Die Nervenendigungen im Nagelbett des Menschen. Arch. mikroskop. Anat. *64:* 173 (1904). – l) Der fibrilläre Bau der Nerven-endapparate in der Haut des Menschen und der *Säugetiere* und die Neuronentheorie. Anat. Anz. *27:* 97 (1905). – m) Zur Frage über den fibrillären Bau der Sehnenspindeln oder der Golgischen Körperchen. Arch. mikroskop. Anat. *67:* 638 (1906). – n) Die Endigungen der sensiblen Nerven in den Augenmuskeln und deren Sehnen beim Menschen und den *Säugetieren.* Arch. mikroskop. Anat. *68:* 501 (1906). – o) Zur Frage über den Bau der Kapseln der Vater-Pacinischen und Herbstschen Körperchen. Folia neurobiol. *4:* 218 (1910). Dogiel und Willanen: Die Beziehung der Nerven zu den Grandryschen Körperchen. Z. wiss. Zool. *67:* 349 (1900). A comprehensive description of the peripheral receptor system was given by Ph. Stöhr in Mikroskopische Anatomie des Menschen, vol. 4, part 1; Nervensystem (Springer, Berlin 1928).

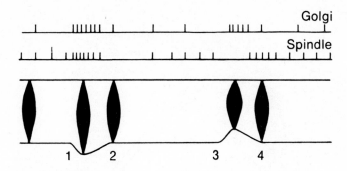

Fig. 7. Response of muscle spindle and of a Golgi tendon organ during stretch and contraction of the muscle. Muscle drawn in black; action potentials shown in upper lines.

Intrafusal fibers of the spindle are arranged in parallel with *extrafusal* muscle fibers. Both are connected to *tendons*. When extrafusal muscle fibers contract, tension in the intrafusal fiber spindle is reduced. Contrariwise passive stretching of the muscle increases its length and tension, including that of the intrafusal fibers which embraces the spindle. The *central* portion of the intrafusal muscle fiber in contrast to its *distal* segments is not contractile. Its nuclei are surrounded by spirally arranged nerve fibers. *Secondary* sensory endings or flower-spray endings are continuous with afferent fibers which end in the spinal cord *posteriorly*. Muscle spindles receive *motor* innervation through fibers from the *ventral* horn of the spinal cord. They terminate in the *motor end plate*. Thus, contraction of distal parts of intrafusal fibers results in stretching of the central 'sensory' part of the intrafusal fiber with consequent stimulation of the sensory endings. Stretching of the muscle as a whole has a like effect.

The design of the *Golgi tendon organ* is simpler than that of muscle spindles. The Golgi tendon organ consists of large myelinated nerve fibers which terminate with a spray of fine endings between bundles of collagenous fibers situated close to junctions of tendons and muscles. Golgi tendon organs are arranged *in series* with extrafusal muscle fibers, in contrast to the arrangement *in parallel* of the intrafusal fibers of the muscle spindle. Thus, whenever the muscle contracts or is stretched passively, the tendon organ is stimulated.

Action potentials can be recorded from single fibers of dorsal roots. There is some spontaneous activity in the receptors when the muscle is at rest, more so from the *spindle* than from the *Golgi tendon organ*. When the muscle is stretched (1), afferent discharges from *both* sense receptors increase. Resting values are re-established when the stretch discontinues (2). Following contraction of the muscle (3) the *Golgi tendon organ* responds by increasing its firing rate while the muscle spindle's electrical discharge ceases. When the muscle relaxes again (4), there is pull on the *spindle* again with the result that it discharges. At the same time the *Golgi tendon organ* does not discharge since no pull is exerted on it. To sum up: *the Golgi tendon organs are tension recorders; the muscle spindles provide information about the length of the muscle* [adapted from Brodal].

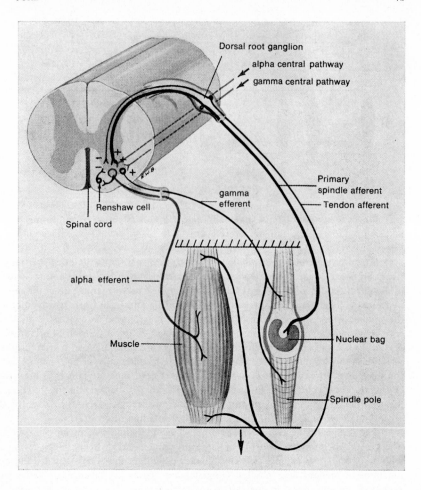

Fig. 8. Diagram of neural basis for stretch reflex. Afferent fibers from muscle spindle and tendon organs and efferent fibers to muscle and spindle (gamma fibers) are shown. Excitation is indicated by plus signs, inhibition by minus signs. The Renshaw cell is an interneuron that provides recurrent inhibition to the active motoneuron pool. Muscle rigidly fixed at upper end and subject to stretch in direction of arrow at lower end [from TUTTLE and SCHOTTELIUS].

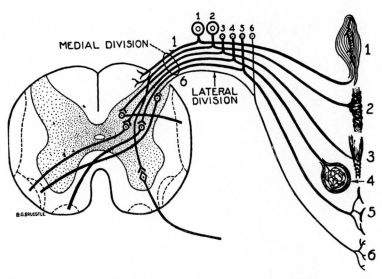

Fig. 9. Diagrammatic cross-section of spinal cord showing principal sites of termination of dorsal root fibres coming from exteroceptors and proprioceptors. 1 and 2 represent large medullated fibres having large dorsal root ganglion cells and passing to the dorsal columns; they arise from Pacinian (1) and muscle spindle endings (2). 3 and 4 terminate on dorsal horn cells that cross and give rise to spino-thalamic and spinocerebellar tracts. 5, a similar cell terminating on neuron that gives rise to ventral spinothalamic tract. 6, a small fibred neuron (pain) terminating in substantia gelatinosa of Rolando giving rise to fibre of ascending spinothalamic tract of opposite side [from JOHN FULTON].

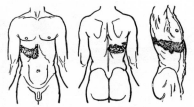

Fig. 10 a. Distribution of the eruption in a case of herpes zoster proved by *post mortem* examination to have been due to inflammation of the ganglion of the seventh posterior root [HENRY HEAD].

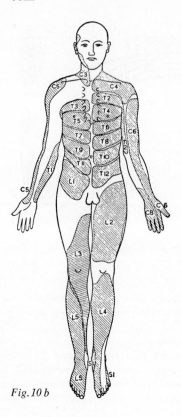

Fig. 10 b

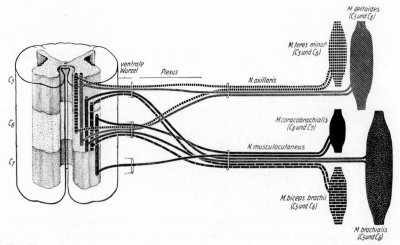

Fig. 11 a

Segmentinnervation der Muskeln des Rumpfes.
[Nach v. Schumacher, Sitz.-Ber. Kais. Akad. Wiss. Wien **117**, 131 (1908) und Villiger, Gehirn und Rückenmark (1930)]

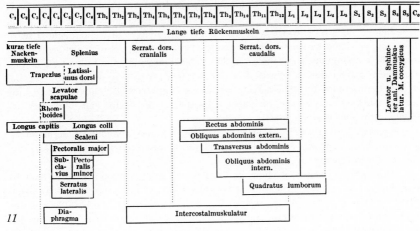

11

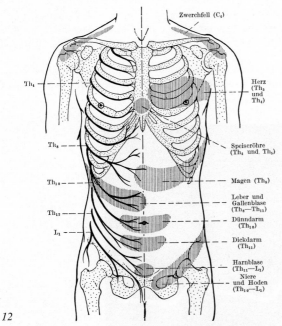

12

Fig.10-12. Three schemata. How exteroceptors are patterned over the body surface of man was shown in 1900 by HENRY HEAD who had studied the distribution of skin eruptions due to inflammation of posterior root ganglia in patients with herpes zoster. He drew the map which details the segmental stratification of dermatomes (fig. 10 a, b). To pursue his investigation HEAD took up residence for two years in a mental hospital where herpes zoster was common in paretic patients.

HEAD'S study followed publication of the results of SHERRINGTON'S experiments with monkeys in 1894: Section of three consecutive dorsal nerve roots above and three below single roots which were kept intact, left areas of skin intact whose sensibility was mediated by single roots. Because of the overlapping of individual dermatomes three contiguous roots were cut above and three below the root whose dermatomal distribution was examined. The isolated skin areas were designated 'essential sensory dermatomes' or 'remaining aesthesias'. SHERRINGTON also emphasized that the sensory supply of the *skin* is divergent from that of the motor and of the sensory supply of the *muscular* system (fig. 11 a, b).

Between 1910 and 1937, DUSSER DE BARENNE conducted studies of 'dermatomes' in monkeys by applying strychnin to dorsal roots thus rendering *segmental dermatomes hyperaesthetic.*

At OTFRID FOERSTER'S clinic in Breslau, we studied dermatomes in man by assessing *antidromic vasodilatation* after stimulation of dorsal roots; and by identifying *remaining cutaneous sensibility* after dorsal root section. Like SHERRINGTON, FOERSTER cut three roots above and three roots below an intact root, explaining that the number of algogenic fibers from a given region sufficed to give to desired therapeutic results wherever the operation was indicated because of pain. In 1948, KEEGAN and GARRETT determined cutaneous dermatomes in patients with prolapsed intravertebral discs causing damage to single posterior roots. '*Algesic segment lines*' thus assessed were synonymous with those dermatomes demonstrated by HENRY HEAD and FOERSTER.

Figure 12 summarizes the findings reported in HENRY HEAD'S doctoral thesis submitted in 1892 on 'Disturbance of Sensation with special reference to pain of *visceral* disease'; it was published in German in 1898.

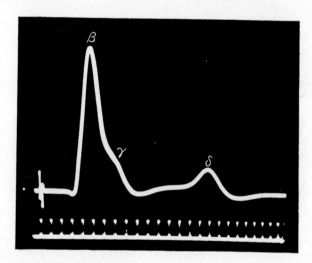

Fig. 14. Text see page 47.

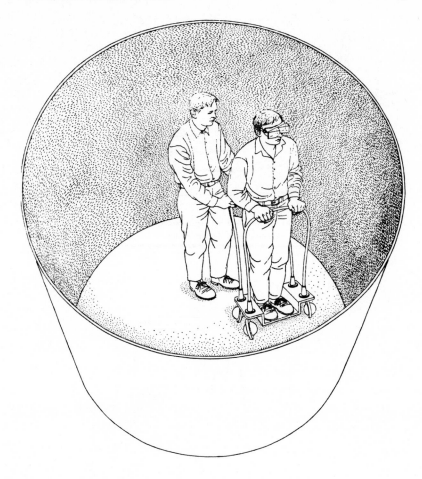

Fig.13. Depicts a man wearing distorting prism goggles. He is transported passively. The purpose of the study was to test the hypothesis that adaptation to visual distortion depends on active movement. Subjects who were restrained from moving freely could not adapt themselves to the situation created by the wearing of prisms, while subjects who walked about did adapt themselves notwithstanding the visual handicap. Movement involving orientation against the ubiquitous pull of gravity is the prerequisite for the resumption of normal functioning of the central nervous system even when vision is distorted or excluded; while severance of this link causes physical and mental deterioration [after HELD].

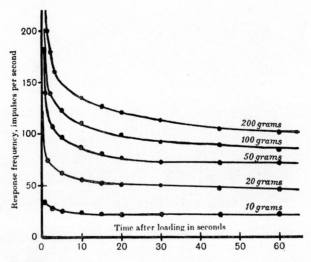

Fig.15

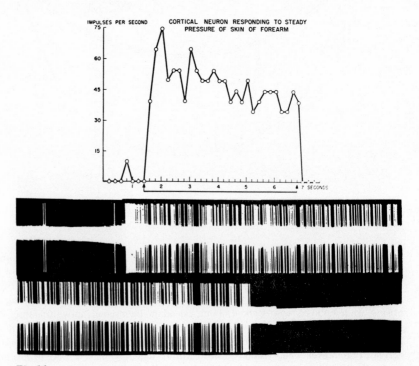

Fig.16

Fig. 14-16. Since the first postulation by Sir CHARLES BELL in 1826 of a 'sixth sense' or 'muscle sense', there has been a continuous discussion of the modality of conveyance of the 'sense of movement'. There were those who assumed that the 'willing of motor action' is experienced 'inside the brain'; while others ascertained that the activated muscle 'informs the brain' through afferent feedback. As is explained in the text, both assumptions were proved valid.

Figure 14 to 16 represent evidence pertaining to the feedback theory, figure 14 showing the action potential of an afferent nerve; figure 15 the differential response 'messages' in the somato-sensory cortex.

JERZY E. ROSE and VERNON B. MOUNTCASTLE'S tracing (fig. 14) p. 44 shows compound action potential of the saphenous nerve of the cat recorded at a distance of 54 mm from the locus of stimulation. Several elevations (denoted by Greek letters) are recorded because, in the nerve, fibers of different sizes are grouped around several peaks. Since the saphenous nerve lacks the largest afferents arising from the muscle stretch receptors no elevation is indicated even though the sizes in the α- and β-groups overlaps. The β- and γ-elevations as denoted here are sometimes referred to as α- and β-peaks. All elevations pertain to A fibers. The elevation due to C fibers is not shown. Time line: 5,000 cps.

MATTHEWS (fig. 15) analyzed the response of muscle spindles to stretch. The abscissa shows time in seconds, the ordinate impulse frequencies per second.

V. B. MOUNTCASTLE and T. P. S. POWELL recorded action potentials of a single cell in the postcentral gyrus of the monkey, *Macacus rhesus*. The neuron is driven by steady pressure applied to the cutaneous receptive field located on the volar surface of the contralateral forearm. Onset and release of pressure indicated by solid bar under the graph, which shows the number of impulses per second plotted at 200-msec intervals (fig. 16).

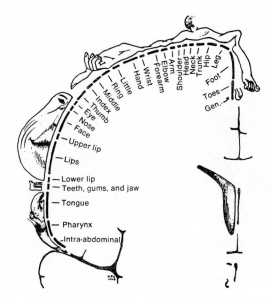

Fig. 17. Text see page 49.

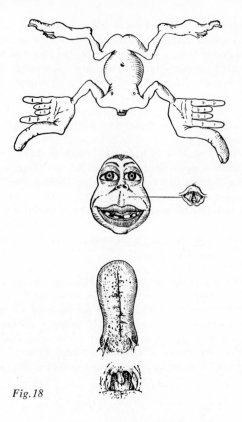

Fig. 18

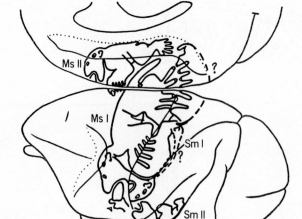

Fig. 19

Fig.17-19. Three homunculi. The idea to present a picture of the sequence and extent of responses elicitable from the sensory-motor strip occurred to PENFIELD in 1937. Figure 18 has been included ever since in textbooks on physiology. It illustrates the fact that the specifically human attributes of *Homo sapiens* are conspicuously dominant. The result is the caricature which reveals the significance of the two features that single out he human species: speech and the tool function of the hand.

Painstaking analyses of motor and sensory parts of the cortex carried out by PENFIELD over many years revealed that motor and sensory homunculi do not completely correspond. Figure 17 shows the latter.

E. D. ADRIAN pointed out that peripheral somatic stimuli evoke responses not only in the main somatosensory cortical area of the brain but also in a small field beneath its lower end. The latter is referred to as *second* somatosensory area, to distinguish it from the former which is referred to as the *first* somatosensory area. At least one more region which responds to peripheral somatic stimuli is situated on the medial surface of the cerebral hemisphere. None of these areas are purely sensory. It has, therefore, been suggested to designate them *sensorimotor cortical areas* [Figure 19 from BRODAL].

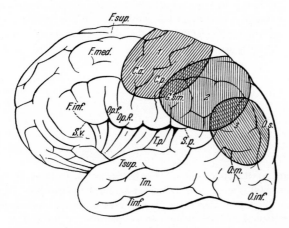

Fig. 20. Cortical regions whose impairment causes syndromes that may include apraxia. Almost invariably the parietal lobe of the left hemisphere is involved. However, the vulnerability of the adjoining portions of the frontal and occipital lobes has been demonstrated [from LIEPMANN].

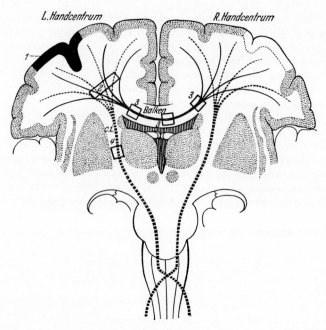

Fig. 21. Lesions of the left cerebral hemisphere may cause bilateral as well as unilateral apraxia. The left half of the brain controls all cognitive and praxic functions. It communicates with the right half through the corpus callosum. LIEPMANN showed how pathways connecting the motor control area for the right hand in the left cerebral cortex with its corresponding opposite area in the right cerebral hemisphere depart from the cortex ('L. Handcentrum') together with those that eventually traverse inner capsule, pons, and pyramidal decussation, but that those going into the corpus callosum are sorted out subcortically [from LIEPMANN].

Fig. 22-24. 'The Support Reaction and the Central Nervous Control of Progression' [JOKL and SCHEPERS]. Patient with left-sided cerebral lesion whose distribution is shown in figure 22. A hemiplegia (fig. 23) was present on the contralateral side together with a homolateral support reaction and impairment of the arm-swinging movement during walking (fig. 24). Both the latter symptoms were due to elimination of cortical control. The condition was unaccompanied by agnosia and apraxia since the cortical injury left the greater part of the left parietal lobe and adjoining areas intact (compare with LIEPMANN'S fig. 21).

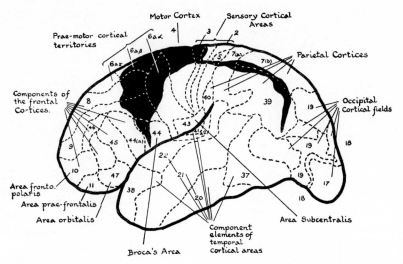

Motor Cortex

Sensory Cortical Areas

Prae-motor cortical territories

6a∝

4

3

2

6aβ

5

7a

7(b)

Parietal Cortices

6aγ

Components of the frontal Cortices.

8

40

39

19

Occipital Cortical fields

1+6

43

19

9

44

44(a)

1+2

19

18

45

22

10

47

21

37

19

17

Area fronto-polaris

11

38

20

18

Area prae-frontalis

Area orbitalis

Area Subcentralis

Broca's Area

Component elements of temporal cortical areas

Fig. 22

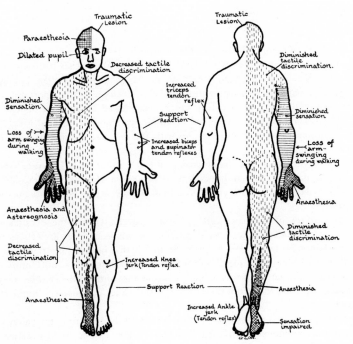

Traumatic Lesion

Traumatic Lesion

Paraesthesia

Diminished tactile discrimination.

Dilated pupil

Decreased tactile discrimination

Increased triceps tendon reflex

Diminished Sensation

Support Reaction

Loss of arm swinging during walking

Increased biceps and supinator tendon reflexes

Diminished sensation

Loss of arm-swinging during walking

Anaesthesia

Anaesthesia and Astereognosis

Diminished tactile discrimination

Decreased tactile discrimination

Increased Knee jerk (Tendon reflex.

Support Reaction

Anaesthesia

Anaesthesia

Increased Ankle jerk (Tendon reflex)

Sensation impaired

Fig. 23

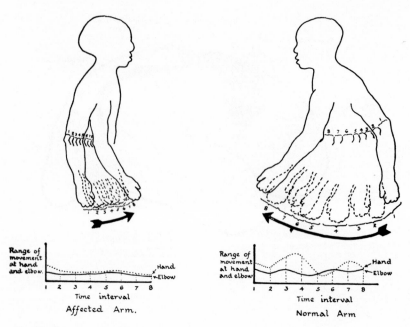

Fig. 24. Text see page 50.

Fig. 25. At the Olympic Village in Melbourne in 1956, I examined Mr. Karoly Takacs, 46 years of age, member of the Hungarian team. Mr. Takacs had represented Hungary in the pistol shooting competition at the Berlin Games in 1936, holding at that time several national and international championships in this sport. In 1938, Mr. Takacs was involved in an accident which necessitated amputation of the right arm midway between elbow and wrist. On his discharge from the hospital Mr. Takacs decided to continue competing by shooting with the left hand which he had never used prior to the amputation.

In 1939, he won the world championship in pistol shooting at Lucerne, Switzerland. At the Olympic Games in London in 1948 and in Helsinki in 1952, he was awarded gold medals for the same event. The 1956 Games in Australia were his fourth Olympic contest.

Several features of this case call for comment. First, that the excellence of Mr. Takacs' athletic performance was so great as to place his pre- as well as post-amputation attainments into a class by itself. Secondly, that initially the training which established his world fame as a pistol shot was confined to the right arm. Thirdly, that the switch-over to the left arm was achieved within the comparatively short time of 8 months. And fourthly, that from then onward Mr. Takacs continued to improve his athletic competence.

This report indicated the extent to which an established neuromotor pattern can be projected into previously untrained skeletomuscular regions. It also raised the question of the dominance of one cerebral hemisphere over the other. Prior to the

25

amputation, Mr. Takacs was right-handed. It would, as he pointed out, not have occurred to him to shoot with his left hand. The postoperative switch involved not only the motor aspect of the skilled performance but also its sensory counterpart, especially in respect of the functional integration of the tactile, proprioceptive and visual components.

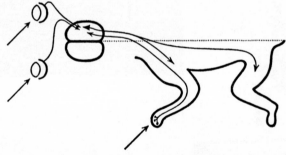

Fig. 26. Crossing of sensory and motor pathways to the cerebral hemispheres in conformity with the crossing of nerve fibers from the nasal side of the eye.

'The visual messages are the most complex, and the most important of all those reaching the human brain. In other animals, hearing and smell may be the dominant senses, but in our brains the olfactory pathways have shrunk and smells play a very small part in our thoughts and decisions. In all the vertebrates, however, the eyes seem to have dictated the general arrangement of the cerebral hemispheres. Owing to the camera structure of the eye an object on the left is focused on the right half of the retina and vice versa. Now the natural arrangement would seem to be that the right half of the retina of either eye should be connected with the right cerebral hemisphere and the left half of the retina with the left hemisphere, and this is in fact accomplished by a sorting-out of the nerve fibres at the optic chiasma before they enter the brain. The result is that an object on the left is signalled by a visual message which is sent to the right cerebral hemisphere. Now, if the object came into contact with the left side of the body or if the left hand were put out to touch it, it would be confusing if the tactile messages were sent to one side of the brain and the visual to the other. What seems to have happened, therefore, is that the crossing-over of the visual message has made it necessary for all the sensory pathways to cross over, so that all the information reaching one side of the brain comes from the opposite side of the body and its surroundings; and as all the information goes across to the opposite hemisphere, the orders for the muscles must also be elaborated in that hemisphere and the messages for them must come back across the midline on their way down to the spinal cord' [from E. D. ADRIAN].

Fig. 27. Blind gymnastic champions. The performances on the horizontal bar and the parallel bars are extremely difficult. Few gymnasts with unimparied vision are capable of executing them. The observations are of interest not only from the point of view of sensory physiology. They also illustrate the exceptional forces of motivation that are at times mobilized in handicapped individuals as a result of their reflection on being inferior [from LORENZEN].

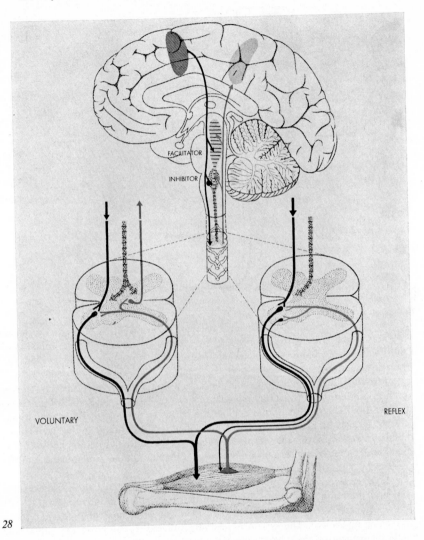

28

Figs. 28 and 29. 'ARAS' (ascending reticular activating system). The *reticular formation,* consisting of the multineuronal, multisynaptic central core of a region which extends from medulla to hypothalamus, receives collaterals from specific sensory pathways. It projects diffusely upon the cortex. Impulses derived from it are brief, discrete, direct and of short latency, in contrast to those via the unspecific ARAS which are persistent, diffuse and of long latency.

Corticofugal pathways converge on the reticular formation of the lower brain stem. Stimulation of widespread cortical areas gives rise to electric potentials in the reticular formation. Impulses originating in the cortex are capable of exciting the ARAS.

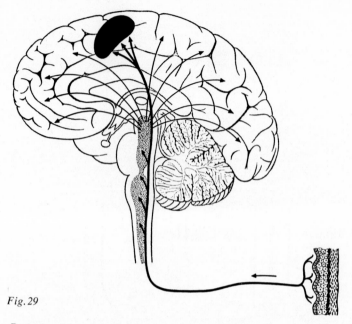

Fig. 29

Reticular formation. A peripheral receptor (lower right) is connected to a sensory area in the brain cortex (upper left) by a pathway extending up the spinal cord. On its way up this pathway branches into the reticular formation which 'awakens' the brain in its entirety.

Movements are modified by the RAS. In voluntary movement afferent nerves conduct impulses from the muscle spindle to a sensory area in the brain. Motor nerves conduct impulses from the motor area of the cortex to the muscle. Both afferent and efferent systems branch into the RAS from which impulses originate that facilitate or inhibit. In reflex movement sensory impulses are passed on to motor nerves in the spinal cord, some of which activate muscle and maintain 'tone' while others sensitize the spindle. The RAS controls both.

Investigations by MAGOUN and collaborators demonstrated the existence of a system in the brain stem whose stimulation induces changes in the electrical activity of the cerebral cortex. These changes are identical with those observed in natural awakening or arousal. The system is medially situated. It comprises the bulbar and midbrain tegmentum, the subthalamus, the hypothalamus, and the ventromedial portion of the thalamus. The medial *afferent* system is distinct from the lateral or lemniscal *corticipetal* system of classical neurology. Its action on the cortex is mediated by thalamic and extrathalamic pathways. The effects of the latter are not confined to specific cortical areas but affect the entire neocortex. The same portion of the brain stem reticular formation facilitates lower motor outflows: Stimulation of the reticular substance lying ventromedially in the caudal part of the bulb inhibits the knee jerk, decerebrate rigidity and movements resulting from stimulation of the motor area of the cerebral cortex. A facilitatory region situated laterally in the reticular formation runs upward into the midbrain tegmentum, the central gray matter and the subthalamus. Stimulation of these regions facilitates the knee jerk and augments responses to stimulation of the motor area. 'The reticulospinal tract', RUCH writes, 'obeys SHERRINGTON'S law of reciprocal innervation'. [Adapted from J. D. FRENCH, G. MO-RUZZI, H. W. MAGOUN, T. C. RUCH, J. F. FULTON and P. BARD].

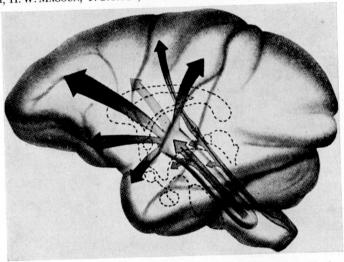

Fig. 30. Ascending reticular activating system (ARAS). The reticular formation, consistang of the multineuronal, multisynaptic central core of the region from medulla to hypothalamus, receives collaterals from specific or classical sensory pathways and projects diffusely upon the cortex. Impulses via specific sensory pathways are brief, discrete, direct and of short latency in contrast to those via the unspecific ARAS which are persistent, diffuse and of long latency [from MAGOUN].

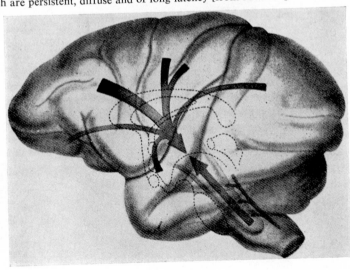

Fig. 31. Corticifugal pathways. Collateral pathways converging on the reticular formation of the lower brain stem. Stimulation of widespread cortical areas gives rise to electric potentials in the reticular formation, hence functional connection is assumed by corticoreticular paths. Afferent impulses from all sources and impulses originating in the cortex are capable of exciting the ARAS, which in turn maintains the cortex and behavior in a state of arousal and alertness [from MAGOUN].

Fig. 32-36. Five drawings presented in 1609, 1653, 1815 and 1938 by KEPLER, FLUDD, GALL and SPURZHEIM and SIGMUND FREUD as well as one published in 1968 purporting to show 'the seven psychic centers of chakras' of Yoga. Each of them pretends to elucidate the attachment of the human mind to the body. They have in common that they are arbitrary in design and that they are not derived from demonstrable facts. KEPLER'S 'Planetary Orbits' and ROBERT FLUDD'S 'A Universe Within the Mind' reflect the undisputed authority held among scientists during the 16th and 17th centuries by astrology (fig. 32 and 33). The 'Physiognomical System' of GALL and SPURZHEIM is not related to astrology but it is likewise a mere product of imagin-

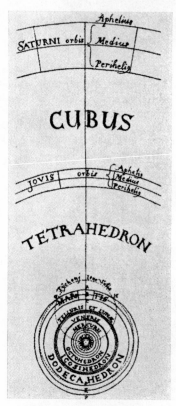

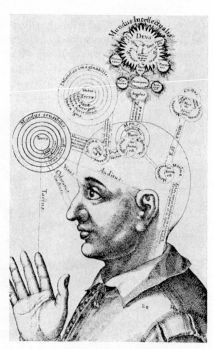

Fig. 32　　　　　*Fig. 33*

ation[6] (fig. 34). FREUD'S diagram belongs in the same category (fig. 35). It is of special interest because prior to devoting his efforts to psychoanalysis, FREUD was a recognized authority on neurological subjects such as aphasia (fig. 38). He as well as WERNICKE were pupils of THEODORE MEYNERT (1833-1892).

The 'seven psychic centers' serve as basis for the teachings of Yoga. They are said to 'raise the feminine kundalini at the base of the spine until it unites with the masculine sahasrata atop the skull' (fig. 36) [from CHUSID].

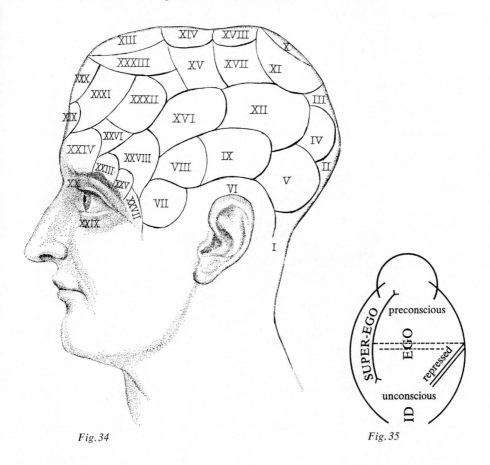

Fig. 34   Fig. 35

6   The regions marked in Roman numerals were said to indicate the following characteristics: organ of amativeness, philoprogenitiveness, inhabitiveness, adhesiveness, combativeness, destructiveness, constructiveness, covetiveness, secretiveness, selflove, approbation, cautiousness, benevolence, veneration, hope, ideality, conscientiousness, firmness or determinateness, individuality.

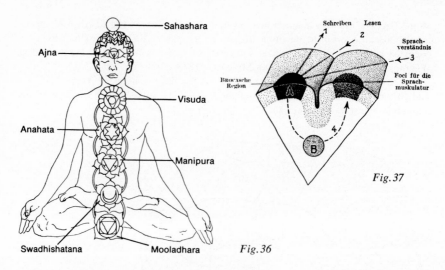

Fig. 37

Fig. 36

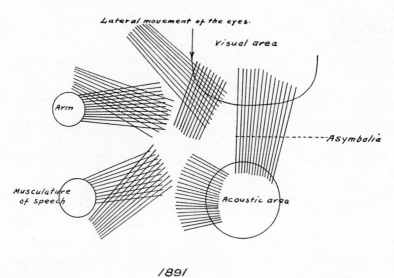

Fig. 38

Fig. 37-40. Four diagrams summarizing neurological observations of patients with aphasia [fig. 39 from WERNICKE, 1874; 38 from FREUD, 1891; 40 from BASTIAN, 1900; and fig. 37 from KLEIST, 1915]. All four diagrams reflect the acceptance of the concept first formulated by WERNICKE that the different clinical manifestations of aphasia are caused by impairment or destruction of 'centers' or of connexion pathways through which they interact ('A', 'B' in KLEIST'S diagram, fig. 37).

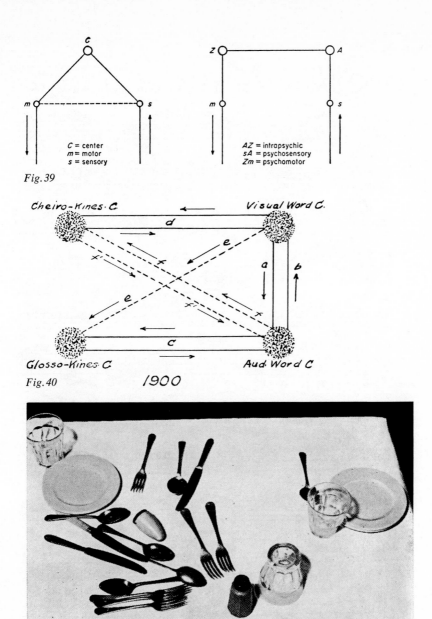

Fig. 39

C = center
m = motor
s = sensory

AZ = intrapsychic
sA = psychosensory
Zm = psychomotor

Cheiro-Kines. C.

Visual Word C.

d

e

a   b

x'   f

e   x'

x

c

Glosso-Kines. C.

Aud. Word C.

Fig. 40   1900

Fig. 41. Constructional apraxia [from MacDonald Critchley]. Constructional apraxia in a patient of MacDonald Critchley with biparietal atrophy. The patient, a woman of 57 years, was instructed to lay a table for a meal for 3 people. Though not paralyzed, the patient was unable to carry out the task with which she was confronted. The cognitive failure caused the apraxia. As J. Lange pointed out, it is impossible to differentiate between the sequelae of agnosia and apraxia.

Fig. 42. 'Images', manuscript page of Sonata for Violin, Johann Sebastian Bach.

Acknowledgements

I am indebted to the following authors from whose writings I have quoted or whose illustrations I have reproduced: E. D. ADRIAN, A. BRODAL, J. CHUSID, MAC- DONALD CRITCHLEY, J. D. FRENCH, JOHN FULTON, W. HAYMAKER, R. HELD, H. W. MAGOUN, B. MATTHEWS, V. B. MOUNTCASTLE, W. PENFIELD, T. C. RUCH, A. SCHOTTE- LIUS, CHARLES SHERRINGTON and E. W. STRAUS.

References

ADRIAN, E. D.: The physical background of perception (Oxford University Press, London 1947).

BARD, PH.: Medical Physiology (Mosby, St. Louis 1956).

BORING, E. G.: Sensation and perception in the history of experimental psychology (Appleton-Century, New York 1942).

BORING, E. G.: A history of experimental psychology (Appleton-Century-Crofts, New York 1950).

BRODAL, A.: Neurological anatomy (Oxford University Press, London 1969).

CHUSID, J. G.: Correlative neuroanatomy and functional neurology (Lange, Los Altos 1970).

CLARA, M.: Das Nervensystem des Menschen (Johann Ambrosius Barth, Leipzig 1959).

CLARKE, E. and O'MALLEY, C. D.: The human brain and spinal cord (University of California Press, Berkeley 1968).

CRITCHLEY, MACD.: The body image in neurology. Lancet *1950:* 335-341.

CRITCHLEY, MACD.: The parietal lobes (Arnold, London 1953).

CRITCHLEY, MACD.: Clinical investigation of disease of the parietal lobes of the brain Med. Clin. N. Amer. *46:* 837-857 (1962).

DE REUCK, A. V. S. and KNIGHT, J.: Myotatic, kinesthetic and vestibular mechanisms. Ciba Symposium (Little, Brown, Boston 1967).

FRENCH, J. D.: The reticular formation. Sci. Amer. (May, 1957).

FULTON, J. F.: Physiology of the nervous system (Oxford University Press, London 1943).

GESCHWIND, N.: Disconnexion syndromes in animals and man. Brain *88:* 237-294, 585-644 (1965).

GESCHWIND, N.: Carl Wernicke, the Breslau school, and the history of aphasia; in Brain function, vol. 3 (University of California Press, Berkeley 1963).

GORDON-TAYLOR, Sir G. and WALLS, E. W.: Sir Charles Bell, his life and times (Living- stone, Edinburgh/London 1958).

HAYMAKER, W. and SCHILLER, F.: The founders of neurology (Thomas, Springfield 1970).

HEAD, H.: Studies in neurology, vol. 1-2 (Oxford University Press, London 1920).

HELD, R.: Plasticity in sensory-motor systems. Sci. Amer. *213:* 84-94 (1965).

HENSEL, H.: Allgemeine Sinnesphysiologie (Springer, Berlin/Heidelberg/New York 1966).

HERRNSTEIN, R. J. and BORING, E. G.: A source book in the history of psychology (Harvard University Press, Cambridge 1965).

JOKL, E.: Neurological case histories of two Olympic athletes. J. Amer. Med. Ass. 14 (1957).

JOKL, E.: Motor functions of the human brain; in 'Biomechanics II', pp. 1-27 (Karger, Basel 1971).

JOKL, E. and BOSHOFF, P. H.: Errors of refraction and visual efficiency. Acta med. orient. 6: 12 (1947).

JOKL, E. and SCHEPERS, G. H. W.: The support reaction and the central nervous control of progression. Sth afr. J. med. Sci. 38: 227-258 (1942).

LANGE, J.: Agnosien und Apraxien; in Handbuch der Neurologie, vol. 6, pp. 807-958 (1936).

LORENZEN, H.: Lehrbuch des Versehrtensports (Enke, Stuttgart 1961).

MAGOUN, H. W.: The waking brain (Thomas, Springfield, Ill., 1972).

MERTON, P. A.: Absence of conscious position sense in the human eye; in BENDER The oculomotor system (Hoeber, New York 1968).

MOUNTCASTLE, V. B.: Touch and kinesthesis; in Handbook of physiology, section I: Neurophysiology, vol. 1, pp. 387-430.

NIELSEN, J. M.: Agnosia, apraxia, aphasia. Their value in cerebral localization (Hoeber, New York 1936).

OLMSTED, J. M. D.: François Magendie (Schuman's, New York 1944).

PENFIELD, W.: The cerebral cortex of man (Macmillan, New York 1952).

PLESSNER, H.: Die Einheit der Sinne (Bonn 1923).

PLÜGGE, H.: Vom Spielraum des Leibes (Müller, Salzburg 1970).

RIESE, W.: A history of neurology (MD Publications, New York 1959).

RUCH, T. C. and FULTON, J. F.: Medical physiology and biophysics (Saunders, Philadelphia/London 1960).

SCHILDER, P.: The image and appearance of the human body (Kegan, London 1935).

SHERRINGTON, Ch.: Man on his nature (Cambridge University Press, London 1940).

SIGERIST, H. E.: Great doctors (Allen & Unwin, London 1933).

STRAUS, E. W.: The upright posture. Psychiat. Quart. 26: 529 (1952).

STRAUS, E. W.: Psychologie der menschlichen Welt (Springer, Berlin/Göttingen/Heidelberg 1960).

STRAUS, E. W.: Phenomenological psychology (Basic Books, New York 1966).

SUNDERLAND, S.: Nerves and nerve injuries (Williams & Wilkins, Baltimore 1968).

TEUBER, H.-L.: Wahrnehmung, Willkürbewegung und Gedächtnis: Grundfragen der Neuropsychologie. Studium Generale 22, pp. 1135-1178 (1969).

TUTTLE, W. W. and SCHOTTELIUS, A.: Textbook of physiology (Mosby, St. Louis 1969).

VINKEN, P. J. and BRUYN, G. W.: Disorders of speech, perception and symbolic behaviour; in Handbook of neurology, vol. 4 (North-Holland, Amsterdam 1969).

WOLSTENHOLME, G. E. W. and KNIGHT, J.: Homeostatic regulators (Churchill, London 1969).

Author's address: Prof. E. JOKL, University of Kentucky, *Lexington, KY-40506 (USA)*

Medicine and Sport, vol. 8: Biomechanics III, pp. 65–72 (Karger, Basel 1973)

Status Report on Biomechanics

J. WARTENWEILER

Eidgenössische Technische Hochschule, Abteilung X für Naturwissenschaften, Zurich

Biomechanics can be defined as the mechanics of living systems; to wit, in biomechanics the structure and function of living systems are investigated by using the knowledge and methods of mechanics.

In *biostatics* problems of static strain and tissue solidity are discussed. In *biodynamics* motion is investigated. *Biokinematics* is concerned with motion as a displacement of body parts, or of the whole body, without taking into consideration the forces that induce motion. *Biokinetics* refers to the forces that are to be found in statics and dynamics.

Which living systems should be the object of our studies? Our interest is principally directed at human beings; but to obtain a better insight into the human system, it is desirable to study structure and functions of animal systems in order to make a comparison, as is done in anatomy and physiology. Living systems are constructions of many organic systems, which will have to be studied for themselves and in their interaction. Biomechanics can, therefore, be classified into that of the locomotive, cardio-vascular, respiratory and other systems.

In the following I will describe biomechanics as it is understood in the International Seminars of the ICSPE. We restrict ourselves to biomechanics of the active and passive locomotive system, as well as to its pilot and regulation processes, mainly with regard to *human beings*.

Fundamental Research

Structure and Function of the Locomotive System

Structure and function of the locomotive system are made comprehensible by means of knowledge in the following fields: anatomy and physiology, anthropology, kybernetics and, of course, mechanics. Anatomy describes the structure and construction of the locomotive system, and gives us information about the rotation points in joints, as well as the leverage and extent of the various groups of muscles. We expect, especially from anthropology, data on body proportions. The indices of Fischer, Bernstein and Dempster about the distribution of mass and the position of the centre of gravity of the various body parts are well-known. But we must carry out our own measurements as soon as exact calculations are involved.

Forces of Motions

In biomechanics we can distinguish between inner and outer forces, i. e. forces originating within the body itself and forces like gravity and resistance which influence the body from the outside. Muscular force is the most important force, about which some details now follow.

Static muscular force is already quite thoroughly investigated. I should like to mention especially the catalogue of HETTINGEN and HOLLMANN [6] in the German periodical Sportarzt und Sportmedizin.

The dissertation of HAVENER [5], under the guidance of GROH, informs us exhaustively about the dependence of the muscular force on the physiological cross-section.

As is well-known, muscular force is also dependent upon the length of muscle in its various contracted phases, respectively; measurements of force will, therefore, have to be related to the length of the muscle.

The correlation between static and dynamic muscular force has been the object of studies since HILL [7], WILKIE [16] and FENN [4] as, for instance, those by ASSMUSSEN [1], BOUISSET [2], VREDENBREGT [13] and my team [14]. Static and dynamic force increase are being intensively studied by VREDENBREGT, HOERLER, NELSON, STOTHART, and others.

Speed, Acceleration
and Frequency of Motion

Speeds of motion – contrary to forces of motion – have not been the object of many studies. Above all, there is no review of maximal speeds classified by joints and body segments. We, for instance, have measured average maximum values of 14–20 rad/sec for rotations in the shoulder joint. ROESLER [10] indicates maximum values of about 40 rad/sec for movements of the hand and 20 rad/sec for forearm movements.

When researching the *movement frequency* the covered distance of the movement will have to be taken into account. We have measured, for instance, average frequencies of 6–7 Hz in forearm swinging with a declination of 10°.

Finally, the *acceleration* values are interesting. We have measured maximum values ranging from 50–100 m/sec² in pulling and pushing motions of the arms. The same motions performed with the legs resulted in maximum values ranging from 30 to 50 m/sec². PERTUZON and BOUISSET [9] indicate acceleration values in the vicinity of 100 rad/sec² for forearm rotation. Such indications are found very seldom in the literature. A more detailed catalogue is entirely wanting.

Course of Movement

The course of movement can be shown by film, photocyclography, stroboscopy and light traces, and can be recorded by characteristic curves which yield diagrams determining the position, speed and acceleration of various body parts. Pressure curves and EMG are also used.

It is the aim of biomechanics to ascertain all characteristic forms of movement and to explain them, if possible, from their causes.

Innervation and Control of Posture and Movement

Posture and movement are controlled by the central nervous system.

I should like to refer herewith to the chapter 'Integrative action of the nervous system' from the proceedings of our 2nd International Seminar in Eindhoven [11]. All regulative mechanisms of body movements can be regarded kybernetically. This is a wide area for future research which, in this field, is still in its initial stage. I am thinking in this connection of the chapter 'Muscular coordination' of the Eindhoven Seminar [12].

Psychological Aspects

Psychological terminology in biomechanics has caused many discussions as criteria are introduced like 'fear' and 'joy', which are primarily subjective.

It is very interesting for us to compare the objectively measurable movement values with subjective experiences.

Applied Science of Biomechanics

Biomechanics principally involves sports and physical education, manual labour, orthopedics and rehabilitation for its practical results.

1. In *sport* it is endeavored to obtain an increase of achievement. *Citius – altius – fortius!* Thus, biomechanics of sports investigates movements mainly with regard to improvements in technique and style.

2. *Physical education* has a greater ambition. Here the movement is evaluated as an expression of human personality and the intellectual, emotional and spiritual background is of very great importance. All these associations can also be researched on the biomechanical level.

The whole development of movements should thus be studied from the new-born to adulthood and should result in a typology of movement.

3. More studies should also be dedicated to the *artistic expressive movement*; be it in everyday gestures, in artistic dancing or in the playing of an instrument.

4. It is not so easy to clarify how up-to-date the study of *manual labour movements* is today. More and more machines are installed for workers and the problem is not so much physical overexertion as the lack of physical exertion.

5. *Human engineering* is a very interesting chapter with all its problems of the relationship between man and machine. There was a time when the machine laid down the rule and man had to adjust himself. Today the trend has been reversed. Biomechanics should help to clarify how machines can be adjusted to man.

6. Furthermore, *orthopedics and rehabilitation* are important practical fields of biomechanics. If artificial limbs should be fitted, the characteristics of moving with the sound ones should be well-known.

Biomechanics will thus have to give data about the pilot and regulative conditions. Knowledge of medicine and biomechanics is presumed in rehabilitation.

Theory of Biomechanics and Biomechanical Models

A theory will have to describe and order systematically the phenomena with which it is concerned.

Motion Categories

When taking into account the mechanical terminology, body movements can be divided into rotations, translations and torsions. The most important movements are rotations in the joint axes which are transverse to the limb axes. Throwing is mainly a rotation of the arm, so are beating and chopping. Translation movements are understood to be especially the pushing and pulling movements of arms and legs. Torque results in the longitudinal twisting of the spine.

Most movements exist in combinations. Walking, for instance, consists of translations of the legs combined with additional rotations in the hip joints, whereby the translations prevail in stepping, and the rotations prevail in the forward swinging of the legs.

There are other classifications of movements, which might be used as the case occurs.

Motoric Factors

A theory of movement performance will most especially deal with the following motoric factors, which determine basically the movements: static force, dynamic force, speed, frequency, equilibrium, precision, rhythm and co-ordination.

Principles of Motion

Every science aims at the summary of its knowledge in rules, laws and principles. In biomechanics it is not yet clear as to what can be defined as a rule, a law or a principle. ZEUNER [17] enumerates 5 laws of motion: the laws of economy, totality, rhythm, tripartition and the right input of force. HOCH-MUTH [8] also defines 5 biomechanical basic laws, which he calls principles: the principle of the initial force, the optimal trajectory of acceleration, the co-ordination of partial impulses, the reaction and the conservation of momentum.

We ourselves have chosen the expression 'kinetic rules' in our endeavour to summarize the conformities of force development in simple swinging movements [15]. We speak of a law of shifted phases, which says that in period-ically shifted phase movements the share of the counter-movement increases when the frequency is increased, opposite to the share of the comovement, which decreases.

As DONSKOI suggested [3], it would be justified to determine principles as inclusively as possible. To this end we should like to define rhythm as being a principle of movement of the first order.

Rhythm as a Principle of Motion

Rhythm belongs to the repeatedly discussed concepts in sports and physical education. It might be seen defined as: 'Rhythm is an ordering of time, a dynamic-energetic ordering, a changing of tension and relaxation, a periodical change.' Other authors, trying to give the psychological aspect, describe rhythm as an experience that escapes objective comprehension as, for instance, fear or joy. We should like to add to this that rhythm can, no doubt, be experienced but that this experience can be attributed to very concrete processes, which are determined by rhythm. As a result, it must be possible to classify rhythm objectively.

If repetition is to be stressed, rhythm could be defined as being periodical, i.e. as a repetition in regular time periods. But rhythm is not only dependent upon repetition. It can also be found in the process itself. Rhythm, then, is a regular variation, for instance a change of position, of speed, of force, of intensity. Sudden changes, interceptions or omissions interrupt rhythm or cancel it. Rhythm can thus be defined as being a regular variation that is repeated at regular periods of time.

Rhythm is most clearly manifested in motions. Here, the principle of rhythm determines the regular increase and decrease of speed. The path of the movement with congestions, i.e. sudden changes of speed, is not rhythmical. Certain requirements are expected from the acceleration or force, respectively, if the variation of speed should be regular. Acceleration – being the most sensitive measurement of rhythm – and its equivalent force must be apportioned, i.e. it may not start suddenly and halt abruptly or even change its sign.

The rhythm of a movement can be represented by a wave. A wave develops, for instance, from the time path diagram of a swinging movement. The amplitude represents here the covered distance of the motion. The wavelength represents the metrum, which decides the beat for the repetition. The frequency amounts to the total number of repetitions per time period. The relation of amplitude to wavelength is a measure for the intensity, i.e. for the force input, with which a movement is being performed. The quality of the rhythm can be seen from the trajectory of the curves: suddenly occurring changes in the acceleration are proof of deviants in movement.

Biomechanical Models

As biomechanics is a science still in its initial stage, we may not yet have too many expectations about mechanical and mathematical models. All-round models for the locomotive system do not exist. On the other hand, concepts of models to represent muscular contraction or kybernetic processes have worked out satisfactorily.

Learning Process

In biomechanics it is necessary to know the learning process, as biomechanics will not only ascertain the course of movements, but it also endeavors to determine optimal performance and is not only studying the movement learning process *per se*, but most especially with regard to the teaching methods of movement.

Instrumentation

A last word on instrumentation, before I conclude. Whereas, in the past, every laboratory of biomechanics produced its own instruments, some of these are on the market today. I should like to bring to your attention the exhibition which is held during this congress, and most especially the plate for measuring reaction forces from Kistler Instrumenten Co. CH-Winterthur. I shall be glad to give all desired information about a pulsating light-emitter device, electronic goniometer and accessory, differentiator, integrator of force and acceleration, which are being produced in my laboratory. As far as I know, no standard models for telemetry are available yet.

To be honest, there is still much research lacking in biomechanics. Some research has been done for years in various scientific disciplines, most of it in physiology. A systematic research of movement, however, has only recently been undertaken. We cannot assert from any movement that we understand its basis and there are almost no fundamentals for a theory of human movement. It is, therefore, of the greatest importance that we meet here in Rome to exchange results of researchers from all over the world.

References

ASMUSSEN, E.: The relation between isometric and dynamic strength in man. Bull. Féder. int. Educ. physique *1:* 9–19 (1967).

BOUISSET, S.; CNOCKAERT, J.C. et PERTUZON, E.: Sur la vitesse maximale de raccourcissement du muscle au cours d'un mouvement monoarticulaire simple. J. Physiol., Paris *58:* 474 (1966).

DONSKOI, D.: Bewegungsprinzipien der Biomechanik im Sport. Medicine and Sport, vol. 2, pp. 150–154 (Karger, Basel 1968).

FENN, W.O. and MARSH, B.S.: Muscular force at different speed of shortening. J. Physiol., Lond. *85:* 277–297 (1935).

HAVENER, M.: Zur Geschichte der Untersuchung über die absolute Muskelkraft. (Universitätsbibliothek, Saarbrücken 1970.)

HETTINGER, T. und HOLLMANN, W.: Dynamometrische Messungen am Muskel. Sportarzt Sportmed. *1:* 18–25 (1969).

HILL, A.V.: The heat of shortening and the dynamic constants of muscle. Proc. roy. Soc. B *126:* 136–195 (1938).

HOCHMUTH, G.: Biomechanik sportlicher Bewegungen (Sportverlag, Berlin 1967).

PERTUZON, E. and BOUISSET, S.: Instantaneous force-velocity relation in human muscle. Medicine and Sport (Karger, Basel 1973).

ROESLER, H.: Messungen zur Dynamik der oberen Extremitäten. Jb. dsch. Ver. Rehab. Behind. *70:* 260–262 (1969).

VREDENBREGT, J. and WARTENWEILER, J.: Biomechanics. 2. Medicine and Sport, vol. 6, pp. 30–55 (Karger, Basel 1971).

VREDENBREGT, J. and WARTENWEILER, J.: Biomechanics. 2. Medicine and Sport, vol. 6, pp. 62–83 (Karger, Basel 1971).

VREDENBREGT, J. and WESTHOFF, J.M.: The dynamic behaviour of the human muscle. Report No. 28 (Institut voor Perceptie Onderzoek, Eindhoven 1960).

WARTENWEILER, J.; STEINEMANN, M. und WETTSTEIN, A.: Über die Kraftentfaltung bei Kontraktionen der Skelettmuskulatur. Verh. schweiz. naturforsch. Ges. *142:* 138–140 (1962).

WARTENWEILER, J. and WETTSTEIN, A.: Basic kinetic rules for simple human movements. Medicine and Sport, vol. 6, pp. 134–145 (Karger, Basel 1971).

WILKIE, D.R.: The relation between force and velocity in human muscle. J. Physiol., Lond. *110:* 249–280 (1950).

ZEUNER, M.: Versuch einer Begründung der Leibeserziehung als Wissenschaft vom Begriff der Bewegung her. Leibesübungen *11:* 3–11 (1959).

Author's address: Dr. J. WARTENWEILER, Eidgenössische Technische Hochschule, Abteilung X für Naturwissenschaften, *CH-8000 Zürich* (Switzerland)

I. Methods

1. Data Processing

Medicine and Sport, vol. 8: Biomechanics III, pp. 74–83 (Karger, Basel 1973)

Poor Man's Graphics
Liberating a Line

R. GARRETT, T. BOARDMAN and GLADYS GARRETT[1]
Purdue University, Lafayette, Ind.

Introduction

In the past few years, knowledge of human movement and techniques for dealing with it have increased significantly. Research is continuing at a rapid rate to improve not only our empirical knowledge of human responses, but to develop mathematical models for predicting responses where empirical data is not available. We are now faced with the question of how we can possibly make use of the overwhelming amount of information that is being generated in our individual research and educational activities. We need a device to store large amounts of information (the results of previous empirical studies) and to analyze rapidly the mathematical models that we either have or are developing. We have such a device in the digital computer. The computer alone is not enough. We need a means of communicating with it in our own languages and especially in the most universal language of all—pictures. Communication in the visual mode is possible through the graphic computer terminal. It is the goal of this paper to discuss the application of these computer devices to the study of biomechanical problems.

There will be discussion in this paper on specific applications of the computer and its graphics terminal currently in progress at Purdue University in the areas of sport, physical education, rehabilitation and human motion. Further discussion will center on some general aspects of digital computers and graphic terminals such as their cost, availability and potential in the field of biomechanics.

Football Strategy Simulation

Sports tend to have a very clear set of rules and procedures which can be easily learned with little or no practical experience. The manual dexterity

1 The authors would like to thank 3 mechanical engineering students, PAT GERARDOT, RICK PUTNAM and JIM RUSSEY, for their programming efforts.

and reflexes required of various sports can be learned and practiced individually even in the absence of a competitive environment. Strategy, however,
cannot be taught—it must be the result of an actual competitive confrontation.
For games such as football, the strategem is difficult to teach and practice
since it requires the personnel, equipment and realistic environment in order
to simulate competitive conditions.

Computer simulation can provide a means of allowing an individual to
somewhat experience the competitive conditions necessary to experiment
with actual strategies. The basic procedure is as follows: the computer is
programmed with the statistics of actual previous football games and with
most of the rules and procedures of the game. Two individuals may compete
against each other by calling plays with the computer determining the out-
come of the plays based on actual statistics.

The computer is capable of prompting the players for a response, evaluating the response based on previous statistics and checking all facets of the
game such as rules, time expended and score. Some of the plays available are
indicated below by a single character:

```
r  =  run,
t  =  tricky run,
f  =  field goal,
p  =  punt,
s  =  short pass,
l  =  long pass.
```

Although this is a somewhat limited example because it was designed for
a standard teletype terminal, it does provide the basis for a more detailed
program involving graphics and the ability to interact with the computer to
adjust items such as defensive and offensive patterns. Application programs
such as these are being developed at Purdue through the undergraduate and
graduate project courses. Almost all applications such as the football program
give the student an excellent experience in modeling, programming, analysis
of data and interacting with other students in areas different from their own
in an interdisciplinary atmosphere.

Anatomical Analysis of Human Movement

One of the most interesting applications of the computer and computer
graphics to biomechanics is its use in kinesiological instruction. The following

Structure of human body (upper portion)

Segments	Bones	Joints	Action capabilities
Trunk	vertebral column (cervical thoracic lumbar sacrum coccyx)	spine	flexion extension lateral flexion rotation
Shoulder girdle	clavicle scapula	shoulder girdle	adduction abduction rotation (up + down) elevation depression
Upper arm	humerus	shoulder joint	adduction abduction flexion extension horizontal flexion horizontal extension rotation (in + out)
Forearm	radius ulna	elbow joint	flexion extension
Hand	Carpals metacarpals phalanges	wrist joint	flexion extension

Fig. 1. Teaching aids in structural kinesiology is a program designed to help students understand the science of human movement. The program gives the student the option of having the subject-matter presented to him or of being tested on it. If the system asks for a response but you are uncertain what the form of the response should be, type in the word 'help'. To get a definition of any of the kinesiological terms, type 'define'. Do you wish to be taught or tested? ? Taught. When you wish to continue, type 'continue'. If you want a term defined, type 'define'. ? Define flexion.

is an attempt to trace through the implementation of one computer-aided instruction system developed at Purdue designed to teach bone-muscle inter-action in the human body.

The first phase of the project was accomplished without graphic output so that a standard teletypewriter could be used by the student. The intent was to provide programmed instruction to supplement class lectures and labora-tories in the anatomical analysis of human movement. Figures 1, 2 and 3

Segments	Bones	Joints	Action capabilities
Pelvis (lower part of trunk)	pelvic girdle	hip joint	flexion extension abduction adduction rotation (in + out)
Thigh	femur	knee joint	flexion extension
Lower leg	tibia fibula		dorsal flexion plantar flexion
Foot	talus calcaneus metatarsals phalanges	ankle joint	adduction abduction eversion inversion

Fig. 2. Flexion-bending; bringing bones together as in elbow joint when hand is drawn to shoulder; exception, shoulder joint-movement of humerus to the front, a forward movement, also considered flexion. When you wish to continue, type 'continue'. If you want a term defined, type 'define'. ? Continue. Structure of human body (lower portion). When you wish to continue, type 'continue'. If you want a term defined, type 'define'. ? Continue. Which joint would you like to examine? (For a list of acceptable names for the joints, type 'help'.) ? Elbow.

Muscle	Bones of attachment	Principal action
Biceps brachii	scapula radius	flexion-forearm
Brachioradialis	humerus radius	flexion-forearm
Brachialis	humerus ulna scapula	flexion-forearm
Triceps	ulna humerus	extension-forearm

Fig. 3. Elbow joint. Which joint would you like to examine? (For a list of acceptable names for the joints, type 'help'.) ?

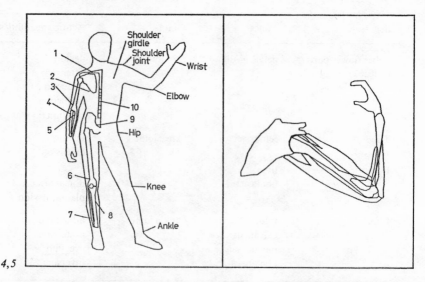

Fig. 4–5. Name the bones labeled above: (1) clavicle: *correct*; (2) humerus: *wrong, this is the scapula*; (3) humerus: *correct*; (4) ulna: *correct*; (5) radius: *correct*; (6) femur: *correct*; (7) fibula: *correct*; (8) tibia: *correct*, and (9) pelvis: *correct*. *You have 8 out of 9 correct.*

depict the terminology of the body structure and the body motion. The student is provided with the chance to be taught or tested by selection of the appropriate mode. A 'define' routine is available that will give the definitions of any of the terms used in the programmed instruction.

It soon became apparent that a nonpictoral, static study of human motion was of value only as a means of learning the vocabulary. The laboratory manual already used in class provided this type of information in even more depth. What was needed was a way for students to visualize living, moving anatomy so that analysis of gross motion of the body could be made readily. Textbooks, even cadavers, leave much to be desired in this most important aspect for physical educators. A computer-graphic display system might possibly prove an answer to this dilemma.

Information on the structure of the body including an illustration of a human figure (fig. 4) and on the terminology of joint action was stored in the system. The student has the options of being taught or tested as he interacts with the computer. In this paper the elbow joint was selected as an illustrative example of interacting with the system. As the student works with the light pen he can elicit the motion capabilities of the joint. Simultaneously he 'sees' the motion, what muscles are causing the motion and the terminology associ-

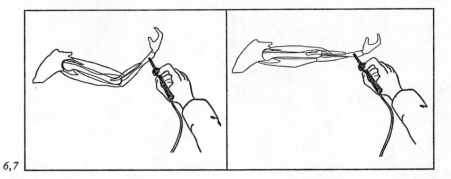

6,7

Fig. 6–7. Description see text.

ated with the action. Figures 5, 6 and 7 depict the actions of flexion and extension of the elbow joint.

This program is being developed specifically for the needs of women majors at Purdue, providing knowledge in gross analysis of movement.

Golf Swing Analysis

Many types of human motion involve the complex interrelation of several body segments over short time intervals. Difficulty often occurs in developing skills in these areas because it is virtually impossible to analyze problems sufficiently to determine which segments were making what contributions. Simulation of these types of motions using a digital computer offers a partial solution to this problem by allowing the student to vary individually the parameters which contribute to the motion and visualize the overall effect.

Perhaps in the hope of personal benefit, the writers have chosen to simulate the golf swing. Our goal is simple: determine all significant factors which contribute to the trajectory of the golf ball over which the golfer has control and permit him to visualize the results of varying these factors. The golf swing is actually an excellent example of complex, rapid human motion as it involves the interrelation of hands, wrists, forearm and upper arm, in addition to the general relation of these segments through the trunk and legs to the ground. Further, the typical swing takes less than 1 sec and results in the delivery of a peak force of 2,000 lb to the golf ball, sending it away at 100 mph.

The following parameters may be varied individually and the resulting motion studied: (1) club-type (No. 1 wood, or No. 5 iron, $\cdots$); (2) hindrance

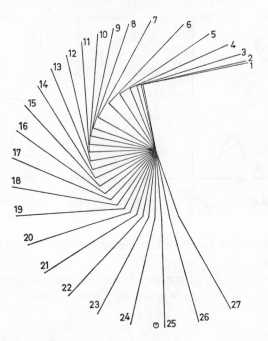

Fig. 8. Stroboscopic diagram. Club: No. 1 wood. Carry: 208 yd. Range: 221 yd. RN: 0.250.

parameter (a measure of the golfer's strength and ability to 'snap' wrists); (3) arm length and moment of inertia; (4) club weight, elasticity, flexibility and length; (5) angle of impact between club head and ball; (6) plane of impact, and (7) the downswing torque (an indication of how hard the golfer swings at the ball).

There are many possible forms of output from the analysis, but of most interest are the carry (distance traveled in the direction of the swing), range (total distance traveled), and the angle representing the degree of slice or hook. Also, a stroboscopic diagram of the downswing showing the dynamic relation of the arms to the wrists is of great interest. A typical output is shown in figure 8.

Speech Articulation Simulation

Another application of computers and graphic terminals is being accomplished in the area of rehabilitation of the physically handicapped. In

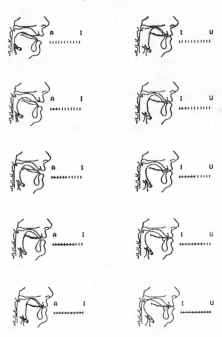

Fig. 9. Description see text.

many cases, a prime ingredient in the treatment of handicapped individuals is a comparison of the normal and abnormal motions; i.e. the ability to say 'this is what you *are* doing and this is what you *should be* doing'. The combination of modern data acquisition and measurement devices with the computer graphics terminal provide this ability.

This approach is being taken in the analysis and treatment of speech defects in conjunction with the Audiology and Speech Science Department at Purdue. Empirical data from normal subjects has been recorded by way of X-ray photography of the basic sounds. From this, a cross-section of the neck-head area can be drawn showing the relation of the tongue, lips, teeth and palate for each type of sound. The computer can be used to interpolate between these individual sounds yielding a 'talking' model of normal speech. A stroboscopic view of this is shown in the progression from *A* to *I* to *U* in figure 9.

The data acquisition procedure may then be repeated for individuals with speech defects so that 2 'talking' models may be viewed, side by side. The patient can then make a detailed, precise comparison between his formulation of sounds and what is considered normal.

Human Motion Simulation

Work is continuing on the motion analysis and simulation as reported in the Transactions of the 2nd International Seminar on Biomechanics entitled *Human motion: simulation and visualization*. The purpose of this research project is to enable the researcher to quantify movement with enough precision to accurately assess individual differences in the movements of small children. Graphical models of the body and body segments may be viewed along with the mathematical models of the motion data.

Availability of Equipment

The computer graphics field (at least graphics at a reasonable cost) is a relatively new field. For that reason we should like to present some facts about what hardware is available, its costs and some hints about what to look for in acquiring and using such devices.

Firstly, most of the applications described in this paper require access to a large-scale computer such as an IBM 360, UNIVAC 1108, CDC 6400 or GE 635. Most larger universities, industries and hospitals have access to such machines; those who do not should be able to access them through commercial time-sharing companies. An important consideration in the use of such large computers is that, for graphics work, transmission rates above 60 characters/sec (preferably 240–480 characters) are advisable. Slower rates often lead to intolerable delays in transferring pictures to the remote graphics terminal.

There are basically 2 types of low-cost ($10,000–20,000) graphics devices available today: the storage tube and refresh-type display terminals. The storage tube terminals are capable of very high resolution, accurate drawing on an 8 × 6 in. screen. As the name implies, the picture and associated text are stored on the screen as drawn and need not be continuously redrawn. The resolution of these devices is typically 170 lines/in. and virtually any amount of drawing may be done on the screen without flicker or blurring.

Storage tubes have two disadvantages. Firstly, they can display only static pictures. In order to make any change on the screen, the entire picture must be erased and redrawn. Secondly, since these devices have no local compute capability or core storage, all analysis and drawing must be done by an external computer. For some of the applications described in this paper, an ARDS storage tube terminal made by Computer Displays Inc. of Boston, Massachusetts, USA, was used.

The refresh-type displays permit dynamic pictures by including a local mini-computer in their configuration. The local mini-computer (generally a 4,000- to 8,000-word, 2 μsec or better cycle time, single register device) is capable of refreshing the image on the screen 30–60 times/sec. This permits selected erase and animated drawings in real time. Further, the mini-computer is often capable of performing many of the calculations involved in the analysis. For example, the entire football game and speech 'talking' model can be programmed in the local computer. The device used at Purdue University is an IMLAC made by IMLAC Corporation of Waltham, Massachusetts USA.

These devices are generally very reliable and no more difficult to operate than a movie camera or video tape system. However, they must be programmed to accomplish the desired analysis and display functions. The programming time for the projects described herein is between 3 and 6 man-months each and requires programmers somewhat familiar with the computer and knowledgeable of the application.

Conclusion

The entire concept of visual communication through computer graphics, enabling solution of entirely new classes of problems, is in tune with the new educational spirit. Students, teachers and researchers of today want to be involved in creativity and decision-making in an atmosphere that offers some degree of immediacy of response and results. The characteristics of this atmosphere must be such that these individuals can easily do productive work in a minimal amount of time. It is the writer's opinion that interaction of man and computer can provide a good foundation for this creative atmosphere.

Authors' address: RICHARD GARRETT, THOMAS BOARDMAN and GLADYS GARRETT, Purdue University, *Lafayette, Ind.* (USA)

Medicine and Sport, vol. 8: Biomechanics III, pp. 84–103 (Karger, Basel 1973)

Interactive Use of Laboratory Computer for Biomechanical Studies [1]

J. H. Milsum, R. E. Kearney and H. H. Kwee

Bio-Medical Engineering Unit, McGill University, Montreal

Biomechanical studies of neuromuscular control systems typically involve complex combinations of electromyographic (EMG) and kinesiological (position, velocity, acceleration and torque) signals. The synthesizing, processing and systems analyzing of these signals can be greatly facilitated by the interactive use of a computer, both on-line for control and display of experiment, and off-line for analysis. Fortunately computers designed with the laboratory environment in mind are now commercially available relatively cheaply ($20,000 upwards) (fig. 1). They have many distinct advantages in this role as compared with any of: (1) a collection of special-purpose devices for particular functions; (2) a conventional analog computer, and (3) a large centralized digital computer, even if the latter is connected to the laboratory by a remote terminal.

The major necessary features in such a laboratory computer typically include the following: (1) adequate speed (order of 1-μsec operations); (2) analog-digital multiplex channels for data input from the analog world of the experiment, and digital-analog conversion output channels; (3) oscilloscope for display of raw and processed results; (4) simple interface equipment for experimenter's interaction with the computer in modifying experimental conditions, parameters, etc., and (5) simple connection between experiment and computer, so that different experiments in the same lab may be serviced.

Although the basic system should be digital, hybridizing with an analog system can have real advantages in exploiting the ability of analog computers to solve differential equations fast. Further, innovations are occurring in

1   Presented at the 3rd International Seminar on Biomechanics, International Council on Sport and Physical Education, Rome, 27.9.71–1.10.71.

Fig. 1. Laboratory computer and experimental equipment.

hybrid computer technology, such as the invention of the hybrid multiplier (one analog input and one digital input with analog output, to be mentioned later) which allow better exploitation of the inherent advantage of the analog computer conferred by its 'parallel' mode of operation.

I. Experimental Control

In this section we are concerned with the interactive mode of experimentation in which the biomechanical experiment is conducted through an interface with the computer, as monitored or controlled by the experimenter. This improves his ability to design appropriate probing signals and to control their initiation according to present neuromuscular condition. We are doing this in our studies of neuromuscular control in the forearm and of the postural control system.

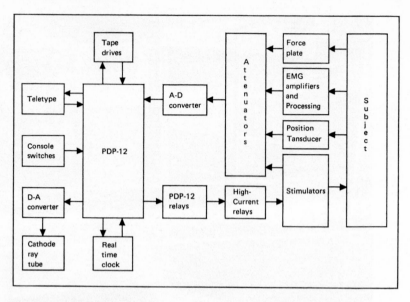

Fig. 2. Schematic of laboratory computer in postural control study.

A. Experimental Protocol

It is, of course, true that physiological experimentation in general, and biomechanical studies in particular, initially involve a lot of 'fishing' in order to zero in on what specific controlled experiment should be made, during which a small-enough number of parameters is changed so that the experimenter can meaningfully analyze the results. However, this 'fishing' is only a preliminary stage and, once completed, it should be possible to define an experimental protocol to be followed. It seems that there is great advantage in committing this experimental protocol to computer control. Among other things, this can tend to eliminate occasional experimental subjectivity which can creep in, although it still allows the experimenter the freedom to have all outputs displayed to him so that he can intervene when it is really necessary. But it tends to discourage him from casually chasing after some new aspect which occurs to him as the experiment proceeds. Of course, it may be necessary to do the latter, but it is usually better that this be in the form of a separate experiment.

B. Experimental Interaction

Given that the relevant data are being displayed online on the oscillo-scope or other display equipment, the experimenter is enabled to control the experiment in an analog way, for example by use of potentiometers on the digital computer control panel. This applies where active human interaction in the experimental protocol is desirable.

C. Incorporation of Hardware

Various special-purpose pieces of hardware, either commercially obtained or specially developed for the particular experiment, may be relatively easily incorporated into a laboratory oriented setup through the A–D and D–A channels. A typical setup is shown in figures 1, 2 and 3, in this case our postural control study [KEARNEY 1971].

D. Identification of Systems State

In many biomechanical studies, it may be important to ensure that the experimental data are obtained subsequent to a certain initial condition only. For example, in studying the postural control system there is a normal pos-tural tremor or sway, and if the response to a disturbing torque is to be studied, then it is important that the subject start from the same initial conditions each time. This can be done relatively easily through the setup described, in the sense that the appropriate state variables (here the position and angular velocity) can be monitored until they fall within the allowable initial condition range, at which point in time the computer can recognize this condition and initiate a stimulus onset.

E. Repeatable Control Stimuli

When the experiment is actually to be conducted by stimulating the sub-ject, it is of great advantage to be able to use repeatable 'random' (pseudo-random) signals, especially in view of the usual necessity for averaging a num-ber of responses in order to enhance signal-noise ratio. There is no difficulty in providing such patterns through the digital computer, whether they be the

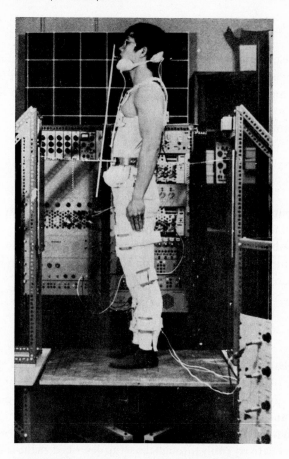

Fig. 3. Subject prepared for postural control experiment.

sums of sinusoids or other particular time patterns. Furthermore, the digital computer is capable of fast decision making in switching from one type of signal input to another, according to such criteria as described in section ID. In our studies of forearm movement control, a particularly useful signal has been a frequency modulated sinusoidal signal, approximately a random signal as far as the human operator perceives it. By using a voltage-controlled generator driven by a pattern stored in computer memory, we can smoothly insert short patches of constant frequency in an otherwise continuous 'noise' signal. Thus, although the subject may sometimes be on the threshold of recognizing that the signal has become a regular sinusoid, it has always blended back into

Fig.4. Human forearm control study.

being noise again before he has time to switch into any preprogrammed or predictive mode of control.

F. 'Negative Time'

In various transient kinesiological responses, it is clearly desirable to have a 'negative time' part of the record, namely that which occurs before the stimulus is actually applied. Now if the stimulus is under computer control as described in section I D, such that the triggering occurs at some non-predetermined instant, this part of the record before the stimulus occurs is effectively

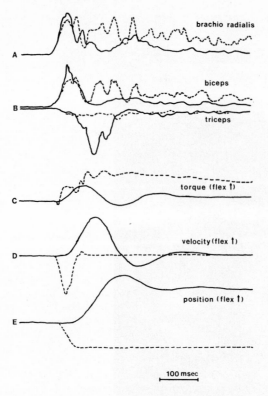

Fig. 5. Voluntary forearm movement, constrained after initation.

in 'negative time'. However, in the computer setup, the records for this 'negative time' can easily be obtained by continually updating with the most recent data, an appropriately long past history in computer storage, until such time as the trigger occurs to locate the pertinent 'zero' time of the particular experimental record. Our study of the forearm movement about the elbow with the setup shown in figure 4 exemplifies this [KWEE, 1971]. Specifically a subject has been asked to perform a maximally fast small-range flexion or extension at an arbitrary time of his own choosing. The records of his movement, both kinesiological and EMG, are to be studied in both constrained and unconstrained conditions (fig. 5).

In the 'constrained' condition, the electrohydraulic actuating system used, '*Swinger*' [KWEE and MILSUM, 1969], is switched under computer control from a particular loading impedance for which the subject has become conditioned, to a very different condition. For example, we typically use a high-compliance spring as load for the subject in the unconstrained condition, and either a rigid clamping or a fast ramp-and-hold movement for the constrained condition. Note that in order to trigger the switch from unconstrained to constrained condition accurately with respect to the subject's unannounced movement we can make use of the fact that the EMG signal starts to rise some 30 msec before any torque can be developed by its muscle. In turn, the increase of this EMG signal above some threshold can be used to trigger through computer control the imposition of the constrained stimuli, following a variable delay programmed by the experimenter.

II. Signal Processing

There is, of course, a wide variety of possible signal processing procedures that may be performed and here we only choose to mention two that have been of particular significance to us.

A. Coherent Averaging

Computers of average transients represented a great step forward for physiological studies when introduced about 10 years ago. Many are now available commercially as separate small oscilloscope-type packages, and include further features such as the ability to compute correlation functions. They all suffer from the two disadvantages that they are useful only for the particular limited number of programs for which they are designed and, secondly, that their averaging process almost always produces growing averages. In contrast, there is no difficulty in programming a general purpose laboratory-type digital computer for nongrowing averaging of successive transients to be super-imposed on each other [DEWHURST, 1967]. Furthermore, in view of the recognized difficulty that averaging of successive transients can in fact hide information which the experimenter ought to have that the system is nonstationary (that is, shifting its parameter values in time), such nonstationarity can always be monitored in the setup we are describing by showing each successive transient singly for comparison with the averaged transient to date. Furthermore, the variability inherent in responses can be checked by displaying such information as the mean absolute deviation of the response against time (see for example, figure 6).

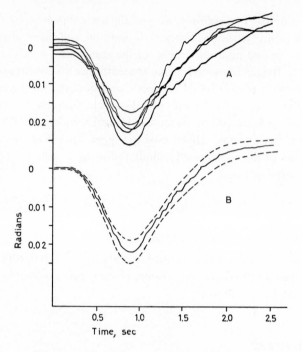

Radians

Fig. 6. Superimposed and averaged responsed in postural control.

B. Digital Filtering

Both the kinesiological and EMG variables typically need filtering to reduce noise. In the kinesiological case, signal-noise characteristics of goniometers and the available filtering, determine whether velocity and acceleration transducers should preferably be installed also. Certainly in our studies with *Swinger* we have found it desirable to incorporate separately position, velocity and acceleration transducers (fig. 7). For EMG variables, the 'noise' consists largely of the individual motor unit contributions, and for kinesiological use the filtering need is for envelope detection to reveal the underlying neural command signals. In this situation, the digital computer provides an unrivaled online ability provided that the data can be stored and processed with at least a small time delay. Specifically digital filters can be programmed which have symmetric impulse response (weighting function) characteristics. These have the highly desirable characteristic that they do not introduce any phase lag in processing the particular signal and, therefore, do not render such

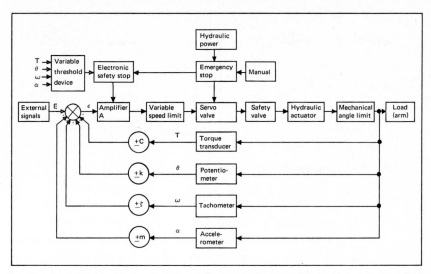

a

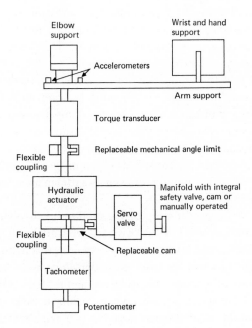

b

Fig. 7. Schematic of human forearm control study.

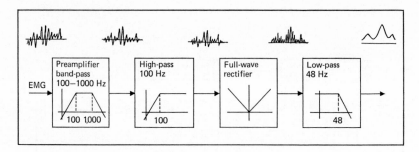

Fig.8. Analog filtering of EMG.

signals possibly misleading in comparing with other signals for time delays, etc. A specific example is in the generation of force-EMG relationships. Here the force signal itself may not need much filtering but the EMG must be rectified and analog filtered for envelope detection in the fairly well agreed way (fig. 8) [KWEE, 1971]. Then symmetrical digital filtering can be designed to be of low-pass or any other desired spectral shape so that the noise can be attenuated from the underlying signal very effectively, and as already noted, without introducing phase lag. Inasmuch as we want a discrimination ability in the order of 1 msec then this unique ability of digital filtering is especially important when trying to infer from the EMG pattern which different central nervous system neural loops may be contributing their effects.

It should also be noted that to obtain accurate force-EMG records in dynamic rather than static conditions, the same analog filtering must be applied to both channels. This problem is overcome by symmetric digital filtering in that the necessary filtering can be accomplished without phase lag in either channel and therefore each becomes directly comparable. Note that if such precautions are not taken then the force-EMG relationship will tend to be changed into a more nonlinear condition, or even away from an actual linear condition [CHENG and MILSUM, 1968].

Figure 9 shows the overall process for a typical digital filtering involving signal-noise improvement [MILSUM, 1971]. Also shown are arbitrary but typical signal wave-forms and spectra, and filter frequency responses throughout the process. The 'analog filter' box will be discussed under *aliassing* below.

A particularly simple class of symmetrical low-pass digital filters (DF) for separating signal from noise is the 3-point one producing a weighted average, y_n, at time $t_0 + nT$, of the original signal x_n and its two neighbouring samples x_{n-1} and x_{n+1}, viz

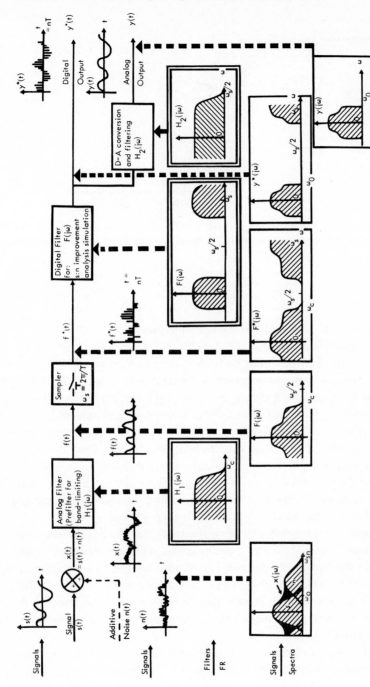

Fig. 9. Generalized schematic for digital filtering of analog signal.

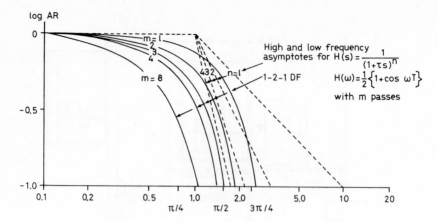

Fig. 10. Frequency responses of 1–2–1 digital filter and low-pass analog filters.

$$y_n = \frac{1}{a+2}(x_{n-1} + a\,x_n + x_{n+1}),\qquad\qquad (1)$$

where t_0 is time 'zero' and T is the sampling time.

The particular case for $a = 2$ yields a DF with the 'best' frequency response characteristic in this class [WILCOX and KIRSNER, 1969] giving approximately a 12 dB/octave cut-off in the frequency range of interest, and a half-power point (3 dB down) at $\omega_b = 1.14/T$ (fig. 10) from WILCOX and KIRSNER [1969]. Note also that this filter is very simple to implement digitally, requiring only adding and shifting operations, and very little computer time. Finally there is the advantage that the experimenter may carry out more and more heavy filtering by repeated passes with this filter until he feels that he has achieved the best discrimination. The figure indicates the filtering result for various values of m, the number of passes, and the above reference gives a graph determining the effective cut-off frequency which results.

The maximum number of successive trials that can be performed in each experiment, under stationary conditions, is often smaller than would be desirable statistically. Under these conditions coherent averaging may still leave much variability (high-frequency 'noise'). Then the ability of the human observer to interpret the results may be facilitated by the low-pass digital filtering (smoothing) just described. All these manipulations can be done 'on-line' if desired.

Multiple passing through a simple filter constitutes multiple self-convolution, and the net result is the equivalent of a single pass through a much more

Given $\quad h_1(mT) = \dfrac{1}{4}\left\{\delta(|m+1|T) + 2\delta(mT) + \delta(|m-1|T)\right\}$

$= \text{'1-2-1'}$

1 2 1	1 2 1	1 2 1	1 2 1	1 2 1
1 2 1	1 2 1	1 2 1	1 2 1	1 2 1
1	2+2	1+4+1	2+2	1
1	4	6	4	1

$h_2(mT) = \dfrac{1}{16}\left\{\delta(|m+2|T) + 4\delta(|m+1|T) + 6\delta(mT) + 4\delta(|m-1|T) + \delta(|m-2|T)\right\}$

$= \text{'1-4-6-4-1'}$

Digital convolution

```
                    1
                  1   1
h1(mT) ──▸       1   2   1            (x1/4)
               1   3   3   1
h2(mT) ──▸    1   4   6   4   1       (x1/16)
            1  5  10  10   5   1
h3(mT) ──▸  1  6  15  20  15   6   1  (x1/64)
          1  7 21  35  35  21  7  1
h4(mT) ──▸ 1 8 28  56  70  56  28 8 1 (x1/256)
```

Pascal's triangles

Fig. 11. Pascal's Triangles and Digital Filter (DF) weighting functions.

complicated DF. Interestingly, for the case a = 2 in equation 1, the equivalent single-pass filter is given directly by the alternate rows of the classical array called Pascal's triangles (fig. 11).

Clearly much computational effort is involved in these higher-order DF, but fortunately the effort can be reduced through a different technique if it becomes costly. For this we pass from the nonrecursive (nonautoregressive) class of DF (in which the above filters belong) to the class of recursive (auto-

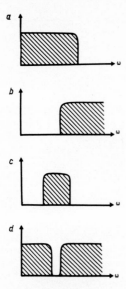

Fig. 12. Some classes of filters. *a.* Low-pass – cutting – off high frequencies, sharply, after some cut-off frequency, ω_c. *b.* High-pass – cutting-off low frequencies, sharply, before some cut-off frequency, ω_c. *c.* Band – pass – a combination of the above so that only a band of frequencies is passed. *d.* Notch rejection – the inverse of band-pass, as for example in rejecting the power line frequency of 60 Hz.

regressive) DF. These provide a very flexible and efficient capability, for example in designing such various classes of DF as: low-pass, high-pass, band-pass and notch-rejection filters (fig. 12). As one example, the recursion formula for a particular 'triangular' low-pass DF is [LYNN, 1970]:

$$y_n = -y_{n-2} + 2y_{n-1} + x_{n-12} - 2x_{n-1} + x_{n+10}. \tag{2}$$

It is called 'triangular' because the weighting function (or averaging process) on the original data points would be of triangular pattern involving the 10 neighbouring data points on either side of the current point. Thus the non-recursive filter would involve 21 terms to be manipulated for each data point (fig. 13), whereas the recursive filter only involves 5, and with small integral coefficients.

The frequency response of this DF is also very interesting. It should first be emphasized that all digital processing of data carries the risk of *aliassing*, or inherently degrading the data unless the sampling rate is adequate for the bandwidth of the signal. Specifically Shannon's *sampling theorem* requires that there be *at least two samples per cycle of the highest frequency present in*

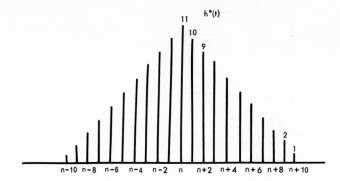

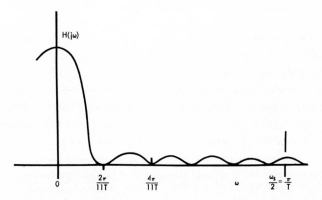

Fig. 13. Weighting function and frequency response of triangular filter.

the signal (plus its noise if of high frequency), if aliassing, which involves a loss of information present in the original signal, is to be avoided. In the frequency response curve of figure 13, the value $\omega_s/2 = \pi/T$ (where T is the sampling time) represents this maximum frequency permitted in the signal (ω_s is called the sampling frequency). Hence, the chosen form of this DF frequency response implies that there is much noise in the range

$$\frac{2\pi}{11T} < \omega < \frac{\pi}{2T}$$

which is to be filtered from the 'true' signal lying mainly in the range

$$0 < \omega < \frac{2\pi}{11T}.$$

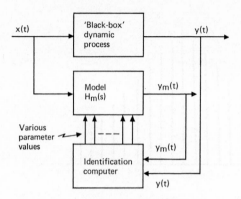

Fig. 14. Conceptual configuration for identification.

It is very important to note that if the desired (or even the maximum permitted) sampling rate does not satisfy the sampling theorem because of high frequencies in the signal (whether or not this be noise), then the only solution is to use analog prefiltering to reduce the high frequencies 'adequately', as implied in figure 9. The criterion for 'adequate' is rather arbitrary here because no real filters have infinite cut-off rates, and hence some residual levels of signal amplitude will persist at high frequencies. In practice if the spectrum can be attenuated to about 1 % or less of its value in the frequency range of importance, then the resulting aliassing will not be significant.

III. Identification of Dynamic Systems

The identification of physiological systems often takes the form of developing mathematical models which describe the system behavior. The model-making process can be simplified and speeded up by the use of a digital computer operating in a hybrid manner with an analog computer.

The analog machine is typically used to implement most of the system model, and in any case that part which can be described by linear differential equations and by those non linearities which are relatively simple to fulfil in an analog fashion (limiting, multiplication, squaring, etc.). Since the analog machine operates in a parallel manner the speed at which this part of the model can be simulated is much greater than possible with digital simulations. Typical solution times for a transient response of a system, of whatever complexity, can be of 1-msec order.

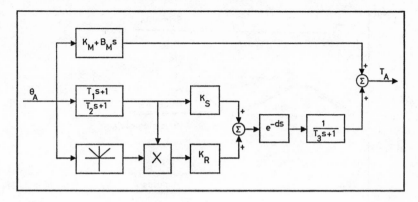

Fig. 15. The gain-control model of the PCS.

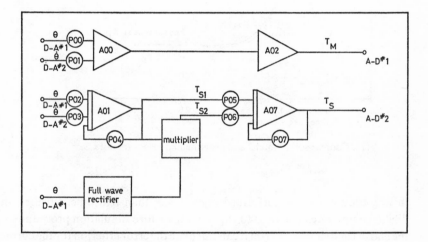

Fig. 16. Analog Patching of figure 15.

On the other hand, the digital computer has a number of important roles to play in this hybrid computation. First the use of A–D and D–A converters makes possible the simulation of functions which are very difficult to implement in a purely analog manner (notably time delays, functions of more than one variable, stiction, etc.). Secondly, if the actual experimental data have been stored digitally, they may be used as inputs for the simulation. Then the model response may be compared directly to the known system response, as already stored digitally (fig. 14). Since it is not necessary to do this on-line

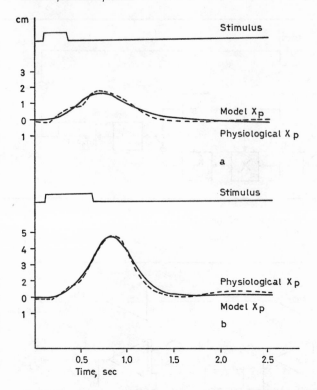

Fig. 17. Comparison of model and experimental responses in PCS.

this procedure may be performed much faster than real time. Further, the digital computer can be used to supervise the entire simulation procedure. At the end of each simulation run it can calculate the error criterion for the current model and then through the use of some type of hill-climbing technique determine the new parameter values to be used. If some type digital attenuator is available (like the hybrid multiplier already mentioned, or servo-set potentiometers) the parameter adjustment can be done digitally and the entire procedure automated. This is especially useful in cases where the number of parameters to be determined is so large that the experimenter loses his 'feel' for the system. Indeed it is impractical, and probably impossible, for a human to control a general hill-climbing procedure involving more than about 5 parameters. Further, the number of runs required in such situations becomes so large that it would be very tedious for the human, but easily implemented automatically.

We have used the manual method of parameter adjustment to best fit a model of the postural control system in an exoskeletally restrained healthy human. This present model includes a gain term which is controlled by magnitude of lean, in order to simulate the observed responses to anterior pulses of torque adequately [KEARNEY, 1971]. Figure 15 shows the block diagram, figure 16 the analog computer patching diagram, and figure 17 the typical matching obtained between model and experimental data, when the parameters are 'best' adjusted. Note that the number of parameters actually adjusted in figure 15 was effectively only 3 or 4. The error criterion was the integrated magnitude of error between experimental and model response normalized with respect to experimental response magnitude.

References

CHENG, M. and MILSUM, J.H.: Dynamic relationship between isometric force and EMG of human muscles, Digest 2nd Canadian Med. Biol. Engin. Conf., (1968).

DEWHURST, D.J.: Neuromuscular control system, Trans. Eng. in Med. and Biol. IEEE *14:* 167–171 (1967).

KEARNEY, R.E.: Modelling the postural control system of the exoskeletally restrained human. M. Eng. thesis, Biomedical Engineering Unit, McGill University, Montreal (1971).

KWEE, H.H.: Neuromuscular control of human forearm movement studied with active dynamic loading. Ph.D. thesis, Biomedical Engineering Unit, McGill University (1971).

KWEE, H.H. and MILSUM, J.H.: A high-performance electro-hydraulic actuator system for neuromuscular control studies. Proc. 8th ICMBE and 22nd ACEMB (1969).

LYNN, P.A.: Autoregressive digital filters Rep. No. 1 (Imperial College of Science and Technology, London 1970).

MILSUM, J.H.: Digital filtering and systems simulation. Unpublished lecture notes, McGill University, Montreal (1971).

WILCOCK, A.H. and KIRSNER, R.L.G.: A digital filter for biological data, Med. biol. Engin. *7:* 653–660 (1969).

Authors' addresses: J.H. MILSUM, Director, Division of Health Systems, Health Sciences Center, University of British Columbia, *Vancouver*; R.E. KEARNEY and H.H. KWEE, Bio-Medical Engineering Unit, McGill University, P.O. Box 6070, *Montreal 101* (Canada)

Medicine and Sport, vol. 8: Biomechanics III, pp. 104–107 (Karger, Basel 1973)

Treatment of Biomechanical Information by Electronic Computer

A. VAIN

Research Laboratory of Muscular Activity, Tartu State University, Tartu

I. Introduction

Several biomechanical data, which characterize body movements in space and time, are used for the analysis of sports technique. These data can be achieved by 2 methods: (1) direct recording of the body movements, and (2) calculation of the unrecordable parameters.

The calculation method permits determining the biomechanics of separate points of the body and also of the whole body. The direct recording method, however, only gives data about the movement of the points of the body that are connected with the apparatus.

The calculation method requires a mathematical model of the human movements. Mathematical models for the analysis of the less complicated movements are described by FISCHER [4] and BERNSTEIN [1], and computer programmes for this analysis are compiled by DUSHKOV [2] and VAIN [5]. The initial data required for the models can be obtained by very complicated methods. That is why the precision of the initial data is usually not high [3]. Hence, the usage of the calculation method requires the determination of the precision of the initial data and the adequacy of the mathematical model.

A. Recording of Primary Data

Kinematography was used to record high-speed movements. A tripod-mounted 35-mm camera was placed perpendicular to the line of motion. Marks were placed on the center of the joints and on the locations of the centers of parts of the body. Electromyography (EMG) was performed with skin electrodes. Accelerogram was taken by means of a mechanical accelerograph which was fixed to the back of the subject. The separation mechanism inserted into camera guaranteed synchronization of all the data on the recording paper in an oscillograph and on the accelerogram. A print was used to note down the coordinates of the locations of the centers of the parts of the body and the axes of the body segments. The results obtained were entered on a table by the following data arrays:

1. The constants comprise the number of the experiment, total number of frames, coefficient of the scale, total mass of the body and number of frames per second.
2. Constants of the accelerometer.
3. Coefficents of the mass of the parts of the body.
4. Lengths of the body parts (segments).
5. Coordinates of the registered points.
6. Data of the accelerometer.
7. The recorded action potential of muscles.

B. Procedure of Computation

A computer programme was composed on the grounds of the mathematical model of body movements.
1. Computation of biomechanical data (table I).
2. The precision of the measurement of initial data is estimated by scattering of the distance between 2 fixed points.
3. To estimate the adequacy of the model (E) and the data registered by the accelerograph (O) we used the formula

$$M = \Sigma \, | \, E_i - O_i \, | \, / (NO_i).$$

4. Recording of the biomechanical data and EMG were made on a magnetic tape. The programme used allowed arrangement of all data in groups of persons and calculation of the biomechanic parameters for time moments, corresponding to every shot of the kinematograms.
5. Statistical procedure of the biomechanical parameters, EMG, physical capacities, anthropometric measurements and sports results were done using general methods.

The described method was used in analyzing some exercises of gymnastics and athletics. Results have shown that the precision of the measurements of initial coordinates was 15.8 ± 2.5 mm and the error (M) of the adequacy of the mathematical model was 10–20%.

Table I

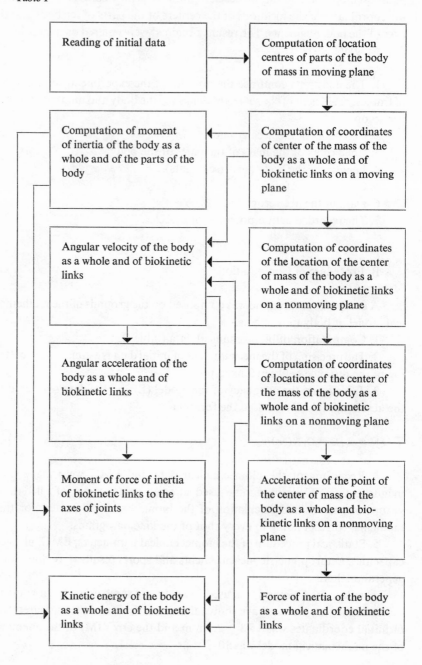

References

1   BERNSTEIN, N. A.: Physiology of movements and physiology of activity (in Russian),
     pp. 39–44 (Medicine, Moscow 1966).
2   DUSHKOV, B. A.: Muscular activity in space flight and in isolated chambers (in Rus-
     sian), pp. 35–44 (Medicine, Moscow 1969).
3   FEDERLE, S.; SCHILLE, D. und TRACHE, L.: Probleme der Messgenauigkeit bei kine-
     matographischen Analysen. Theorie und Praxis der Körperkultur, vol. 18 p. 152 (Sport-
     verlag, Berlin 1969).
4   FISCHER, O.: Theoretische Grundlagen für eine Mechanik der lebenden Körper (mit
     speziellen Anwendungen auf den Menschen sowie auf einige Bewegungsvorgänge an
     Maschinen) (Teubner, Leipzig 1906).
5   VAIN, A.: Biomechanics of a free moving body. Diss. Tartu (in Estonian) (1969).

     Author's address: Dr. A. VAIN, Tartu State University, Research Laboratory of Mus-
cular Activity, Ylikooli 18, 202 400 *Tartu, ESSR* (USSR)

Medicine and Sport, vol. 8: Biomechanics III, pp. 108–115 (Karger, Basel 1973)

Data Modeling Techniques
in Cinematographic-Biomechanical Analysis

Carol J. Widule and D. C. Gossard

Department of Physical Education for Women, Purdue University, West Lafayette, Ind., and Department of Mechanical Engineering, Massachusetts Institute of Technology, Cambridge, Mass.

I. Introduction

Measurement from film, as in other forms of measurement, is subject to error. Because error is magnified when deriving estimates of velocity and acceleration from the basic positional data, attention should be given to either obtaining reliable estimates of position or to obtaining reliable estimates of the derivatives. Three procedures for obtaining estimates of motion characteristics based on kinematographic data are reviewed.

A. Averaging [6, 9]

From a film of a subject executing a standing long jump, the position of the center of gravity of the body was determined for 18 frames during the flight phase – take-off to landing (fig. 1). Using the take-off position and each subsequent experimental point to calculate the initial angle of projection and the initial velocity produced differing results. The true path of the center of gravity of the body, during flight can be estimated using the average initial angle and average initial velocity and is shown in figure 2 in relation to the experimental points. The application of this procedure is limited to modeling the motion of a gravity accelerated projectile.

B. Least Squares [2, 7, 9]

Least squares is a method for determining the constants in the equation of a curve based upon the principle that the line of best fit is obtained when

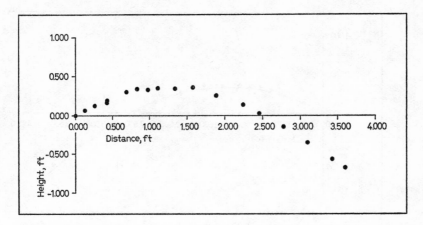

Fig. 1. Measured positions of the center of gravity of the body during flight of a standing long jump. [from fig. 1 in WIDULE and GOSSARD, [9].]

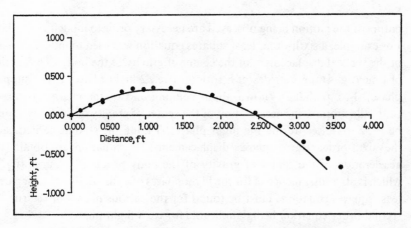

Fig. 2. Trajectory of the center of gravity of the body in the standing long jump – measured (*) *versus* average theoretical (—). [from fig. 2 in WIDULE and GOSSARD, [9].]

the sum of the squares of the deviations from the line and the experimental points is a minimum.

The method of least squares may be used to model any motion of the body. In the case of gravity accelerated motion, a second-order least squares equation is appropriate (as shown in fig. 3 in relation to the same experimental points calculated for fig. 1). However, judgments concerning the

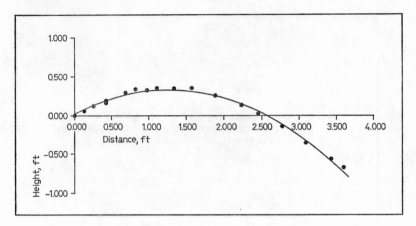

Fig. 3. Method of least squares (—) applied to measured positions (*) of the center of gravity of the body. [from fig. 4 in WIDULE and GOSSARD (9)].

nature of the motion being observed are necessary prior to modeling the data. For example, a sixth-order least squares equation was used to model the path of the vertical displacement of the center of gravity of the body as a function of time (fig. 4) for 2 strides of a sprinter (i.e., 2 flight phases and 1 support phase). From what is known about running and of the characteristics of projectile motion, the acceleration of the center of gravity of the body should be − 32 ft/sec/sec during the flight phases. Taking the second derivative of the sixth-order least squares displacement-time equation to obtain the acceleration of the center of gravity of the body produced a curve (fig. 5) which is obviously incorrect for the flight phases of a run. In this case, separate least squares equations must be found for the various phases of the run – a second-order equation being applicable for each flight phase.

C. Finite Differences [1,3,5,8,9]

The finite difference method is used to approximate the slope of a line which is tangent at a point on a curve. If, for example, the experimental data are expressed in the form of displacement with respect to time, the finite difference method permits an evaluation of velocity and acceleration at a point. This evaluation is based on the values of points symmetrical about a given point. Depending upon the degree of precision desired, any number of

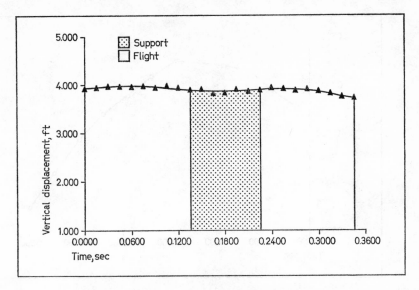

Fig.4. Vertical displacement of the center of gravity of the body as a function of time for a sprint run – measured (+) *versus* sixth-order least squares equation (—).

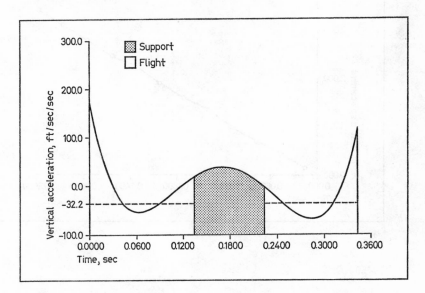

Fig.5. Vertical acceleration for a sprint run based on the second derivative of the vertical displacement-time least squares equation (—) *versus* actual acceleration of center of gravity of the body the during flight (–––).

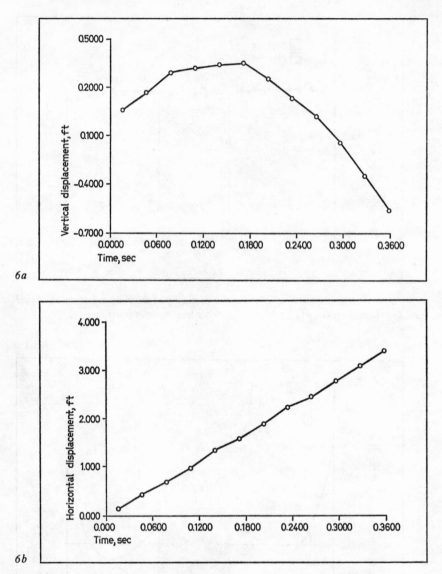

Fig.6. Application of the first central finite difference equation for velocity (*c* and *d*) to standing long jump displacement data (*a* and *b*). [from fig. 6 in WIDULE and GOSSARD (9)].

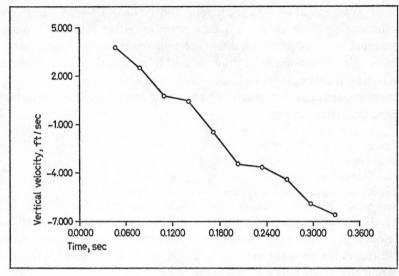

6c

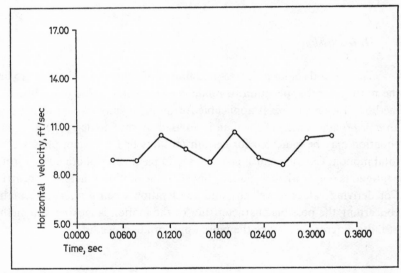

6d

Fig. 6c–d

points symmetrical about a given point may be used. Usually, a first central difference equation involving 1 data point on either side of the point in question, or a second central difference equation involving 2 data points on either side of the point in question, is used to approximate the derivative. If y_i is defined as the point in question, numerical values of a function $y(t)$ may be determined at uniform time intervals, h time units apart, from the following finite difference equations.

First central:

Velocity: $y'_i = (y_{i+1} - y_{i-1})/2h$.

Acceleration: $y''_i = (y_{i-1} - 2y_i + y_{i+1})/h^2$.

Second central:

Velocity: $y'_i = (y_{i-2} - 8y_{i-1} + 8y_{i+1} - y_{i+2})/12h$.

Acceleration: $y''_i = (-y_{i-2} + 16y_{i-1} - 30y_i + 16y_{i+1} - y_{i+2})/12h^2$.

In figure 6 the application of the first central finite difference equation for velocity (c and d) to the standing long jump displacement data (a and b) is illustrated. Six positions were omitted from the data illustrated in figure 1 to permit equal time intervals between each position.

II. Discussion

The methods of least squares and finite differences would appear to have the most general application to modeling kinematographic data. While the method of least squares is applicable for modeling any motion of the body, an *a priori* assessment of the data is advised. A specific degree least squares equation may be indicated for certain motions or for certain phases of the total motion. Care must also be exercised in passing a smooth curve through contiguous points when a sudden change in the motion is actually occurring. For deriving values of velocity and acceleration when *a priori* information concerning the possible characteristics of the motion is lacking, the method of finite differences may be the more appropriate procedure.

Summary

Kinematographic measurement, as in other forms of measurement, seldom provides an exact description of reality. Because error is magnified in the process of taking a derivative, particular attention should be given to the procedures for obtaining estimates of velocity and acceleration from displacement data obtained from film. Three procedures for modeling kinematographic data are reviewed: (1) averaging; (2) least squares, and (3) finite difference method for differentiation.

References

1   CRANDALL, S.H.: Engineering analysis, pp.154–159 (McGraw-Hill, New York 1956).
2   DILLMAN, C.J.: Kinetic analysis of running. Proc. CIC Symp. on Biomechanics, Bloomington, Indiana 1970, pp.137–165 (The Athletic Institute, Chicago 1971).
3   GARRETT, G.E.: Methodology for assessing individuality. A computerized approach; Ph.D. diss. Purdue (1970).
4   GARRETT, G.E.; REED, W.S.; WIDULE, C.J., and GARRETT, R.E.: Human motion. Simulation and visualization. Medicine and Sport, vol.6, Biomechanics II, (Karger, Basel 1971).
5   GARRETT, G.E.; REED, W.S.; WIDULE, C.J., and GARRETT, R.E.: Human movement via computer graphics. Convention Amer. Assoc. Health Phys. Educ. and Rec., Boston 1969.
6   GARRETT, R.E.; WIDULE, C.J., and GARRETT, G.E.: Computer-aided analysis of human motion. Kinesiol. Rev. 1–4 (1968).
7   PLAGENHOEF, S.C.: Computer programs for obtaining kinetic data on human movement. J. Biomech. *1*: 221–234 (1968).
8   SHAFFER, B.W. and KRAUSE, I.: Refinement of finite difference calculations in kinematic analysis. Annual Meet. Amer. Soc. of Mech. Eng. Paper No. 59-A-53 (1959).
9   WIDULE, C.J. and GOSSARD, D.C.: Data modeling techniques in cinematographic research. Res. Quart. Amer. Ass. Hlth Phys. Educ. *42*: 103–111 (1971).

Authors' addresses: Dr. CAROL J. WIDULE, Department of Physical Education for Women, Memorial Gymnasium, Purdue University, *West Lafayette, IN 47907*; Dr. DAVID C. GOSSARD, Department of Mechanical Engineering, Massachusetts Institute of Technology, *Cambridge, MA 02142* (USA)

Medicine and Sport, vol. 8: Biomechanics III, pp. 116–119 (Karger, Basel 1973)

Computer Simulation of Springboard Diving

Doris I. Miller

School of Physical Education, University of Saskatchewan, Saskatoon

The computer simulation technique of studying human motion in sport has not been widely used to date. It is anticipated, however, that this method will be employed with increasing frequency as its capabilities and potential are more fully appreciated. Admittedly, the computer cannot replicate all facets of human performance exactly. Although a number of simplifying assumptions must be incorporated into any workable simulation model, valuable insights concerning the operation of the system can still be gained. The construction of the model itself may provide a deeper understanding of the mechanics of the performance since all aspects must be carefully defined and translated into mathematical terms. Once an adequate and valid model is available, the particular sports skill may be investigated under strictly controlled conditions and the influence and interrelationships of the movement components can be studied. In this paper, a digital computer model capable of simulating the flight portion of a dive will be described and examples of the application of this model will be cited.

Description of the Model

The construction of the simulation model of a diver required that all phases of the action be expressed numerically. Therefore, to facilitate this process, it was decided to restrict the model to the flight or free-fall phase of nontwisting dives performed in the pike and layout positions.

The first step in the construction was to obtain a mathematical description of the body of the diver. This was accomplished by modifying Hanavan's computerized model [1964] into a 4-segment representation consisting of a combined head and trunk, legs, right arm and left arm. The masses, centers of mass, principal moments of inertia and lengths of each of these rigid, geometrically shaped segments were defined by a series of anthropometric measurements of the diver's body. This feature made it possible to provide a somewhat 'personalized' representation of a specific diver.

The principle of the conservation of angular momentum, in conjunction with the body segment parameter data, was utilized to determine the orientation of the diver. Since air resistance was considered negligible, the angular momentum of the diver about his mass center remained unchanged throughout the flight phase of the dive. The complexity of the equations of motion derived from this relationship was a function of the number of body segments. For this reason, the body of the diver was represented by only a 4-link system.

The path of the diver's mass center was established by the application of Newton's laws of motion. This information was related to the location of the mass center within the body to determine the position of the diver throughout the flight.

The equations governing the diver's body segment parameters and his translation and rotation during a nontwisting dive were programmed for solution on a digital computer. Input to this program included the anthropometric measurements of the diver, the initial conditions of the dive (take-off velocity and trunk angle at take-off) and its boundary values (the displacement and velocity of the arms and legs, and the angular momentum of the diver with respect to his mass center). The computer then processed this information and supplied numerical output completely defining the orientation and location of the diver as a function of time. These data, in turn, were channeled into a computer graphics program which provided a 3-dimensional representation of the simulated dive.

Applications of the Simulation Model

This computer model, which has been validated, can be utilized to invesetigate the influence of any of the input variables upon the total performanceof the dive. The interrelationships of these variables can also be studied. It would be possible, for example, to examine the effect of different body proportions upon a dive. A preliminary investigation with 8 university athletes, 2 basketball players and 6 divers, provided quantitative data on their moments of inertia about the horizontal spinning axis through their centers of mass. Two quasi-rigid layout positions were used. In the first, the arms were extended above the head; in the second, the arms remained at the sides. The model showed that the moments of inertia of the basketball players were 25–92% greater than those of the divers (table I). While it would be generally acknowledged that the divers would have a smaller resistance to angular acceleration, quantitative evidence of the magnitude of this resistance and its effect upon performance is not often available.

The simulation model could also assist in providing answers to such questions as: 'How much would the performance of a dive be effected if the action of the arms was altered?' Figure 1 shows the results of 2 different patterns of lateral arm displacement in a forward dive layout. In the first case,

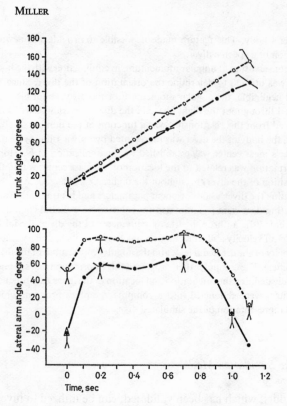

Fig. 1. Influence of arm action upon trunk angle in a forward dive layout. ○-----○ = dive 1; ●——● = dive 2.

the arms were moved to a position approximately 90° from vertical. In the second, they were displaced only about 65°. All other facets of the dives were the same. The simulations indicated that the greater lateral arm displacement was accompanied by an increased rotation of the trunk as compared to the more restricted arm action. By the end of 0.5 sec, the difference in trunk angles was 12.38° and this increased to 22.76° at the completion of one second of the flight.

Conclusion

The model which has been described is capable of providing both numerical information on simulated dives as well as pictorial evidence through the use of 3-dimensional computer graphics routines. The influence of any of the input variables can be investigated either

Table I. Moments of inertia of the body about the spinning axis in 2 quasi-rigid layout positions

Subject No.	Height, in	Weight, lb	Varsity sport	Moment of inertia (slug-feet-feet)	
				position 1[1]	position 2[2]
1	75.94	196.25	basketball	16.01	10.90
2	79.34	194.80	basketball	17.98	12.15
3	70.17	162.00	diving	11.74	8.02
4	71.33	165.07	diving	12.84	8.61
5	65.80	153.20	diving	9.73	6.64
6	68.81	153.33	diving	9.57	6.33
7	69.63	144.80	diving	10.14	6.98
8	68.36	147.37	diving	10.57	7.30

1 Body in an erect position with the arms extended beyond the head.
2 Body in an erect position with the arms by the sides.

singly or in combination. This would include body proportions, angular displacements and velocities of the arms and legs, angular momentum of the system, take-off angle and take-off velocity. The potential of the computer simulation type of research can thus be appreciated. The greatest requirement of the diving model at present is for more descriptive data to define the approximate ranges of the input variables for different types of dives. These data would provide a sound basis for simulation studies of springboard diving.

References

HANAVAN, E.P.: A mathematical model of the human body. AMRL technical report No. 64–102 (Wright-Patterson Air Force Base, Ohio 1964).

MILLER, D.I.: A computer simulation model of the airborne phase of diving. Physical education diss., Pennsylvania State University (1970).

Author's address: Dr. DORIS I. MILLER, School of Physical Education, University of Saskatchewan, *Saskatoon* (Canada)

2. Motion Systems and Models

Medicine and Sport, vol. 8: Biomechanics III, pp. 120–123 (Karger, Basel 1973)

Biomechanics as the Science of the Coordination of the Movements of Man

L. W. Chkhaidze

Sportcommittee, Moscow

Biomechanics was formerly regarded as one of the methods for studying the way in which living beings (principally man) conform to the laws of mechanics when performing voluntary movements. This method was applied in the nineteenth and early twentieth century by all researchers investigating the movements of man (Marey, Fick, Demery, Braune, Fischer, Steinhausen, Drill). As a result, the works of these and other researchers contained nothing but bare facts [Braune and Fischer], or dealt with the solution of purely practical problems [Taylor, Gilbert]. All this was undoubtedly useful (for example, when searching for more rational methods of executing this or that movement, e.g. the labour movements of man); it failed, however, to bring about a 'revolution' in the development of biomechanics, which for a long time remained an accessory, confined to accumulating factual material for other, more established sciences, such as anatomy, physiology, the theory of physical training, methods of perfecting labour movements, etc. The above circumstance was also conditioned by the fact that, at that time, the subject of biomechanics was rather indeterminate.

This situation prevailed until the thirties of the current century; more exactly, until the Soviet school of biomechanics [Bernstein, 1936], had firmly established the following 2 facts.

1. The acceleration of moving sections of the body executing voluntary movements is invariably in functional dependence upon the activity of the central nervous system.

2. A relevant analysis (e.g., of parametric graphs of the acceleration of body sections) enables the researcher to determine this or that form of the coordinating activity of the central nervous system, e.g. either as a direct spontaneous innervational impulse accounting for the movement of the

section in space, or as a response to reaction forces caused by external factors.

This seems to prove that biomechanics of the movements of man and animals does not merely reflect the coordinating activity of the central nervous system, but it enables the researcher to form a judgement as to the principles and the structure of movement coordination as a whole.

Consequently, as far back as 1940, the Soviet school of biomechanics had clear-cut views on movement coordination in man, views which anticipated by a number of years the statements and views of the now universally recognized cybernetics.

According to these views, movement coordination should be regarded as an hierarchical system of ring coordination in which every ring controls a definite part of a movement; the higher this ring, the more important part of a movement is regulated by it. Thus, the highest ring controls the activity of definite muscular synergidae (there may be several such rings: N. A. BERN-STEIN counted no less than 5).

It is self-evident that each of the rings acts according to a certain programme, the formation of which is the most important part of movement training, while concrete central impulses reach the muscular periphery through general direct channels of communication (the impulses of the higher ring being differentiated by those of the lower one).

The most important part of coordination begins after the excitation of the relevant muscular fibres sets the osteo-muscular periphery in motion: external receptors accumulate information as to how the target-aimed part of the movement in question is proceeding, while the proprio-receptors accumulate information on the fulfilment of the programme of muscular interaction. This information is independently fed back to the analyzing mechanisms which compare the factual execution of the given movement with that which had been programmed and, if necessary, make the required corrections.

N. A. BERNSTEIN and his school succeeded in proving their theory relating to the system of movement coordination as described above, not only through numerous facts of a purely biomechanical order and differential equations of movement, but also through the philogenesis and ontogenesis of man's central nervous system, as well as through the evolution of coordinated movement.

The practical value of this theory is indubitable: it helps solve the cardinal problem – how to proceed in evolving the principles of training in unfamiliar skills, whether in work or in sports. It is sufficient to form programmes in all the rings of coordination and to 'launch' these rings. In the highest ring, this programme is formed rather easily with the participation of external receptors (through demonstration, reading, description and the like); as for the inner receptors, it is necessary that the movement be executed in this or that form, otherwise the osteo-muscular periphery, connected as it is with the system of proprio-receptors, will not be brought into operation. This usually constitutes the major difficulty in any form of training in motor skills, as the first execution of the movement in question is effected without any pre-set programme, but in conformity with the biomechanical expedience of the movement being mastered.

The quest for a way out of this impasse is still going on; the Soviet school of biomechanics has discovered methods of obtaining information as to how the biomechanical part of the movement is being executed that enables the researcher to control the dynamic resultativeness of the movement practically immediately upon execution; this, in its turn, serves to determine the correctness of the initial execution of the programme of the internal rings and to introduce the necessary corrections, thus helping to considerably shorten the time required for training in motor skills and achieving markedly better results.

Thus, the general theory of movement coordination evolved on the basis of biomechanical laws has not only proved to be valid, but has mapped out a path for further research.

At the same time, a historical review of the development of allied sciences dealing with the principles of movement coordination shows that these sciences have not yet created a theory explaining all the details of movement (mechanical, mathematical, anatomical, physiological, etc.). Anatomy has not even attempted to create such a science, limiting itself exclusively to the anatomical characteristics of movements of body sections and resulting displacements; physiologists have also failed to propose a general theory of the coordination of voluntary movements, disregarding, as they usually did, almost all of mechanics and, in particular, the mathematical foundations of movement. It is only biomechanics that is able to provide a satisfactory answer to all the questions that have arisen, because it takes everything into account: the structure of the motor apparatus, physiological functions, the laws of mechanics and the mathematical interrelationships involved.

In the light of what has been stated above, we would suggest that at present the general theory of movement coordination in man (and in the higher animals) should be evolved by the combined efforts of several sciences: the structure of the motor apparatus – by anatomy; the vital functions – by physiology; the general principles of control – by cybernetics, and so forth; but biomechanics alone is in a position to combine all these data into an integral whole on the basis of mathematics and mechanics.

In conclusion, we would like to observe that such a theory is immediately connected with not only purely applied disciplines, such as methods of training in this or that motor skill, but also with many other problems. For example, psychology turns out to be directly dependent upon the activity of the higher rings, medicine is dependent on the lower ones, methods of movement training upon the formation of programmes, and the evolutions of the animal world may be followed through the philogenesis of movement coordination, etc.

It follows that biomechanics is one of the contemporary exact sciences that came into being at the junction of several others; but it will finally come into its own if its subject is clearly defined – which it can only be as movement coordination.

Author's address: Prof. L.W. CHKHAIDZE, Zkatertny Perevlox 4, Sportcommittee, *69 Moscow* (USSR)

Medicine and Sport, vol. 8: Biomechanics III, pp. 124–128 (Karger, Basel 1973)

Control Exercised over the Reconstruction of the Movements' System

D. DONSKOI

Central Institute for Physical Culture, Moscow

Modern biomechanics has emerged as a science which treats active human and animal movements and positions. The subject-matter of bio-mechanics should relate in principle to the *origin* and all *manifestations* of mechanical movements in animals. A vast number of movements which serve to make motor actions is combined into systems of movements. Hence, while studying motor actions, it is always necessary to take into account some *system* and *structural features* of movements [DONSKOI, 1968].

The major task of biological mechanics in human movements is to examine mechanical and biological factors which cause movements, some specific features of movements and their efficiency. Eventually, the aim of biomechanical experiments will be to improve human motor actions.

Movements in sport can be improved by the successful solution of the following problems. These include: (1) some *requirements* to execute move-ments in a rational manner and to improve athletes' techniques (these require-ments depend on some biomechanical specific features with due regard to psychological factors), (2) some concrete problems solved to improve various functions of the motor apparatus, which is a *biomechanical system*. This can be done on the basis of requirements used to improve athletes' techniques, and (3) *reconstruction* of the movements system by improving the potential of the motor apparatus.

The logical aspects of improved movements are quite elementary, but the experiments staged to improve athletes' movements entail a number of difficulties. As in any complicated controllable system it is desirable to reduce, on the one hand, the number of the characteristics to be checked and con-trolled and maintain the relative autonomy of subsystems and, on the other hand, it is rather difficult to decide on the characteristics and deter-

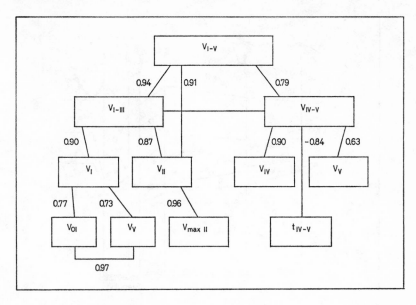

Fig.1. Kinematic correlational model of the ski sliding step. V = phase velocity of the leg.

mine them in quantitative terms according to their actual variability and relationship.

The system of movements can be improved only after laying its foundations. Training results in the formation of the major structure of movements system which is a sufficiently consistent and objective interaction of system components [DONSKOI, 1961]. The structure of this system determines the progress of inherent processes, their interaction with the environment, emergence of new properties and some potential for further development of the system. The system of movements can be improved further by its reconstruction, which is intended to differentiate further between various components, establish more intricate interrelations within their structure [BERNSTEIN, 1967] and change the role of forces in the movements system as well as their value, position and time of application.

To control the reconstruction of the system of movements it is necessary, first and foremost, to obtain some information on the *basic characteristics* used to improve various movements and on the interrelations among movement components. While discovering some structural interrelations, researchers now develop new *models* for the system of movements. These models prove helpful in evolving a *hypothesis* on the mechanism of movements which

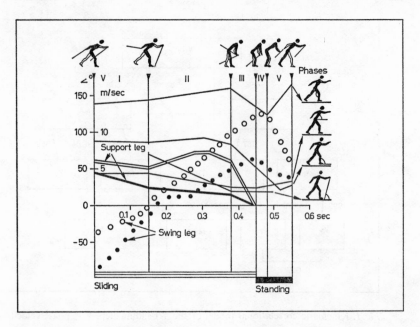

Fig. 2. Standard for rational techniques of the ski sliding step and its efficiency. (The characteristics to be checked are indicated by thin lines and those to be controlled, i.e. the speed of the leg, are indicated by thick lines and circles.)

is an aggregate of motor processes with their mechanical and biological nature. The validity of this hypothesis is verified in training when athletes improve their physical abilities and professional skills.

While experimenting, we applied a *system analysis* which helped us determine some components of the movements system including phases of movements and elementary actions. To this purpose we took advantage of quantitative biomechanical characteristics which made it possible to discern the boundaries of phases as momenta showing some considerable changes in movements and in their particular aims. At the same time we also carried out a *system synthesis* which was intended to determine some structural inter-actions in the system of movements including intercommunication between various subsystems and their relative autonomy. To this purpose we compared the characteristics of various components in the system of movements and their relationship with the total result of the action.

In cycle locomotion (ski-racing) Soviet researchers have developed a structural model of a sliding step. Figure 1 shows only part of this model (speed correlation according to step phases) while the whole model includes

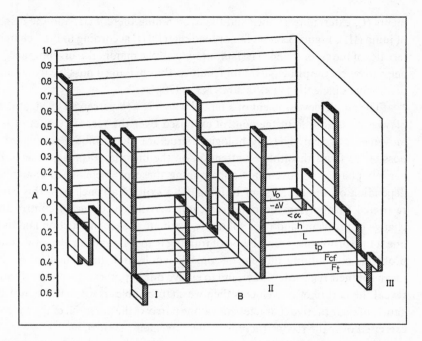

Fig. 3. Rotated factor loading (orthogonal rotation). The structure of the second step in the triple jump. A = factor-loading. B = characteristics.

the correlation of space and time characteristics. Figure 1 clearly shows the leading role of the maximum speed in the second phase (sliding with the help of ski-sticks) $V_{max}II$. In accordance with this structure GROSS and DONSKOI [1971] established some requirements for the system of movements. On the basis of these requirements these researchers have managed, through retrieval, sampling and retention of information, to develop a modified structure of the movements system. Figure 2 shows a number of requirements for the rational system of movements in dynamics (phases I–V), according to some of the controllable space characteristics. This figure also shows the changing dynamics of the controllable characteristics which, in this case, is the speed gathered by the supporting and swinging legs of a top-class and a poorly-trained skier, respectively. The trajectory made by the legs of the top-class skier is shown in figure 2 by the open lines and circles (= and ○) while that of the poorly-trained skier is depicted by a heavy line and solid circles (– and ●).

In noncycle locomotion (triple jump) our collaborator, Y. VERKHOSHAN-SKY, established with the help of factor analysis the kinetic and dynamic structure of movements, and some values of such factors as speed of movements (I), external resistance and the overcoming thereof (II) and the height of jump (III), Figure 3 shows factor loadings (I–III) according to the 8 characteristics of the second step. The factor model of the major take-off components helps to reveal the principal motor faculties, thus making it possible to develop them more efficiently [Verkhoshansky, 1970].

The ample development of a model which is well perceived by an athlete plays a decisive role in the control exercised over the reconstruction of the movements system. The psychological structure of the motor act makes it possible to focus the athlete's attention on the most important problems. In various phases of movements and training these problems keep changing, depending on the progress of training. The skill of a coach is in this case, restricted to applying the objective characteristics in order to control the change of problems and their solutions, estimate the reconstruction and outline future problems. This method of training helps one gain an idea of various problems confronting movements on the whole and, in particular, of the means of their realization. If we fail to evolve our hypothesis theoretically and reveal the nature of movements then we shall be able to see that the investigation of quantitative characteristics alone narrows the potential of the control exercised over the reconstruction of the movements system.

Thus, system and structural research methods in modern biomechanics provide an ample understanding of complicated motor activity and of the control exercised over its improvement.

References

Bernstein, N.: The co-ordination and regulation of movements (Pergamon Press, London 1967).
Donskoi, D.: Biomechanik der Körperübungen (Sportverlag, Berlin 1961).
Donskoi, D.: Bewegungsprinzipien der Biomechanik im Sport. 1st Int. Seminar of Biomechanics, Zurich (Karger, Basel 1968).
Gross, H. i Donskoi, D.: Tekhnika lyzhnika gonschika (Ski-racing techniques), Russian ed. (FiS, Moskva 1971).
Verkhoshansky, Y.: Osnovy spetsialnoi silovoi podgotovki v sporte (Special muscle strength training in sport), Russian ed. (FiS, Moskva 1970).

Author's address: Prof. D. Donskoi, Central Institute for Physical Culture, ul. Kazakova 18, *Moskow K-64* (USSR)

Medicine and Sport, vol. 8: Biomechanics III, pp. 129–133 (Karger, Basel 1973)

The Significance of the Stabilizing Function in the Process of Controlling the Muscle Groups of Upper Extremities

K. Fidelus

Academy of Physical Education, Warsaw

The aim of this work is to formulate and to verify the hypotheses with regard to some principles on the muscle control by the nervous system.

Basic Definitions and Hypotheses

We shall introduce the following definitions concerning the muscle-joint system. By *muscle acton* we understand the muscle, its part or head whose fibers are turned in the same direction in relation to the axis of rotation of the joints. The muscle fibers of the acton perform the same functions. By *function of the muscle acton* we understand the positive and negative components of the torques, which an acton can develop about the axes of rotation of the joints, above which it passes. From the point of view of control, the functions of the muscle acton can be divided into stabilizing and motive. By *motive functions* of the muscle acton we understand the components of the torque, acting upon these degrees of freedom of the joints, which are effected by all forces independant from the nervous system, i.e. mainly external and inertia forces. By *stabilizing functions* of the muscle acton we understand the components of the torque which are balanced by the functions of the antagonistic muscle actons: in other words, the components of forces dependent on the control process of the nervous system [2].

Let us assume

$$\overline{M}_m = \overline{M}'_a + \overline{M}''_a - \overline{M}_s.$$

Where $\overline{M}_m$ = sum of torques developed by all the muscles, $\overline{M}'_a = \overline{M}_e$ = sum of the torques developed by the antagonistic actons in relation to the balanced external torque $\overline{M}_e$, $\overline{M}'_a = \overline{M}''_s$ = sum of the torques developed by antagonistic muscle actons, balancing the torques of synergists $\overline{M}_s$. M'_a values balancing M_e perform the rôle of the motive functions. M_s values which demand development of the additional torques of the M''_a muscles, stand for stabilizing functions.

By *geometric parameters* of the muscle actons we understand the physiological cross-section area, the arm of the force of the acton and the length of its fibers. These last two parameters change as the angles change of the joints over which the actons pass. The geometrical parameters of the actons have essential influence on the magnitude of developed torques [1, 2]. The greater the geometric parameters of the muscle actons, at the same magnitude of the nervous stimulus, the greater is the torque. An adequate distribution of stimulation of different actons can economize the muscle effort, ensuring at the same time necessary value of the torque, developed in the joint. Our hypotheses assume the optimization of the control program according to the criterion of minimization of the energy losses. This may occur when the magnitude of the stimulation of the muscle acton is the greater. The less stabilizing functions the muscle acton is developing in a given motion, the greater the values of the geometric parameters of the given muscle. These two hypotheses do not change the fact that the magnitude of the stimulation is proportional to the force developed by a muscle.

Results of Investigation

Investigations of the muscles of upper extremities were carried out according to the method described [1–4]. The value of the muscle tension was determined by means of the coefficient U_i/U_{imax}, where U_i is instantenous

Table I. Correlation coefficients for investigations of various joints

Muscles			M_{max}		$0.5\,M_{max}$		$0.25\,M_{max}$	
			PG	n_{stab}	PG	n_{stab}	PG	n_{stab}
Joints	elbow	extensors	0.5	0.5	−0.5	1.0	−0.5	1.0
		flexors	−0.6	0.9	0.2	0.7	−0.1	0.9
	wrist	extensors	0.4	0.6	0.2	0.8	0.4	0.8
		flexors	−0.8	0.8	−0.8	0.8	−0.8	0.8

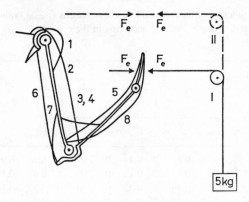

Fig. 1. Position of investigated muscles of the upper extremity. 1 = m. deltoideus, pars anterior (DPA); 2 = m. biceps brachii, caput breve (BBCB); 3 = m. brachialis (B); 4 = m. pronator teres (PT); 5 = m. flexor digitorum superficialis (FDS); 6 = m. triceps brachii, caput longum (TBC long); 7 = m. triceps brachii, caput laterale (TBC lat.); 8 = m. extensor digitorum (ED). F_e = external loading, 5 kg; I = position beneath the axis of rotation of the shoulder joint; II = position above the axis of rotation of the shoulder joint.

value of the integrated voltage of EMG of i-th muscle, and U_{imax} is maximal value of U_i under conditions of a given experiment. Value of geometric parameters of muscle actons and the number of their functions were determined on cadavers, models and living individuals. The rank coefficient of the correlation of Spearman (ρ) was used for determining the dependence between the state of stimulation and investigated parameters of the muscle.

The first investigations deal with separate measurements of U_i for a group of flexors and extensors of the elbow and wrist joint, while developing the maximal torques M_{max}, $0.5\ M_{max}$, and $0.25\ M_{max}$ of the external loading. The value of the correlation coefficient is given in table I. These data bring us to the conclusion that the values of the geometric parameters (PG) of actons do not effect essentially the magnitude of the selective stimulation of the muscles by the nervous system. But the dependence can be observed between the magnitude of the selective stimulation of the muscle and its number of stabilizing functions (n_{stab}).

In further investigations we tried to prove the influence of the number of stabilizing functions on the level of stimulation of the muscle. The investigated person held a 5-kg weight in his hand in a standing position. The arm was hanging loose down the trunk and the forearm was flexed at the elbow joint at an angle of 90°. Six variants were investigated with a change of pronation

Table II. Number of acton functions (nf) their stabilizing functions (n_{stab}), values U_i/U_{imax} and the rank correlation coefficients (ρ) occurring in the course of pushing and pulling of the 5-kg weight

Muscles	nf	No. of exercise												ρ
		1		2		3		4		5		6		
		n_{stab}	U	n_{stab}	U	n_{stab}	U	n_{stab}	U	n_{stab}	U	n_{stab}	U	
DPA	3	1	0.2	0	0.4	1	0.3	0	0.4	0	0.6	1	0.1	0.900
BBCB	4	3	0.12	2	0.12	2	0.2	2	0.12	0	1.0	1	0.92	0.886
B	1	1	0.27	1	0.13	1	0.36	1	0.23	0	0.64	0	0.32	0.800
PT	2	1	0.5	1	0.3	2	0.05	2	0.05	0	1.0	0	0.8	0.972
FDS	6	3	0.36	3	0.32	6	0.1	6	0.05	1	0.82	1	0.9	0.947
TBC long.	3	1	0.25	2	0.2	0	0.5	1	0.2	3	0.1	3	0.05	0.943
TBC lat.	1	0	0.67	0	0.55	0	1.0	0	0.61	1	0.1	1	0.05	0.843
ED	7	6	0.31	6	0.5	2	1.0	2	1.0	7	0.25	7	0.25	0.986

DPA = m. deltoideus, pars anterior; BBCB = m. biceps brachii, caput breve; B = m. brachialis; PT = m. pronator teres; FDS = m. flexor digitorum superficialis; TBC long. = triceps brachii, caput longum; TBC lat. = m. triceps brachii, caput laterale; ED = m. extensor digitorum.

and supination of the forearm. The tension of the following flexors was investigated: m. biceps brachii, caput breve (BBCB), and m. pronator teres (PT) as a synergists of flexion and antagonists of pronation and supination of the forearm. The value of the correlation coefficient for BBCB was 0.843 and for PT 0.90. Both values bring us to the conclusion that there is a significant correlation among correlated values. The third set of investigations concerned the horizontal pushing or pulling of the 5-kg weight with the arm. The schematic view of the investigation stand is presented in figure 1. Uni- and multifunctional muscles were investigated at the constant external loading force, developing variable torques in statics. The following variants were investigated.

1. Pushing of the weight with pronating hand taking the position above the axis of rotation of the shoulder joint.

2. As in item 1, but with pushing performed beneath the axis of rotation of the shoulder joint.

3. Hand in supination, pushing the weight with the hand held above the axis of the shoulder joint.

4. As in item 3, but with pushing performed beneath the axis of the shoulder joint.

5. Hand in supination, pulling performed with hand held above the axis of rotation of the shoulder joint.

6. As in item 5, but pulling performed beneath the axis of rotation of the shoulder joint.

The results of investigation are given in table II. All correlation coefficients are statistically significant.

Conclusion

In the course of investigation of one joint, the change of muscle tension was found, measured by means of EMG, though the external torque was the same for a given joint. Further investigations brought us to the conclusion that the selective change of stimulation of muscles can be effected by geometric as well as by structural parameters of muscles.

By analyzing the functions of muscles in neighbouring joints and their relation to external forces the idea of motive and stabilizing functions was introduced. Two hypotheses concerning the criterion of variable magnitude of muscle stimulation by the nervous system were verified by means of experiments and statistical correlation analysis.

The less stabilizing functions (which caused an additional action of antagonistic muscles) that develop in a given muscle, the greater is the stimulation of the muscle from the nervous system. It is in accordance with a criterion of minimization of energy output and it decides the increase of the energetic efficiency of man. It is likely that the process of learning skills, in the course of which the stimulation of some muscles diminishes, depends on the same principle.

References

1   FIDELUS, K.: Some biomechanical principles of muscle cooperation in the upper extremities. Biomechanics, vol. I. 1st Int. Seminar, Zurich 1967, pp. 172–177 (Karger, Basel 1968).

2   FIDELUS, K.: Biomechaniczne parametry układu ruchu kończyn górnych człowieka. Roczniki Naukowe AWF (PWN, Warsaw, 1971).

3   MORECKI, A.; FIDELUS, K. i EKIEL, J.: Bionika ruchu (PWN, Warsaw 1971).

4   Морецкий. А.; Зкель, Ю.; Фиделюс, К., И Назарчук, К.: Исследование взаимного участия мышц в движениях верхних конечностей. Биофизика 13: 306–312 (1968).

Author's address: Dr. K. FIDELUS, Academy of Physical Education, Marymoncka Street 34, *Warsaw* (Poland)

Medicine and Sport, vol. 8: Biomechanics III, pp. 134–137 (Karger, Basel 1973)

Mechanical Functioning System of the Human Body

P. Jenik and W. Rohmert

Lehrstuhl und Institut für Arbeitswissenschaft, Technische Hochschule Darmstadt

The organising principle for various forms of mechanical behaviour of the human body is the relation between load and strain. There are different kinds of load to which the corresponding kinds of strain correlate. The mechanical load factors are to be classified in kinematic and dynamic and in qualitative and quantitative factors. By combining the individual mechanical load factors it is possible to define the kind of movements, of mechanical load, the type of movement and, finally, an individual case of movement.

In a comprehensive study [Jenik, 1972] we have dealt with biomechanics of work movements of the arm. For over 1,000 individual cases of movements the course of mechanical strain parameters during a movement cycle was stated by the computer: angular velocity and acceleration, static and driving momenta in the shoulder and elbow joint, mechanical power and static and dynamic work.

A brief example of free horizontal swinging movements of the straight arm which were accomplished with different weights manipulated (0–1.5 kp), amplitudes (10–50 cm) and frequencies (25–250 movements/min) can be reported here. The investigations were performed by Stier [1959]. The static load by the weight was eliminated.

The measured data show the total energy comsumption, E, as a function of the frequency of movements, F, for individual kinds of movement. For all 50 investigated cases the mentioned function shows figure 1 and, schematically, figure 2. In all cases E contains 2 components: a static one, S, and a dynamic one, D, because also during a dynamic work a static component of strain is produced simultaneously and indivisibly. By dividing the value of E by F the specific energy consumption $e = E:F$ is achieved, (fig. 3). This division refers also to both components $s = S:F$ and $d = D:F$ (schematically in fig. 4). The dynamic component, d, increases exponentially and the

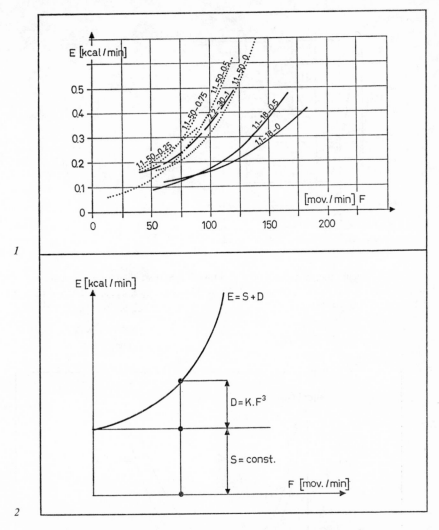

Fig.1. Total energy consumption as a function of frequency for a different kind of movement (mov.).

Fig.2. Total energy consumption, schematically.

static one, s, decreases hyperbolically. The resulting curves of e show in all cases the same character with an evident minimum.

For the investigated movements with sinusoidal moving patterns the dynamic work produced during a movement is proportional to the square of

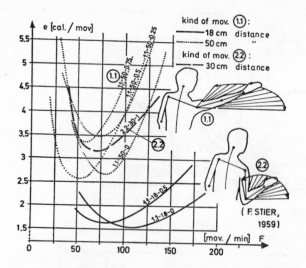

Fig. 3. Specific energy consumption as a function of frequency for a different kind of movement (mov.). Kind of movement 1.1: — = 18 cm distance, ... = 50 cm distance; 2.2: — — = 30 cm distance.

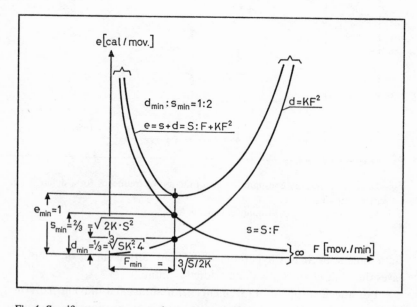

Fig. 4. Specific energy consumption, schematically.

the frequency and the power to its cube. The position of the minimum of e is determined by the proportion $d_{min} : s_{min} = 1 : 2$ (fig. 4).

By means of only 2 points of E it is possible to determinate arithmetically both constants K and S, the form of curves E and e as well as the minimum of e. A satisfactory correlation between empirical and analytical results was found.

The results allow the following generalisation: the quantitative relation between energy and frequency can be formulated by simple expressions which include factors that explain the causality of the mechanical processes and allow a satisfactorily reliable prediction of the energy consumption. The biomechanical analysis enables the formulation of a logically and causally based hypothesis before beginning the investigation and the selection of load factors. Therefore, the experiments can be planned, projected and arranged much more easily and reliably. The discussion of empirical data is supported by objective facts and false results of measuring can be easily identified. Multiple measuring of fewer points guaranties a greater accuracy of results than the same number of measurements in more cases.

In summary, it shall be emphasised that biomechanics by mechanical analysis introduces a new deductive and objective instrument which allows the prediction of the mechanical behaviour of the human body based on the causal relations and in the form of arithmetic functions.

References

JENIK, P.: Biomechanische Analyse ausgewählter Arbeitsbewegungen des Armes. Beuth-Vertrieb GmbH, Berlin-Köln-Frankfurt/M. (1972).

STIER, F.: Die Geschwindigkeit von Armbewegungen. Int. Z. angew. Physiol. *18*: 82–100 (1959).

Authors' address: Dipl.-Ing. PREMYSL JENIK, IAD, and Prof. Dr.-Ing. WALTER ROHMERT, Direktor des Lehrstuhls und Instituts für Arbeitswissenschaft, Technische Hochschule Darmstadt (IAD), Petersenstr. 18, *D-61 Darmstadt* (FRG)

Medicine and Sport, vol. 8: Biomechanics III, pp. 138–142 (Karger, Basel 1973)

Optimization of Human Motions

H. Hatze

Department of Physical Education, University of Stellenbosch, Stellenbosch

A general mathematical method of obtaining optimal human motions, i.e. motions yielding a maximal performance under given constraining conditions and for a given individual is discussed. It is shown that the function to be minimized is, with high probability for all types of motions, the total energy expended.

Introduction

If an individual repeats a specific motion under similar environmental conditions a certain number of times, the motion pattern will change in a particular way. For the healthy individual we may assume that this adaptation goes in the direction of optimizing the motion in question.

The problem being dealt with here is the following: What is optimized in such an adaptation process, and is it possible to determine mathematically the optimal motion for a certain kind of motion, for a given individual and under specific environmental conditions?

In other words: is it, for instance, possible to determine the optimal motion (that is, the motion yielding the maximal performance) of a shot-putter, whose anthropological parameters such as muscle strength, circumferences of joint movements, length, location of centres of gravities, masses and moments of inertia of the limbs are given or can be predicted to change to certain values after specific training? Such an optimal motion may, of course, be found by the individual himself by trial and error in a long series of repetitions but this depends on favourable conditions such as sophisticated training methods, a broad range of previous experience with similar types of motions, and so on. Moreover, the more complex the motion which is to be optimized, the less is the probability of achieving the goal by the method of trial and error in a reasonable period of time. Other examples of such an optimization process are the construction of optimal artificial limbs and to find the most economic work motions. Considerations of this kind have been carried out before: the classic work of Braune and Fischer [1, 2] in 1889 and the recent contributions of Nubar and Contini [3] and of Beckett and Chang [4].

Nubar and Contini [3] contend that the individual will determine his motion in such a way as to minimize muscular effort consistent with given constraints, where muscular effort for one joint is defined as the square of the moment of the joint muscle force times the duration Δt.

A very interesting work is that of Beckett and Chang [4]. Their analysis of the kinematics of gait is based on the assumption that the individual will determine his walking motion in such a manner as to minimize the energy expended. The results of their mathematical predictions, although simplified to 2-dimensional motion, is in good agreement with Ralston's experimental results [5].

Mathematical Optimization of Motions

To tackle the problem in a more general way, we have to simulate the human body by using a mathematical model.

Assuming that the human body can be divided into a number of segments which may be regarded as rigid bodies and are connected by joints of possibly different types and which constitute, as a whole, an open-chain system, the configuration of this system relative to a chosen earth-fixed reference frame and at any instant t can be described by a certain number of functions $q_r(t)$ which are called the time-functions of the generalized coordinates of the system.

Dividing the human body into 23 segments is both reasonably accurate and yet not too complicated. This number varies with different authors, ranging from about 13 to about 52, depending on the view adopted and on the purpose of the model. The configuration of the adopted model can then be described by 60 generalized coordinates, of which 3 are linear, and which determine the position of the midpoint of the line joining the hip-centres relative to the reference frame, and 57 which are angular, and determine the orientation of the segments relative to one another, or relative to the reference frame in the case of the segment pelvis.

Any motion of this system is then completely defined by the set

$$\{q_r(t), \tau : t \in \tau, r = 1, ..., 60\},$$

that is, by the 60 time-functions of the generalized coordinates and by the time-interval during which the motion takes place. There is the possibility of a more sophisticated and compact definition of human motion but we are not concerned with this question here.

A quantitative description of the motion of the partly conservative dynamical system, the human body, involves the setting up of the equations of motion which can be done by using the Lagrange equations. If this is carried out we obtain at most 60 second-order differential equations of the type:

$$\sum_{i=1}^{23}\left\{M_i\left[\left(\frac{\partial f_i}{\partial \dot{q}_r}\right)^{\cdot}-\frac{\partial f_i}{\partial q_r}\right]\right\}+[I_i]\left\{\dot{\omega}_i\frac{\partial \omega_i}{\partial \dot{q}_r}+\omega_i\left[\left(\frac{\partial \omega_i}{\partial \dot{q}_r}\right)^{\cdot}-\frac{\partial \omega_i}{\partial q_r}\right]\right\}=F_{q_r}^L+F_{q_r}^E+F_{q_r}^M$$

which must be solved simultaneously. If there are equations of constraint defined on the system, i.e. if some of the coordinates q_r are not independent, the number of differential equations reduces accordingly.

The meaning of the symbols in the above equations is: M_i, P_i, $[I_i]$, ω_i: mass, position vector of the centre of mass, matrix of the moments of inertia, angular velocity vector of the i-th segment. The terms $f_i = \frac{1}{2}v_i^2 + v_i \cdot u_i - [o\,o\,g] \cdot P_i$ contain scalar products of velocities and position vectors of the i-th segment.

The generalized forces F_{q_r} on the right side of the equations may be either actual forces (if q_r is a linear coordinate) or torques (if q_r is an angular coordinate). $F_{q_r}^E$ denotes external forces (torques) corresponding to the coordinate q_r while $F_{q_r}^L$ denotes internal forces (torques) due to limitations of joint movements. In general, the $F_{q_r}^L$ are rather complicated functions, being of the form

$$F_{q_r}^L = F_{q_r}^L(q_{r-3}, q_{r-2}, q_{r-1}, q_{r+1}, q_{r+2}, q_{r+3}, q_{r+4}, q_{r+5}).$$

In other words, the limitation of a joint movement in a certain direction depends in general on the position of the joint in question as well as on the position of the adjacent joints.

By discussing the remaining generalized force, $F_{q_r}^M$, we arrive at the core of the problem. The $F_{q_r}^M$ are the muscle torques and it is only through these functions that we are able to influence the motion of the body. Hence, any optimization of some function F can only be achieved by alterations of the time-functions of the muscle torques.

There are indications that the optimization of the adaptation process in repeated motions consists of minimizing the total energy expended. This is at least with high probability true for walking, as the work of BECKETT and CHANG [4] shows. But there is other important evidence as well. HILL [6] found that when stimulated muscle is stretched a certain amount beyond its *in situ* length, some energy disappears and he concluded that a biochemical process, in a living muscle cell, can be reversed by application of external mechanical work. Although there is no direct proof yet to this assertion [7, 8], recent work of MORRISON [9] seems to provide very strong evidence supporting

HILL's theory. MORRISON also concluded, and is strongly supported by ELFT-
MAN [10], that the biarticular function of some muscle groups, as for instance
the hamstrings, provides a considerable energy saving.

Now, these biarticular muscle groups are of course also active in fast
movements and it is known that most motions of the human body develop a
motion pattern with distinct stretch-shorten cycles of the working muscle
groups.

Hence, there appears to be very strong evidence that minimizing the
energy expended is at least a major principle in the adaptation process of any
motion type. The question of whether it is the only principle and if it holds
true for all individuals under all circumstances cannot be answered at this
stage, and research in this direction is in progress.

On the grounds of these considerations our main problem can now be
stated as follows.

*Given the dynamical mathematical model of the body of some individual
and given the constraints under which a certain motion has to be carried out,
what should be the functions $Fq_r^M(t)$, $(t \ \varepsilon \ \tau)$, i.e. the muscle torques, so that
$q_r(t)$ are such that the energy expended during this motion (the optimal motion)
is less than the energy expended during any other motion in a certain 'neigh-
bourhood' of this motion.*

This problem is obviously a problem of optimal control theory and a
solution seems possible by using the techniques of the calculus of variations.

The first step in directly attacking this problem is to build a general model
of human muscle which allows for all possible states of different types of
muscles in the body and which predicts correctly the energetic situation for
each possible state.

This is at present being investigated.

References

1 BRAUNE, W. und FISCHER, O.: Über den Schwerpunkt des menschlichen Körpers, mit
   Rücksicht auf die Ausrüstung des deutschen Infanteristen. Abh. math.-phys. kl. K.
   Saechs. Ges. Wiss. *26*: 000–000 (1889).
2 FISCHER, O.: Theoretische Grundlagen für eine Mechanik der Lebenden Körper mit
   Speziellen Anwendungen auf den Menschen, sowie auf einige Bewegungsvorgänge an
   Maschinen (Teubner, Leipzig 1906).
3 NUBAR, Y. and CONTINI, R.: A minimal principle in biomechanics. Bull. Math. Biophys.
   23: 377–391 (1961).
4 BECKETT, R. and CHANG, K.: An evaluation of the kinematics of gait by minimum
   energy, J. Biomech. *1:* 147–159 (1968).

5   RALSTON, H.J.: Energy-speed relations and optimal speed during level walking. Int. Z. angew. Physiol. *17:* 277–283 (1958).
6   HILL, A.V.: Production and absorption of work by muscle. Science *131:* 897–903 (1960).
7   MOMMAERTS, W.F.H.M.: Muscular contraction. Physiol. Rev. *49:* 427–508 (1969).
8   SANDOW, A.: Skeletal muscle. Annu. Rev. Physiol. *32:* 87–138 (1970).
9   MORRISON, J.B.: The mechanics of muscle function in locomotion. J. Biomech. *3:* 431–451 (1970).
10 ELFTMAN, H.: Biomechanics of muscle. J. Bone Jt Surg. *48*(A/2): 363–377 (1966).

Author's address: Dr. H. HATZE, Department of Physical Education, University of Stellenbosch, *Stellenbosch* (South Africa)

Medicine and Sport, vol. 8: Biomechanics III, pp. 143–145 (Karger, Basel 1973)

Some Objective Characteristics of Efficiency in Sport Techniques

V. Zatsiorsky

Central Institute for Physical Culture, Moscow

The aim of this paper is to give a systematic classification of some of the methods currently used in research work and sport practice for efficiency estimation of sport techniques.

I find it possible to single out 2 groups of objective efficiency characteristics in sport techniques. These include: (1) heterogeneous characteristics which depend on many factors such as athletes' strength, speed and endurance, and (2) homogenous characteristics which do not depend on the motor qualities such as strength, speed and endurance of athletes.

With regard to the heterogeneous characteristics of techniques used in modern sport practice, these can be divided into 2 groups. This division depends on some criteria which may be: (1) some first-rate techniques determined on the basis of biomechanical analysis, i.e. with the help of some 'biomechanical standard', and (2) movement techniques of some top-class athletes.

In the first case we compare the movement performed by an athlete with the first-rate standard and estimate the efficiency of sport techniques by the degree of their likeness. In high jumps, for example, we can see that the distance between the normal center of gravity of the athlete's body and the bar when the athlete clears it can serve as a certain characteristic of this technical sportmanship. In swimming the breast stroke, this characteristic is the degree of deviation from the swimmer's regular speed, etc. In some cases it is necessary to take into account some individual specific features of the athlete's build and, in particular, the total dimensions of his body. For instance, in weight-lifting when the athlete snatches and lifts up the bar to his chest it is reasonable, from the biomechanical point, to lift the bar at a strictly definite height which is high enough for the athlete to make a squat. Experiments prove that it is not advisable to lift the bar too high. The observations data shows that the lifting height of the bar depends on 2 factors, i.e. on the body height of the athletes and on their sportsability. The shorter the athlete and the better techniques he displays, the lower he has to lift the bar to make his movements more successful. Therefore, if we know the body height of the athlete, then the height of bar-lifting in successful attempts may serve as a characteristic of his technical readiness.

Another group of heterogenous characteristics used to estimate sport techniques is intended to compare the movement techniques of top- and low-class athletes. The traits which make it possible to compare the movement techniques of top- and low-class athletes can be justifiably called discriminative traits. To discern these traits it is necessary to use the method of discriminative analysis. The use of discriminative traits in sport techniques can be accounted for by the fact that now it is hardly possible to carry out a total biomechanical analysis of most of the complicated movements, and theoretical considerations often fail to determine the significance of some of the separate components of sport techniques such as joint angle, the distance between the feet, etc., for the efficiency of movements.

Let us consider the homogeneous characteristics used to estimate the efficiency of sport techniques [ZATSIORSKY et al., 1964; and others]. The idea of this method is to compare the actual performance of the athlete with the performance the athlete is able to achieve according to his inborn motor abilities such as strength, speed and endurance. In this case, we estimate the efficiency of sport techniques by analysing how well the athlete has taken advantage of his physical ability. In this estimation we proceed from the relationship existing between the following 3 characteristics. These include: (1) sport performance; (2) development of motor abilities (strength, speed and endurance), and (3) efficiency of techniques.

In practice, this can be done by comparing various performances of athletes in: (1) a technically complicated movement (as a rule, this is the movement with which the athlete competes), and (2) technically more elementary movements requiring the development of the same motor abilities as sport movements do.

To cite an elementary example of this comparison: some trampoline acrobats ranging from beginners to top-class athletes made: (1) elementary jumps, and (2) somersaults (table I). The time of their performance was registered in flight.

The elementary jump is quite simple and the height of flight in this case chiefly depends on some strength and speed abilities of athletes. While per-

Table I. The average time in flight spent by trampoline acrobats of various qualifications while making elementary jumps and somersaults (sec)

Rating	Elementary jump, sec	Somersault, sec	Coefficient of use in percentage
III	1.2	0.9	75
II	1.3	1.1	84.6
I	1.4	1.2	85.7
Soviet national team	1.5	1.5	100

forming a somersault jump the athlete should make the best use of these abilities (100% is an ideal). The data in the table shows that only top-class athletes succeed in doing so. These athletes have better 'motor abilities' and their 'coefficient of use' is also higher.

In the example given above we can see the main idea of the methods discussed in this paper. These examples show that technical readiness may be estimated by the difference between the actual performace of the athlete and the performance he is capable of, taking into account his speed and strength abilities. The predicted performance of the athlete is usually defined with the help of regression equations (in this sense the example described above, in which direct comparison was made between actual and predicted performances, proves to be rather an exception).

The homogeneous characteristics of the efficiency of sport techniques have been estimated in some of the events. These include: (1) the 110-metre hurdle compared with the performance in the 100-metre event; (2) javelin-throwing compared with the performance shown in throwing the shot weighing 3 kg, and (3) crawl-swimming compared with the force displayed by the athlete for a unit of the water resistance, etc.

While estimating the degree of the athlete's technical readiness by the regressions equation it is necessary to take into account the fact that all considerations on the athlete's technical sportsability are, in this case, of relative character. We determine whether the athlete's technique is good or bad by comparing it with the average sport techniques which are typical for these athletes ('good', in this case, implies 'better than the average' and 'bad' 'below the average').

Author's address: Prof. Dr. V. ZATSIORSKY, Central Institute for Physical Culture, *Moscow* (USSR)

Medicine and Sport, vol. 8: Biomechanics III, pp. 146–150 (Karger, Basel 1973)

Clumsiness and Stature
A Study of Similarity

ELSBETH HOERLER

Eidgenössische Technische Hochschule, Zürich

Grasshoppers and fleas jump high; cats move with striking elegance. Elephants can hardly jump; one of the possible reasons for the extinction of the dinosaurs was – in the opinion of zoologists – their clumsiness [1], and the big blue wales did not perish, because they live in water where they do not have to support their own weight. An attempt is made to explain why big creatures are generally clumsier than small ones.

Since there is a general relationship between clumsiness and the ability to jump, the effect of body size on the height of jumping is studied in a theoretical way. 'Height of jumping' means here the vertical displacement of the centre of gravity during flight.

To facilitate matters, a very simple model is considered (fig. 1).

Simplifications

1. Only geometrically similar creatures will be considered (Dia. 1). This has two reasons: (a) the size of the body is not sharply defined for different body forms (is it height or largest dimension?), and (b) the height of jumping of existing creatures is not an exact function of the body size (in however 'body size' might be defined). An elephant is indeed bigger and jumps less than a grasshopper, but this comparison does not apply to a horse and a tortoise. In reality there exists only a *general* relationship between body size and height of jumping.

2. The vertical jump without a swinging movement will be considered, and it is assumed that the angles of the joints at the starting position are the same for all bodies (Dia. 2). (In reality, the bigger creatures will probably bend the knee less.)

Existing creatures ⟶ geometrically similar creatures

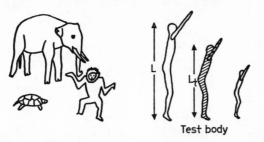

Test body

Fig. 1. A body is characterized by the factor of proportionality, λ: $\lambda = \dfrac{L}{L_t} \cdot L$ = body size.

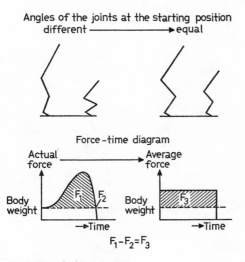

Fig. 2. Description see text.

3. A constant (average) muscular force in the vertical direction is assumed (Dia. 2). (It is ignored that during the phase of pushing off, the force exerted by big creatures is nearer maximum, and that consequently they get more tired during this phase.)

Calculation

1. *General.* The energy equation gives the relationship between the height of jumping, the displacement of the centre of gravity during the phase of pushing off and the quotient: muscular force-body weight (fig. 3).

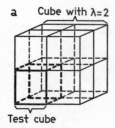

Fig.3. Energy equation: $A(K - G) = \dfrac{MV^2}{2} = GH \cdot K$ = average vertical muscular force, G = body weight, M = body mass, V = take-off velocity. $H = A\left(\dfrac{K}{G} - 1\right)$.

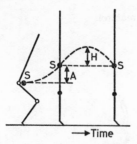

a   Cube with λ=2

Test cube

b

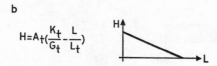

$H = A_t\left(\dfrac{K_t}{G_t} - \dfrac{L}{L_t}\right)$

Fig.4. The figure (a) shows that areas increase with the second power of the factor of proportionality, λ, volumes with the third. $A = \lambda A_t$. $K = \lambda^2 K_t$ (proportional to the cross-sectional area of the muscles). $G = \lambda^3 G_t$ (proportional to the volume). Substitution (b):

$$H = A\left(\frac{K}{G} - 1\right) = \lambda A_t\left(\frac{\lambda^2 K_t}{\lambda^3 G_t} - 1\right) = A_t\left(\frac{K_t}{G_t} - \lambda\right).$$

In figure 4, it is taken into account that the muscular force increases with the second power of the body size, the body weight with the third power. In this way, a general relationship between the body size, L, and the high of jumping, H, is obtained.

This relationship shows that the bigger the creature, the less the height of jumping; that there is a size limit (if the creatures size exceeds this limit, it cannot jump and, once fallen, it would not be able to raise itself to its legs);

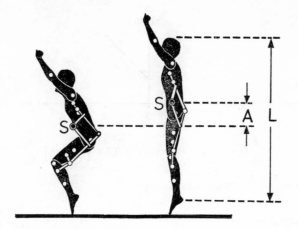

Fig.5. Description see text.

and that the distance between the lowest and the highest position of the centre of gravity (A + H) does not depend on the body size.

2. *Numerical*. The model in figure 5 shows that the displacement of the centre of gravity during the phase of pushing off is about the eighth part of the body size. (That the joints of feet in this model cannot be moved is of no great importance here.)

Applying the main formula (fig. 5) to the jumping test-body a numerical relationship is obtained between the body size and the height of jumping (fig. 6). This relationship shows that the limit, for the size of a human, is 4 m, and that a mini-human body would not be able to jump more than 0.5 m. These results are, of course, only approximate.

The High Jump in Sports

For the high jump in sports, taller people are generally better fitted than smaller ones because their centre of gravity is already higher at the start. The centre of gravity in a body standing upright and on tiptoe is about 0.6 times the body size above the ground. The height of this high jump is approximately that of the culmination point of the centre of gravity above the ground. But this is only correct as long as the sportsman can raise his legs quick enough (or can move horizontally) during the rising time. It can be easily shown that

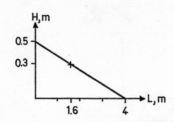 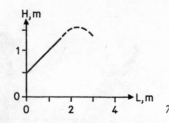

6   7

Fig.6. $L_t = 1.6\,\text{m}.$ $A_t = \dfrac{1}{8}\,L_t = 0.2\,\text{m}.$ $H_t = 0.3\,\text{m} = 0.2 \times \left(\dfrac{K_t}{G_t} - 1\right)\dfrac{K_t}{G_t} = 2.5.$

$$H = 0.2\left(2.5 - \frac{L}{8}\right).\quad H = \frac{1}{2} - \frac{L}{8}.\quad \text{H and L in metres.}$$

Fig.7. The high jump in sports (approximated). $H_{sp} = 0.6\,L + H = 0.6\,L + \dfrac{1}{2} - \dfrac{L}{8}.$
$H_{sp} = \dfrac{1}{2} + \dfrac{L}{2}$ (not correct for very big bodies). H_{sp} and L in metres.

this is not possible for bodies exceeding a certain size. This means that for the high jump in sports there exists an optimum body size (fig. 7).

In conclusion, the main result of this paper is that – within the model discribed – big creatures, during the flight phase of jumping, cannot raise their centre of gravity as high as small ones and that with respect to size there is a limit that cannot be exceeded by any living creature.

Reference

1   HABER, H.: Das mathematische Kabinett (Deutsche Verlagsanstalt, Stuttgart 1967).

Author's address: Miss ELSBETH HOERLER, Eidgenössische Technische Hochschule, Abt. X für Naturwissenschaften, Kurse für Turnen und Sport, *CH-8000 Zürich* (Switzerland)

Medicine and Sport, vol. 8: Biomechanics III, pp. 151–155 (Karger, Basel 1973)

Heredity *versus* Environment in Biomechanics

A. Venerando[1] and M. Milani-Comparetti[2]

Biomechanics combines the unchangeable laws of mechanics with the variability of living beings. Movement in man results from the application of muscular forces to the lever systems of the skeleton. The relative insertions of the muscles, the size of the bones, the position and type of the joints are many structural variables affecting biomechanical performance. The intensity of muscular forces, their balance, the speed of contraction and relaxation, etc., are functional variables, equally affecting biomechanical performance.

In man (as in any living being) a wide range of variability is allowed for both structural and functional traits, diversifying each individual within the limits of the species. Within these limits each individual receives, at conception, a given assortment of genes (his 'genotype') which, interacting with environmental influences, will lead his organism to its full development (his 'phenotype'). Some traits in the individual phenotype are almost fully controlled by the genotype, others are almost fully controlled by environmental influences, with a wide intermediate range.

Biomechanical studies may indicate which traits, alone or variously combined, are conditioning factors for each type of performance. But there immediately arises the problem of whether such traits are inherited or acquired: in the first case, in fact, those who have inherited the trait are found only through selection, while in the second case the desired trait can generally be developed by training.

It is absolutely pointless to try and obtain by training alone, however intense and protracted, the possession of a given genetically determined trait (such as long legs) in an individual who lacks it. It is equally pointless to go through the long procedure of previous selection when seeking the bearers of a trait that is controlled by environmental influences, since it can be obtained much more easily (with the exclusion of a few unfit individuals) by proper training.

It is, therefore, very important to be able to discriminate between inherited and environmentally-induced traits. For this purpose, the Institute of Sports Medicine of the Italian National Olympic Committee and the Gregor

1   CONI Istituto di Medicina dello Sport, Rome
2   Istituto Mendel, Rome

Mendel Institute of Medical Genetics and Twin Research (founded and directed by Prof. LUIGI GEDDA) jointly initiated a series of studies.

The research method adopted is the classical expression of the 'twin method', formulations of which are widely discussed and presented in the publications of Prof. GEDDA and his school [1, 2].

The twin material for the initial study was drawn at random (according to present specifications) from the Italian Twin Register of the Mendel Institute (a computerized register operated with the support of the Italian National Research Council, listing data for some 15,000 twin pairs). The specifications were that the sample was to include only male twins without previous experience of intense physical activity, divided in 2 age groups: 10–13 years and 14–17 years (subsequently identified respectively as '10 years' and '16 years'). At least 10 pairs were to be monozygotic (MZ) and 10 dizygotic (DZ) in each age group. A total of 48 pairs was selected, and zygosity was tested by means of blood groups, dermatoglyphics and 'equivocity' tests, as well as the comparison of 5 stable anthropological traits. Two pairs were excluded because zygosity could not be established with sufficient reliability. The final sample included 11 MZ and 12 DZ pairs in the lower age group, and 13 MZ and 10 DZ pairs in the higher.

Each twin was tested separately for the various structural and functional traits in the laboratories of the Institute of Sports Medicine, without previous knowledge of zygosity (which was established at the Mendel Institute). All experimental data were then recorded and distributed according to age and zygosity. Some data were analyzed separately with the cooperation of undergraduate students in both institutes, and were presented at previous scientific meetings [3, 4].

In view of the size of the experimental material and for purposes of standardization it was decided to analyze all quantitative data by means of a computer, and the required computer program was written in Fortran language by Dr. G. TATARELLI.

The results of computer analysis are listed, in table I, for each of 72 traits or indices. The results are expressed, according to twin study methodology, in terms of Holzinger's index of heredity ($\hat{H}$) with the corresponding ranges of probable error (P_E). $\hat{H}$ values approaching unity are interpreted as indication of outright genetic control in the corresponding traits and, in fact, the percentage of genetic control for a given trait can be taken to correspond to the numerical value of Holzinger's index.

In table I a plus sign identifies those traits for which the probable error exceeds 0.5, since the significance of the corresponding $\hat{H}$ values is obviously

Table I. Holzinger's index of heredity (Ĥ); values at 10 and 16 years of age

	10 years, Ĥ ± P_E	16 years, Ĥ ± P_E
1. Stature	°° 0.9041 ± 0.1661	°° 0.8504 ± 0.2531
2. Sitting vertex height	°° 0.8875 ± 0.1828	− 0.0050 ± 1.8278 +
3. Acromial height	°° 0.8991 ± 0.1794	°° 0.8329 ± 0.2590
4. Radial height	°° 0.9545 ± 0.0811	°° 0.9438 ± 0.0965
5. Trocantherion height	°° 0.9093 ± 0.1611	°° 0.8886 ± 0.1557
6. Iliospinal height	°° 0.9083 ± 0.1647	°° 0.8433 ± 0.2265
7. Dactylion height	− 1.2747 ± 3.2659	0.2734 ± 1.2871 +
8. Tibial height	(°) 0.6365 ± 0.6457 +	(°°) 0.7714 ± 0.3256
9. Chest breadth	0.4291 ± 0.9543 +	(°) 0.5453 ± 0.6965 +
10. Bi-acromial breadth	(°°) 0.7632 ± 0.3930	°° 0.9493 ± 0.0851
11. Bi-iliac breadth	(°°) 0.6852 ± 0.5186 +	(°) 0.6348 ± 0.5320 +
12. Bi-condilean diameter	− 0.1245 ± 1.8106	(°) 0.5297 ± 0.7059 +
13. Insp. mesosternal chest girth	0.2191 ± 1.3742 +	°° 0.8933 ± 0.1579
14. Exp. mesosternal chest girth	− 1.3043 ± 4.0013 +	°° 0.9277 ± 0.1047
15. Max. circumference right arm	(°) 0.5507 ± 0.7204 +	°° 0.8373 ± 0.2327
16. Max. circumference left arm	(°°) 0.7138 ± 0.4632	°° 0.8785 ± 0.1780
17. Left leg circumference	°° 0.9792 ± 0.0377	°° 0.8757 ± 0.1794
18. Right leg circumference	(°°) 0.8274 ± 0.3241	°° 0.8987 ± 0.1566
19. Subscapular fold	°° 0.9652 ± 0.0550	(°) 0.6109 ± 0.5139 +
20. Lumbo-abdominal fold	0.2855 ± 1.0064 +	0.2602 ± 1.1064 +
21. Para-umbilical fold	0.4890 ± 0.7694 +	°° 0.8993 ± 0.1539
22. Left arm dorsal fold	°° 0.8367 ± 0.2188	(°°) 0.7273 ± 0.3391
23. Left arm ventral fold	(°) 0.5637 ± 0.5634 +	°° 0.8951 ± 0.1356
24. Left leg medial fold	(°°) 0.8490 ± 0.2453	(°°) 0.6834 ± 0.4780
25. Left leg lateral fold	− 0.4189 ± 2.1701 +	(°) 0.5436 ± 0.5591 +
26. Weight	(°°) 0.6771 ± 0.5758 +	(°) 0.5986 ± 0.6496 +
27. Right hand dynamometry	0.4319 ± 0.9600 +	(°) 0.5607 ± 0.7016 +
28. Left hand dynamometry	°° 0.8926 ± 0.1756	0.4433 ± 0.8883 +
29. Right arm dynamometry	0.4159 ± 1.0324 +	(°°) 0.7990 ± 0.3422
30. Left arm dynamometry	0.2961 ± 1.2771 +	(°) 0.6469 ± 0.6071 +
31. Back extensor dynamometry	0.4381 ± 0.8930 +	(°°) 0.7414 ± 0.4417
32. Leg extensor dynamometry	(°°) 0.6727 ± 0.5218 +	(°°) 0.7803 ± 0.3541
33. Total dynamometric index	(°°) 0.7570 ± 0.4878	°° 0.8669 ± 0.2289
34. Vital capacity	(°) 0.5910 ± 0.7277 +	°° 0.9122 ± 0.1392
35. MVV	(°) 0.5255 ± 0.8182 +	°° 0.8903 ± 0.1639
36. FEV$_{1.0}$	°° 0.8712 ± 0.2260	°° 0.8629 ± 0.2171
37. Respiratory rate	0.0694 ± 1.0917 +	(°) 0.6517 ± 0.4839
38. Respiratory rate, max. work	− 0.4209 ± 1.5734 +	
39. Respiratory rate 3 min post-work	− 0.4090 ± 1.9962 +	− 1.1773 ± 2.9028 +
40. Heart rate (lying down)	− 0.2823 ± 1.8810 +	0.3837 ± 1.5641 +
41. Heart rate (standing)	0.1802 ± 1.2728 +	− 0.0683 ± 1.6145 +
42. Heart rate, max. work	°° 0.9085 ± 0.1450	(°) 0.5147 ± 0.4635

Table I (continued)

	10 years, $\hat{H} \pm P_E$	16 years, $\hat{H} \pm P_E$
43. Heart rate, 3 min post-work	$0.0370 \pm 1.4646^+$	$^- 1.9290 \pm 4.9922^+$
44. $\dot{V}_{O_2}$	$0.4325 \pm 0.9294^+$	$0.3732 \pm 0.7666^+$
45. $(\dot{V}_{O_2})$ 160	$^{\circ\circ} 0.9391 \pm 0.0997$	$^- 0.7967 \pm 2.8832^+$
46. $\dot{V}_{O_2}$, 3 min post-work	$0.4872 \pm 0.5810^+$	$0.2829 \pm 0.9452^+$
47. $\dot{V}_{O_2}/kg/min$	$0.4458 \pm 0.8379^+$	$(^{\circ})0.5348 \pm 0.5513^+$
48. $\dot{V}_{O_2}/kg/min$, max. work	$(^{\circ})0.6595 \pm 0.5794^+$	$^- 0.7732 \pm 2.3545^+$
49. $\dot{V}_{O_2}/kg/min$, 3 min post-work	$^- 3.2276 \pm 6.5775^+$	$0.3440 \pm 0.8723^+$
50. LPI	$^- 0.2791 \pm 1.9787^+$	$0.1204 \pm 0.9442^+$
51. LPI, max. work	$^{\circ\circ} 0.9746 \pm 0.0430$	$0.0230 \pm 1.5413^+$
52. LPI, 3 min post-work	$^- 0.3033 \pm 1.5735^+$	$^- 0.0742 \pm 1.6529^+$
53. RE_{O_2}	$^- 5.1369 \pm 8.8057^+$	$0.3295 \pm 0.9301^+$
54. RE_{O_2}, max. work	$^{\circ\circ} 0.9655 \pm 0.0597$	$^- 1.5340 \pm 3.5067^+$
55. RE_{O_2}, 3 min post-work	$0.3966 \pm 0.6746^+$	$^- 0.8559 \pm 3.1833$
56. $\dot{V}_{CO_2}$	$(^{\circ})0.5993 \pm 0.5728^+$	$^{\circ\circ} 0.9883 \pm 0.0196$
57. $\dot{V}_{CO_2}$, max. work	$(^{\circ\circ})0.6711 \pm 0.4962$	$^- 0.2275 \pm 2.0617^+$
58. $\dot{V}_{CO_2}$, 3 min post-work	$(^{\circ})0.5489 \pm 2.3138^+$	$(^{\circ})0.7151 \pm 0.3619$
59. R	$(^{\circ})0.6067 \pm 0.4897$	$0.5000 \pm 0.5014^+$
60. R, max. work	$0.3248 \pm 0.9365^+$	$^- 0.5837 \pm 2.0506^+$
61. R, 3 min post-work	$0.1331 \pm 1.0665^+$	$0.4959 \pm 0.6554^+$
62. $\dot{V}$ BTPS	$(^{\circ})0.7012 \pm 0.4452$	$^- 0.6717 \pm 2.2695^+$
63. $\dot{V}$ BTPS, max. work	$^{\circ\circ} 0.8302 \pm 0.2682$	$^- 0.1182 \pm 1.6869^+$
64. $\dot{V}$ BTPS, 3 min post-work	$0.0219 \pm 1.4135^+$	$0.4986 \pm 0.5201^+$
65. T-wave, V_6 (mm) (1)	$^- 0.0766 \pm 1.4398^+$	$^- 0.0832 \pm 1.5021^+$
66. T-wave, V_1 (mm) (1)	$0.3375 \pm 0.6778^+$	$^{\circ\circ} 0.8875 \pm 0.1708$
67. Q-T, cm/sec (1)	$^- 1.2364 \pm 3.3157^+$	$^- 0.2912 \pm 1.9555^+$
68. Q-T, calc. (1)	$(^{\circ})0.6103 \pm 0.5173^+$	$0.3113 \pm 0.8989^+$
69. QRS, cm/sec (1)	0.1703 ± 0.9604	$^- 1.1684 \pm 2.7305^+$
70. P-R, D_2 (cm/sec) (1)	$0.4733 \pm 0.6389^+$	$0.3679 \pm 0.7557^+$
71. P-wave, D_2 (cm/sec) (1)	$0.0098 \pm 1.3018^+$	$0.3731 \pm 0.6375^+$
72. Electric axis, degrees (1)	$(^{\circ\circ})0.7480 \pm 0.3421$	$0.1216 \pm 1.3515^+$

MVV = maximum voluntary ventilation (litres/min). $FEV_{1.0}$ = forced expiratory volume in one second (litres). LPI = Leistungspulsindex (oxygen pulse) (ml/beat). RE_{O_2} = Respiratory (ventilatory) equivalent for oxygen. R = gas exchange ratio (respiratory quotient). BTPS = body temperature and pressure, saturated with vapour. (1) = electrocardiographic date.

doubtful. It is worth noting that all negative $\hat{H}$ values (no negatives are normally expected to occur, according to the formulation of the index) correspond to the larger values of probable error.

One ° sign marks all $\hat{H}$ values exceeding 0.5; two such signs mark all $\hat{H}$ values exceeding 0.66; yet whenever the index value minus the probable error drops below 0.50 these signs are in parentheses. Thus, the traits identified by single or double nonparenthesized ° signs are, in our sample, those for which genetic control is more obvious.

At this point the pattern of genetic or environmental control over the different traits tends to become clear, with the higher values of the index of inheritance generally corresponding to skeletal traits, while more variability is found in functional traits.

Several findings seem to require further verification and explanation, often concerning the methodological and experimental approach as related to relevant differences in $\hat{H}$ values between age groups. We mention, for example, No. 2, 7, 12–14, 45, 51, 54, 63 and 66. Some error is likely to have been made in transcription. In fact No. 38 has one entire age group missing, and some such error may affect other data; yet the necessary checks were impossible at this time.

Further studies are obviously needed, and in fact we are the first to stress the limitations of our sample. Yet we do believe that our contribution, representing a novel approach in the study of biomechanics, may provide some useful information for our colleagues and possibly stimulate scientific cooperation.

References

1   GEDDA, L.: Twins in history and science (Thomas, Springfield 1961).
2   GEDDA, L.: De genetica medica (Mendel Institute, Rome 1963).
3   VENERANDO, A. and MILANI-COMPARETTI, M.: Twin studies in sport and physical performance, Acta genet. med. gemellol. *19:* 1–2, 80–82 (1970).
4   VENERANDO, A.; MILANI-COMPARETTI, M.; DAL MONTE, A.; BRACCI, C.; CALDARONE, G.; GRECO, M.; GAMBULI, N.; PIOVANO, G.; SANTILLI, G., and TATARELLI, R.: Influenza dell'eredità e dell'ambiente su alcuni parametri funzionali legati all'attività fisico-sportiva. Med. sport *23:* 6, 181–184 (1970).

Author's address: Prof. ANTONIO VENERANDO, CONI Istituto di Medicina dello Sport, *Rome* (Italy)

Medicine and Sport, vol. 8: Biomechanics III, pp. 156–157 (Karger, Basel 1973)

Studies of the Growth of the General Centre of Gravity Height of School Children

A. Iliescu, Eugenia Portarescu, Dora Gavrilescu and
Georgeta Bragarea

Institute of Physical Education and Sport,
Faculty of Functional Anatomy and Biomechanics, Bucharest

Since 1966, we have been studying the general centre of gravity of the body (GCG) of school children of several public schools and lyceums in Bucharest and Petrosani (a mountain region of average altitude). We have put the development of the GCG height into correlation with a series of physical development parameters as, for example, body height, body weight, lower extremity length, etc.

Until now we have effected more than 12,000 measurements with 7- to 18-year-old pupils, of whom 60% were girls and 40% were boys. We have used the method of Prof. Dr. E. Repciuc of Bucharest [8] to determine the GCG height. This method uses a special brancard, which is perfectly balanced and is 2 m long and 0.5 m wide. Furthermore, weighing scales were used to take the body weight as well as a steel antropometre and a special support for the Brancard.

The GCG height was determined by the formula:

$$d = \frac{P - T}{G} \times L,$$

where d = the GCG height, P = the weight of both body and brancard, T = the weight of the brancard alone, G = the total body weight, L = the length of the brancard, i.e. 2 m.

This method gives satisfactory results with an approximation of 2–3%, which is caused by the balancing, by the brancard, the masses of which cannot be distributed equally, and by the various positions of the subjects on the Brancard.

We have choosen this method for its simpleness, as the methods of collective investigations with school children are restricted.

Data Progressing

1. The GCG height increases between 7 and 18 years by about 19 cm in boys and 14 cm in girls.

2. During the growth development there are periods of acceleration of the GCG height; with boys, between 12 and 13 years and between 15 and 17 years, and with girls, between 11 and 13 years, followed by a standstill between 13 and 17 years, followed by another growth period.

3. A comparison of the GCG height and the length of the lower extremities shows:

1. With boys, a growth parallel of the GCG height and the length of the lower extremities, which signifies – in our opinion – the proportional development of the masses of the body segments.

2. With girls, an increased growth of the lower extremities between 7 and 11 years, followed by a standstill. After 16 years a dissociation appears; whereas the lower extremities grow gradually, the GCG height is at a standstill and even decreases.

The determination of the GCG height of school children is of great importance from the practical point of view. The body weight, which is regularly taken, gives unsatisfactorily results when comparing the physical development with physical education and sport. It is, in our opinion, necessary to take into account the development of the partial masses of the body (in particular the pelvis) with the help of the GCG curve.

We would propose to introduce the value of the GCG height as another index of periodical medical controls, as it has proven to be equal to other physical development indices and could very well be used as a control criterion.

References
1 ABE, J.: Body proportions. Res. J. phys. Educ. (1960).
2 ANDRONESCU, A.: Anatomia copilului (Ed. Med., Bucharest 1966).
3 DONSKOI, D. D.: Biomecanica exercitiilor fizice (Ed. C.F.S., Moscow 1961).
4 HOCHMUTH, G.: Biomechanik (Sportverlag, Berlin 1967).
5 ILIESCU, A.: Biomecanica (Ed. C.N.E.F.S., Bucharest 1968).
6 IONESCU, A. N.: Auxologie (Ed. Med., Bucharest 1968).
7 NEMESSURI, M.: Functionelle sportanatomie (Akad. Kiado, Budapest 1963).
8 REPCIUC, E.: Anatomie omului (Ed. Med., Bucharest 1962).
9 TITTEL, K.: Beschreibende und funktionelle Anatomie (Sportver ag, Berlin 1965).

Authors' address: Dr. A. ILIESCU, Dr. E. PORTARESCU, Dr. D. GAVRILESCU and Dr. G. BRAGAREA, Institute of Physical Education and Sport, Faculty of Functional Anatomy and Biomechanics, *Bucharest* (Rumania)

3. Technique of Motion Studies

Medicine and Sport, vol. 8: Biomechanics III, pp. 158–164 (Karger, Basel 1973)

A Torque-Producing Stimulator for the Study of Muscular Response to Variable Forces

A. Berthoz and S. Metral

Laboratoire de Physiologie du Travail, CNRS, and Laboratoire de Biophysique, UER Broussais, Paris

In biomechanical studies related to muscle contraction a variety of mechanical stimulations has been used. In man it is particularly relevant to apply forces and this has generally been done by attaching weights to the limbs. This method is, however, very limited as it is difficult to vary the applied force according to the needs of research program or rehabilitation procedure. We have developed a new method which allows the use of variable forces. This method is based on electromagnetic coupling and has been tried in man [Metral et al., 1967; Berthoz and Metral, 1970] and in the cat [Berthoz et al., 1971] for the study of muscle contraction. The purpose of the present paper is essentially to give an outline of the method and of some of the results obtained so far.

I. Description of the Device

The input force is applied by means of an electromagnetic clutch controlled by a function generator (fig. 1). The function generator provides a constant electrical voltage, V, upon which is superimposed a modulation whose wave-form can be modified according to the needs (ramp, sinusoid, triangle, random input). The output of the function generator is connected with the terminals of an electromagnetic clutch. This clutch provides a variable torque link between 2 pulleys, P_1 and P_2. P_1 is driven by a motor at constant speed, P_2 is coupled to P_1 by the electromagnetic powder contained in the clutch, this electromagnetic powder is activated by a magnetic field proportional to the function generator output applied to the terminals. Thus a variable torque is transmitted through this pulley to a metal rod, attached to

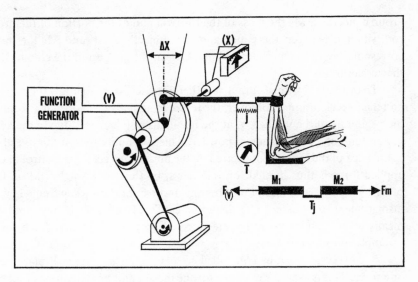

Fig.1. Electromagnetic clutch and experimental setting for human force response research (for abbreviations see text).

the subject limb or to the muscle under study. This allows the tangential component of the force exerted on P_2 to be transmitted to the subject.

II. Measurement of Mechanical Parameters

When no displacement occurs, the force input applied to the subject is known, either from the electrical voltage (v) applied to terminals of the clutch or from the value T_j given by a strain gauge located on the metal rod (fig. 1).

Yet this is true only in isometric conditions, for when a force is applied to the limb, the latter usually moves. The force of inertia then has to be taken into account. The strain gauge, mechanically similar to a stiff spring, transmits the force input completely. Its stretch is not significant compared to the displacement of the pulley and limb. The forces exerted on either side of the strain gauge are then equal at every moment and can be written:

$$T_j = F_m + m_2 \ddot{X} = F - m_1 \ddot{X}, \tag{1}$$

in which F is the force input provided by the clutch assembly. This force always tended to extend the forearm. F_m is the contraction force of the muscle group concerned, M_2 the mass of the forearm and of the portion of the drive rod, situated between the gauge and the subject's wrist, and M_1 the mass corresponding to the mechanical link between the gauge and the clutch. $\ddot{X}$ is the acceleration.

To evaluate F with accuracy it is therefore necessary to know the masses and their acceleration. Moreover, the force, F, actually developed lags behind the beginning of the voltage, v, at the terminals of the clutch, due to the activation time of electromagnetic powder. This is a limitation on the use of v, as an index of the force input exerted on the muscle. An accurate control of the applied force is thus limited, on one hand, by this phase angle and on the other by the forces of inertia that become more important when frequency or displacement are higher. The displacement X of the limb is measured by means of an angular potentiometer located on the axis of the clutch drive assembly output pulley.

Such a system can be controlled so as to impose either a displacement or a force. This is even easier with a torque motor, and this later type of motor has been used recently in experiments with patients by WALSH [1970].

III. Mathematical Description of the System

Another demonstration of equation (1) is by application of the principle of conservation of mechanical energy. This principle can be written:

$$(F - F_m)\, dX = \frac{1}{2}(m_1 + m_2)\, \dot{X}^2, \tag{2}$$

which, when derivated, gives the same equation as equation (1).
Equation (1) gives:

$$\ddot{X} = \frac{F - T_j}{m_1},$$

and

$$F_m = F\left(1 - \frac{m_1 + m_2}{m_1}\right) + T_j \frac{m_1 + m_2}{m_1}$$

and

$$F_m = -F\frac{m_2}{m_1} + T_j\left(1 + \frac{m_2}{m_1}\right)$$

if

$$\frac{m_2}{m_1} = \lambda$$

$$\boxed{F_m = -F\lambda + T_j(1 + \lambda)} \ ,$$

if m_2 can be made equal to m_1

$$F_m = -F + 2T_j$$

so, theoretically, one should be able to comput the value of the force developed by the muscle. In fact, it can be generally useful to consider only the part of the system containing F_m, m_1 and T_j. Then the basic equation is:

$$\boxed{T_j = F_m + m_2 \ddot{X}} \ .$$

In the case where a sinusoidal input of frequency will be used, and if the system under study is compared to a visco-elastic system of mass, m_2, stiffness, k, and viscous damping, c:

then

$$T = T_0 \ \sin \ \omega t \qquad\qquad \omega = 2\pi f$$

$$X = X_0 \ \sin \ (\omega t + \Phi).$$

A classical result from physics of vibration is

$$X_0 = \frac{T_0}{\sqrt{(c\omega)^2 + (k - m_2\omega^2)^2}} \qquad tg\,\Phi = \frac{c\omega}{k - m_2\omega^2} \ .$$

This result shows that if the damping, c, is very low the system will have a resonance frequency for:

$$2\pi f = \sqrt{\frac{k}{m}} \ .$$

In other words, another advantage of that device is that all the classical descriptions used in physics for mechanical systems can be directly applied. The possibility to change the input will also allow the use of more sophisticated techniques of analysis as spectral density or correlation analysis following application of random signals.

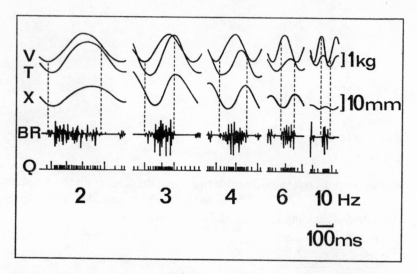

Fig.2. Time course of electromyogram and mechanical parameters for various frequencies of sinusoidal stimulation. V = input voltage of function generator; T = tension measured by strain gauge; X = forearm displacement; BR = brachioradialis EMG; Q = integrated EMG. Upward deflection of curves corresponds to an increase of V and T and to a forearm displacements X in direction of stretch (flexion). Dotted line shows phase relationship between EMG and other parameters.

IV. Application of the Method

This method can be used in studies on normal healthy man as well as in patients or for developing muscle strength studies. It can be used in animal research for the study of elementary properties of muscular control of movement. Finally, one can imagine some rehabilitation procedures based on this type of device.

Figure 2 illustrates an example of recordings obtained while submitting sinusoidal forces to a subject. EMG of biceps brachii and mechanical parameters have been recorded in the experimental situation described by figure 1. When transient forces are applied (fig.3) one can vary the final (constant) level of the force as well as the rate of onset ($\frac{dF}{dt}$), thus providing a powerful tool for the study of the dynamic response of the neuromuscular system.

The recordings have been obtained in healthy men but it is also possible to use the method with patients. An example is given in figure 4. The recording

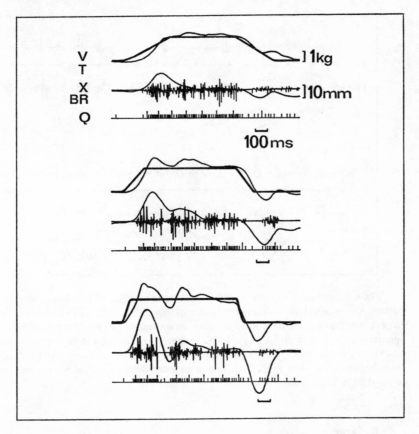

Fig.3. Effect of application of a force ramp of constant amplitude (2 kg). V = input voltage of function generator; T = tension recorded from strain gauge; X = displacement; BR = brachioradialis EMG; Q = integrated EMG. Upper, middle and lower group of recordings were obtained from the same subject by increasing the rate of change of applied force.

was made on a parkinsonian patient whose spontaneous tremor was at 3 cps; whatever the frequency of applied oscillation was, the EMG activity at 5 cps is not modified, while in the healthy subject this EMG activity is synchronous to the applied force. The interpretation of this independance between the frequency of parkinsonian tremor and the frequency of an applied oscillation force strongly supports the current ideas of a central origin for this tremor; it was also shown that these patients did not have a maximum oscillation at 4–5 Hz as healthy subjects do. The maximum was around 2 Hz.

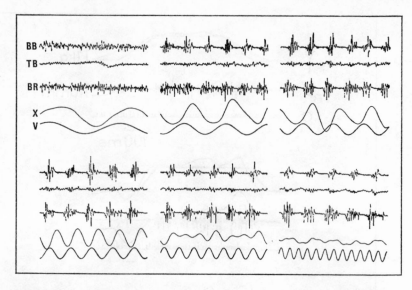

Fig. 4. Recording of movements of the forearm and muscular activity in a parkinsonian patient. A sinusoidal force was applied at different frequencies as described in figure 1. Peak to peak, amplitude of the force was 1 kg, mean value was about 1 kg. This patient shows a spontaneous tremor at about 5 cps. EMG of biceps brachii (BB); triceps brachii (TB); brachioradialis (BR) and recording of output displacement of the forearm (X) and input from the function generator (V). Each sample has a duration of 1 sec and corresponds from left to right to 1.5, 2.5, 3.5, 5 and 12 Hz.

References

BERTHOZ, A. and METRAL, S.: Behavior of a muscular group subjected to a sinusoidal and trapezoidal variation of force. J. appl. Physiol. *29*(3): 378–384 (1970).

BERTHOZ, A.; ROBERTS, W.J., and ROSENTHAL, N.P.: Dynamics characteristics of the stretch reflex using force inputs. J. Neurophysiol. *34*: 412–619 (1971).

METRAL, S.; BERTHOZ, A. et SCHERRER, J.: Effects de l'application d'une force sinusoidale sur la contraction d'un muscle chez l'homme. Rev. neurol. *118*: 563–567 (1968).

WALSH, E.G.: Oscillatory transients in the response to an abrupt change of force at the wrist in parkinsonism. J. Physiol., Lond. *209*: 33–34 (1970).

Author's address: Dr. A. BERTHOZ and Dr. S. METRAL, Laboratoire de Physiologie du Travail, CNRS, *Paris* (France)

Medicine and Sport, vol. 8: Biomechanics III, pp. 165–171 (Karger, Basel 1973)

The Joint Force and Moment Analysis of all Body Segments when Performing a Nonsymmetrical, Three-Dimensional Motion

S. PLAGENHOEF

University of Massachusetts, Amherst, Mass.

The kinetic analysis of human motion includes the determination of joint forces and moments of force. This entails the calculation of displacement, velocity and acceleration of all the body segments that can be defined as 'rigid'. This is defined as a body part that does not change its shape other than the small changes that occur due to muscle mass readjustments.

The anatomical data presented by DEMPSTER [2] has been the most widely used, but modifications of his data may be used for application to living material. Accepting his specific gravity data for the different body parts, the weight of each segment may be determined for the living human by using the water displacement method. With recent data supplied by CLAUSER et al. [1], the center of gravity of the living may be determined by resubmerging the body segments to the correct level of volume displacement. This correct level is between 46.5 and 42.6% of the total volume as the segment is submerged from the distal end. The center of gravity is, therefore, distal to the plane of mid-volume. The mean data as listed by CLAUSER et al. are listed in table I.

Table I

Segment	Volume from distal end to center of mass, %
Whole leg	42.6
Thigh	46.5
Calf and foot	46.4
Calf	45.1
Upper arm	45.6
Forearm and hand	45.2
Forearm	43.7

The moment of inertia of a living body segment cannot be determined, so the following method is suggested as the most reliable technique for analyzing a human motion. The moments of inertia as determined by DEMPSTER [2] are used to calculate the radius of gyration. The percentage difference between these data and his cadaver centers of gravity is then applied to the living human by determining the radius of gyration of the living relative to the measured center of gravity [see reference 3, p. 21, for radius of gyration data]. An example of a calculation to determine the radius of gyration on the living is as follows: the distance from the joint center (elbow) to the center of gravity of the forearm, according to DEMPSTER, is 43% of the forearm length. The calculated radius of gyration of the forearm is 52.6% determined from DEMPSTER's moments of inertia. The length of the forearm of a living person measured 30 cm, and the center of gravity was found to be 13.5 cm from the elbow or 45% of the length. The 9.6-percent cadaver difference added to 45% places the radius of gyration for the living segment at 54.6%, or 16.38 cm from the elbow. The corrected data would be more accurate for use than the cadaver data.

The anatomical data, velocity and acceleration data are used to determine joint forces and moments of force. The formulae needed are available in PLAGENHOEF's work [3] on pp. 28–57. The relative motion method of analysis was chosen over the absolute method so that the relative importance of the motion of each segment could be determined.

A symmetrical, planar motion was ideal for analysis because a single link system includes all parts of the body. However, a nonsymmetrical, nonplanar motion requires a complex analysis in order to determine the forces due to motion of all the body segments, as well as to retain the relative motion method of analysis. It is important to analyze all the links of the body even though there is no relative motion at a joint. This is due to the fact that muscle contraction is needed at this joint to stabilize it, and this effects the muscle action at other joints.

To obtain the forces due to motion of all segments the following method may be used. Separate frames of whole-body motion as recorded by motion pictures were analyzed. A link system was drawn which included all the segments of the body. A primary link system was selected that would best give the information desired. The forces due to the body segments outside of the selected primary chain were introduced as external forces at the appropriate points. Figure 1a–d represents the procedure used for a primary chain that includes the hip to hip link. The following steps are taken to analyze the whole-body motion.

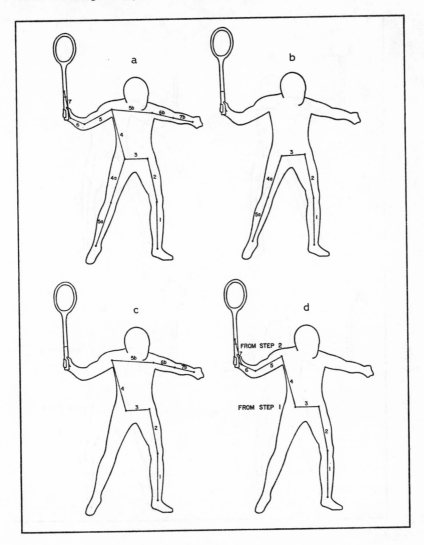

Fig.1. a Total link system (hip rotation). *b* External forces. Step 1. Use Fx, Fy, Mo of segment 4a, apply at end of segment 3 in step 3, *c* External forces. Step 2. Use Fx, Fy, Mo of segment 5b, apply at end of segment 4 in step 3. *d* Final link system for analysis. Step 3, Fx = Horizontal Force; Fy = Vertical Force; Mo = Joint Moment of Force.

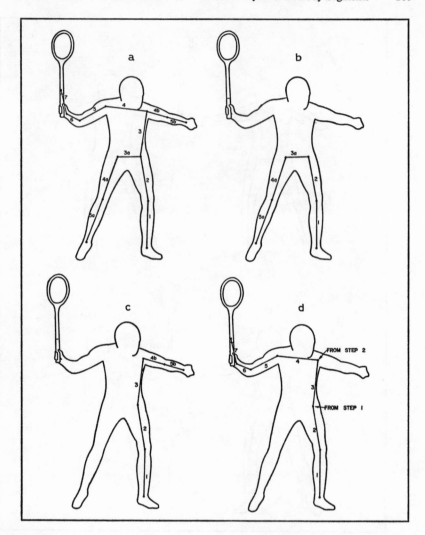

Fig. 2. a Total link system (shoulder rotation. *b* External forces. Step 1. Use Fx, Fy, Mo of segment 3a, apply at end of segment 2 in step 3. *c* External forces. Step 2. Use Fx, Fy, Mo of segment 4b, apply at end of segment 3 in step 3. *d* Final link system for analysis. Step 3.

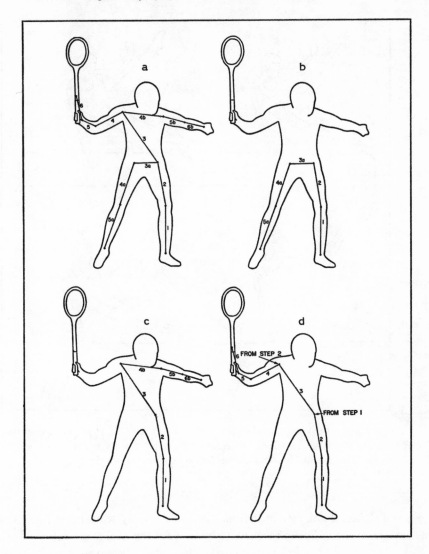

Fig. 3. a Total link system (diagonal, combined hip and shoulder rotation). *b* External forces. Step 1. Use Fx, Fy, Mo of segment 3a, apply at end of segment 2 in step 3. *c* External forces. Step 2. Use Fx, Fy, Mo of segment 4b, apply at end of segment 3 in step 3. *d* Final link system for analysis. Step 3.

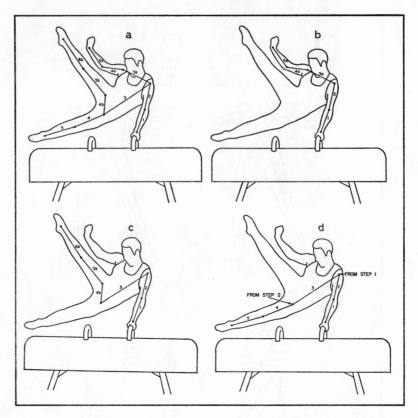

Fig 4. a Total link system. *b* External forces. Step 1. Use Fx, Fy, Mo of segment 3a, apply at end of segment 2 in step 3. *c* External forces. Step 2. Use Fx, Fy, Mo of segment 4b, apply at end of segment 3 in step 3. *d* Final link system for analysis. Step 3.

1. Select a link system that includes all the body segments (fig. 1 a) and includes a selected primary chain to be used for the final analysis (fig. 1 d).

2. Select the two link systems that include the body segments not included in the primary chain (fig. 1 b, c).

3. Determine the forces and moments of body segment 4a to be inserted at the end of segment 3 (step 1).

4. Determine the forces and moments of body segment 5b to be inserted at the end of segment 4 (step 2).

5. Apply the forces and moments obtained from steps 1 and 2 at the appropriate points and analyze the final link system (fig. 1 d) (step 3).

Different link systems may be chosen, depending on the path taken when moving from legs to arms through the trunk. Figure 2a–d shows a link system chosen to obtain information on the motion of the shoulders relative to the hips as they rotate about the vertical axis. Figure 3a–d shows a link system chosen to obtain information pertaining to the combined motion of the hips and shoulders. The same procedure is followed for analyzing the whole motion as shown in figure 1a–d. Figure 4a–d shows how the procedure would be followed if the hand were the fixed point rather than the foot.

The anatomical data needed for this type of analysis are only partially available. Until cadaver dissections are made available, it is necessary to construct a model using DEMPSTER's mass distribution data [2] to obtain any desired unknown data [see reference 3, pp. 20–27]. However, the method of analysis presented here makes it possible to obtain the body segment forces and moments for the whole body during any 3-dimensional motion.

References

1   CLAUSER, C.E.; MCCONVILLE, J.T., and YOUNG, J.W.: Weight, volume, and center of mass of segments of the human body. Wright Air Development Center Report, AMRL-TR-69-70 (Wright-Patterson Air Force Base, Ohio 1969).
2   DEMPSTER, W.T.: Space requirements of the seated operator. WADC Technical Report (1955).
3   PLAGENHOEF, S.: Patterns of human motion (Prentice-Hall, Englewood Cliffs 1971).

Author's address: Prof.S.PLAGENHOEF, University of Massachusetts, *Amherst, Mass.* (USA)

Medicine and Sport, vol. 8: Biomechanics III, pp. 172–174 (Karger, Basel 1973)

An Estimate of Tension Exerted by Hip Extensors During Two Different Balance Poses on the Right Lower Extremity

ALICE L. O'CONNELL

Sargent College of Allied Health Professions, Boston University, Boston, Mass.

Line drawings were made from film of the subject in each of 2 poses. Subsequently, the pelvis and bones of the weight-supporting extremity were included in each drawing (fig. 1, 2), and force diagrams completed in each. The weight of the body supported on the head of the right femur was determined by summing segmental weights, and found to be 111 lb. Location of the center of gravity of the same mass was then made by summing the force moments around the hip axis exerted by each body segment (segmental method for locating the center of gravity). The line of gravity in pose I fell 0.44 in. anterior to the hip axis, but 6.7 in. anterior to the axis in pose II. When the line of force of the hamstring muscles was added to pose I it made an angle of 99° with a horizontal through the hip axis, and crossed the horizontal 3.4 in. posterior to that axis. As both poses form an equilibrium situation, the sum of the moments around the hip axis in each must be equal to zero.

In pose I:

$$3.4 \text{ in} \times VC_H - 0.44 \text{ in} \times 111 \text{ lb} = 0,$$

$$VC_H = \frac{0.44 \text{ in} \times 111 \text{ lb}}{3.4 \text{ in}}$$

$$= 14.36 \text{ lb}.$$

Total muscle force of the hamstrings (TMF_H) in this pose will equal the vertical component of the hamstrings divided by the sine of 99°:

$$TMF_H = \frac{14.36 \text{ lb}}{0.98769}$$

$$= 14.5 \text{ lb}.$$

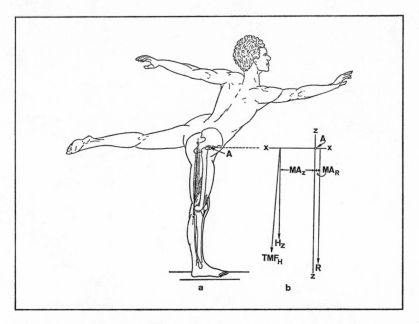

Fig. 1. a Line-drawing of pose I with pelvis, femur, tibia and fibula, and biceps femoris included. *b* Force diagram of both gravitational and hamstring forces acting at the hip. A = hip axis; x–x = horizontal line through hip axis; z–z = vertical line through hip axis; TMF_H = total muscle force exerted by hamstrings; H_z = Z or vertical component of TMF_H; MA_z = moment arm of Z-component; R = resultant of segmental weights supported by right hip joint; MA_R = moment arm of resultant.

In pose II, on the other hand, the line of hamstring force crossed the horizontal through the hip axis at 3.7 in. posterior, but at the angle of 84° (fig. 2b).

In pose II:

$$3.7 \text{ in} \times VC_H - 6.7 \text{ in} \times 111 \text{ lb} = 0,$$

$$VC_H = \frac{6.7 \text{ in} \times 111 \text{ lb}}{3.7 \text{ in}}$$

$$= 206.6 \text{ lb}.$$

TMF_H will equal vertical component divided by the sine of 84°:

$$TMF_H = \frac{206{,}6 \text{ lb}}{0.99452}$$

$$= 218.73 \text{ lb}.$$

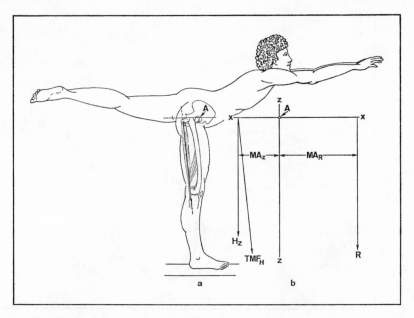

Fig. 2. a Line-drawing of pose II with pelvis, femur, tibia and fibula, and biceps femoris included. *b* Force diagram of both gravitational and hamstring forces acting at the hip. A = hip axis; x–x = horizontal line through hip axis; z–z = vertical line through hip axis; TMF_H = total muscle force of hamstrings; H_z = Z or vertical component of TMF_H; MA_z = moment arm of Z-component; R = resultant of segmental weights supported by right hip joint; MA_R = moment arm of resultant.

While the solution to this problem is based on a number of assumptions and approximations, it at least serves to provide an estimate of the comparative amounts of muscle force exerted to support the body in each of the 2 poses. This procedure can, of course, be used to calculate similar forces acting on other joints, and it can be carried further to determine the resulting pressures on the articulating surfaces.

Reference

O'CONNELL, A. L. and GARDNER, E. B.: Understanding the scientific bases of human movement (Williams & Wilkins, Baltimore, July 1972)

Author's address: Dr. ALICE L. O'CONNELL, Sargent College of Allied Health Professions, Boston University, *Boston, MA 02215* (USA)

Medicine and Sport, vol. 8: Biomechanics III, pp. 175–180 (Karger, Basel 1973)

A Method of Three-Dimensional Analysis of Twisting Movements

W. Duquet, J. Borms and M. Hebbelinck

Navorsingslaboratorium HILO, Vrije Universiteit Brussel

Conventional cinematographic analysis usually is made on images in a single plane. For many physical performances, at least one body segment leaves this plane. If the particular segmental movement forms an essential and integral part of the whole body movement, analysis in only one plane gives a distorted pattern of what actually takes place.

This paper describes a technique for analysis of synchronous frames of movie film from 2 cameras, one mounted at the side of a subject giving an image in the vertical plane and the other mounted overhead giving an image in the horizontal plane. By a procedure of graphically manipulating projected anatomical reference points to a common plane, it is possible to determine actual positions in space and obtain dimensions and angles and, from these, derive covered distances, linear velocities and accelerations as well as angular velocities and accelerations.

As shown in figure 1, synchronous projections of anatomically designated points A, B are projected as A^vB^v in the vertical plane and A^hB^h in the horizontal plane. In order to manipulate these points graphically, our projections on paper are made with the horizontal plane located directly below the vertical plane. In effect, we have rotated the horizontal plane 90° with respect to the vertical plane so $A^vA'A^h$ and $B^vB'B^h$ are straight lines. Because reference points are synchronous, it should be noted, as in figure 2, that the lines joining the points in each plane $A^vA'A^h$ and $B^vB'B^h$ are parallel.

To find the real magnitude of the distance A–B, regardless if this is a distance between points from one frame or between points on different frames, we must conceptually move one of the points so it is in the same

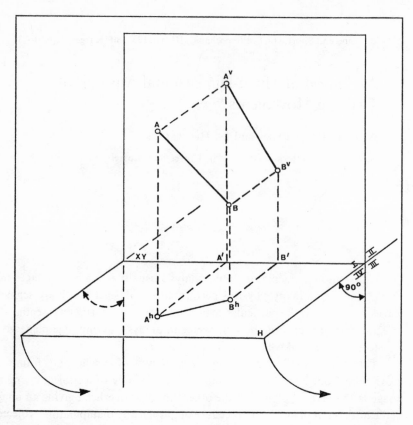

Fig. 1. The arbitrary distance AB and its projections A^vB^v and A^hB^h. The horizontal plane is rotated 90° downwards in order to obtain figure 2.

frontal plane as the other and view the real magnitude A^vB^v from the side view[1].

This is accomplished diagrammatically as shown in figure 3 by using a vertical line, Sch, as an axis and relocating B in a horizontal plane as shown by movement of B^h to B_1^h which, correspondingly, is viewed from the side as moving B^v to B_1^v. It should be noted that $B^hB_1^h$ is an arc of a circle in the horizontal plane with the diagrammatic center at Sch^h; as viewed from the side, this movement is seen as the horizontal line $B^vB_1^v$. The distance $A^vB_1^v$ represents the actual distance between points A and B.

1   In some analyses it is simpler to locate points in the same horizontal plane and view from the overhead position.

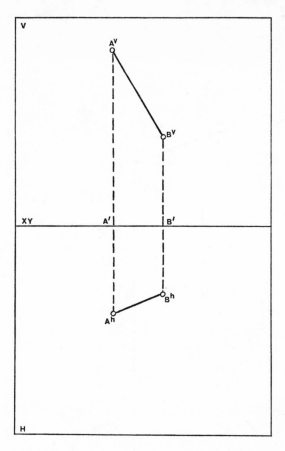

Fig. 2. The projections A^vB^v and A^hB^h determine the real spatial position of any arbitrary line AB.

Method of Three-Dimensional Determination of a Real Angle

An arbitrary angle formed by straight lines SH and HK, intersecting at H, is represented by the vertical and horizontal projections $S^vH^vK^v$ and $S^hH^hK^h$ (fig. 4).

In order to determine the real magnitude of angle $S\hat{H}K$, its projection will be rotated 3 times until it stands perpendicular to one of the points of view (overhead or side view).

First rotation. S is rotated to a position as far from us as H ($S^h \rightarrow S^h_1$). S^v becomes S^v_1, K^h moves to K^h_1 and K^v_1. (Conditions: angle $K^h\hat{H}K^h_1 =$ angle

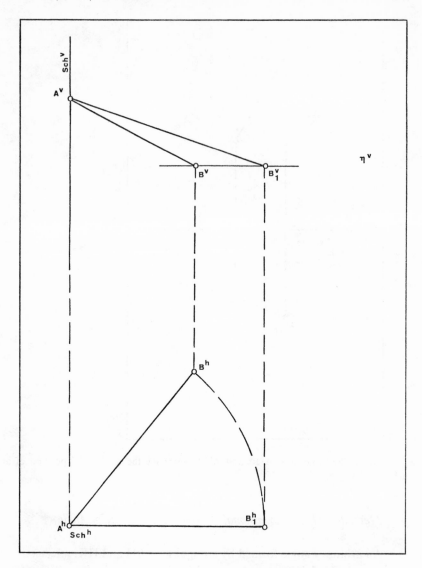

Fig. 3. Example of the determination of the real magnitude of an arbitrary distance AB.

$S^h \hat{H}^h S_1^h$; $K^v K_1^v$ and $S^v S_1^v$ are horizontal.) The axis of rotation was a vertical straight line through H.

Second rotation. S_1 is rotated to a position vertical under H: axis is the headline through H. S_1^v becomes S_2^v, while K_1 must rotate to conserve the angle

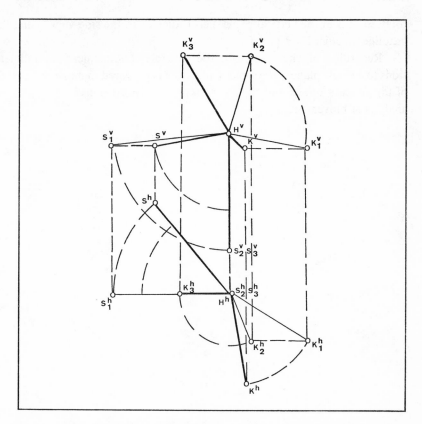

Fig.4. Example of the determination of the real magnitude of an arbitrary angle SĤK.

SĤK: $K_1^v \rightarrow K_2^v$ (condition: angle $K_1^v \hat{H}^v K_2^v$ = angle $S_1^v \hat{H}^v S_2^v$) and K_1^h goes to K_2^h (perpendicular to the horizontal point of view).

Third rotation. $S_2 H K_2$ is displaced to $S_3 H K_3$, so that the plane determined by $S_3 H K_3$ will stand perpendicular to the horizontal point of view. Using the vertical line through HS_2 as axis, K_2^h is rotated to the position K_3^h, where K stands as far from us as H and S. K_2^v moves to K_3^v (condition: $K_2^v K_3^v$ is horizontal); S_3 takes the same spatial position as S_2 because S_2 is a part of the axis. The angle $K_3^v \hat{H}^v S_3^v$ is the real magnitude of the angle KĤS, since the points K, H and S have been projected so they are equidistant from our point of view. The angle KĤS has not changed during the different rotations since $K^h \hat{H}^h K_1^h$ = $S^h \hat{H}^h S_1^h$ and $K_1^v \hat{H}^v K_2^v = S_1^v \hat{H}^v S_2^v$.

We might note that it would also have been possible to start with a rotation of S in the vertical plane, perpendicular to the side view, using the headline through H as an axis.

Regardless of where we begin, the principle of adjustment of all dimensions to a single plane is the same. Unless one is prepared to make these kinds of adjustments, one cannot avoid serious perspective error in cinematographic analysis of movement.

Author's address: W. DUQUET, Navorsingslaboratorium van het HILO, Vrije Universiteit Brussel, Hegerlaan 28, *B-1050 Brussels* (Belgium)

Medicine and Sport, vol. 8: Biomechanics III, pp. 181–184 (Karger, Basel 1973)

An Opto-Electronic Technique for Analysis of Angular Movements

D. L. MITCHELSON

Institute for Consumer Ergonomics, University of Technology, Loughborough

A search of the literature of biomechanics reveals a lack of instrumentation for making on-line recordings of the angular displacement, velocity and acceleration of many limb and body segments, and there is a particular need for a means of recording the relative and absolute angular parameters of non-adjacent body segments.

It was, therefore, decided at the Department of Ergonomics and Cybernetics of Loughborough University of Technology, to attempt to fill this need by developing a technique based on the use of polarised light.

Polarised Light Technique

During the course of developing the apparatus, it was found that the technique was similar in some respects to that reported by REED and REYNOLDS [1969] and GRIEVE [1969].

The principles of operation are illustrated by figure 1. Light emitted from a DC light source is polarised by transmission through a disc of linearly polarising filter. The disc is made to rotate in excess of 125 revolutions/sec. The light is received by a transducer which consists of a pair of matched photo-cells in front of each of which is mounted a piece of polarising filter with the planes of polarisation of the 2 filters fixed at 90° to each other. The outputs from the 2 photo-cells are connected in opposition, consequently any non-polarised light incident on the transducer is rejected. The signal produced by the polarised light whose plane of polarisation is rotating, is sinusoidal and of a frequency twice that of the revolution rate of the disc.

The elapsed time between the occurrence of a reference pulse produced

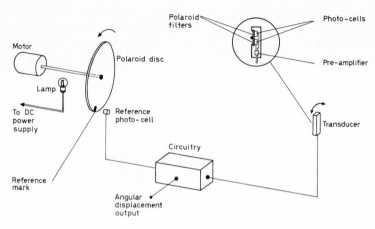

Fig. 1. Schematic diagram of angle sensing apparatus.

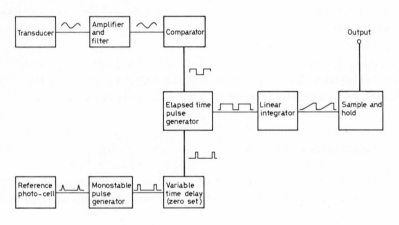

Fig. 2. Block diagram of angular displacement circuitry.

when a mark on the edge of the disc passes in front of a fixed photo-cell and the time at which the sinusoidal signal next crosses zero is measured in the main angle sensing circuitry which is illustrated in block form in figure 2. The elapsed time is electronically converted to a DC output voltage which is proportional to the angular displacement of the transducer from the fixed angular reference determined by the position of the reference photo-cell.

This technique differs from that of REED and REYNOLDS [1969], in that in their apparatus the DC output is obtained by detecting the phase difference between the sinusoidal signals received by a reference photo-cell and the subject-mounted photo-cells. This phase difference is then converted to a square wave pulse train whose duty cycle is proportional to the angular displacement. Subsequent integration and filtering produces the DC output. Unfortunately this process of smoothing inevitably leads to considerable decrease in the frequency response of the apparatus. There is also marked deviation from linearity when the angular displacement approaches 0 or 90°.

In the technique herein described, the DC output is obtained by allowing a linear integrator to ramp up during the elapsed time between the reference pulse and the following zero crossing of the sinusoidal signal. The voltage level which the ramp achieves at the end of the time period is fed to a sample-and-hold circuit which holds the output at that DC level until the following sampling period. The output level is thus updated to the exact value of the angular displacement of the transducer once in each revolution of the disc. When fast angular movements occur the output then appears as a 'staircase' in which the correct angular displacement is represented at the start of each step.

This technique means that the frequency response is the maximum which may be realised for a given revolution rate of the disc.

The output is found to be linear to within 0.3° over the range 0–180°. If the transducer rotates through more than 180°, the output simply returns to the zero level and continues to rise linearly with angle as before.

Additional circuitry has been built which samples the angular displacement output and records the difference between the levels occurring during consecutive sampling periods. In this way, angular velocity output is obtained by what is an electronic finite differences method. Two such angular velocity circuits connected in series provide an angular acceleration output.

Photo-cell transducers may be mounted on all of the relevant limb and body segments of the subject under study to produce simultaneous angular displacement, velocity and acceleration outputs. At present, the outputs are recorded on an ultra-violet paper chart recorder. It is anticipated, however, that outputs will be recorded on analogue magnetic tape and subsequently digitised on entry to a digital computer. Much of the interpretation of the data can then be handled very quickly by the computer. The chief disadvantages of this system lie in the usual problems of line-of-sight common to the film techniques. Also, errors may occasionally be introduced due to certain geometric factors.

If Φ is the angular displacement of a transducer rotated within a plane normal to the line of sight and Ψ and Θ are rotations performed in sequence about the two other axes orthogonal to the line of sight, then the angle recorded by the apparatus, Ω, is related to Φ, Θ and Ψ by the expression:

$$\text{Tan } \Omega = \frac{\tan\Phi \cos\Psi}{\cos\Theta} + \text{Sin}\Psi \tan\Theta.$$

Thus, provided either Ψ or Θ is zero, the error will be zero. In practice, this condition can often be met without imposing special restrictions on the movements under study.

Experimental Applications

This technique has been used successfully at Loughborough to analyse limb and body movements in relation to work skills. It has been possible, for the first time, to examine in detail some interesting relationships in the co-ordination of movements for such skills. Space limitations unfortunately do not permit presentation of the details of these studies.

Future Developments

It may be fruitful to use this opto-electronic technique to provide immediate feedback to subjects, via a suitable display, so that they can monitor particular parameters of their movements and so improve motor skills.

Plans are going forward at Loughborough for the development of opto-electronic techniques for recording linear movements within a 3-dimensional Cartesian co-ordinate system. This will provide a natural complement to the apparatus for recording angular movements. Such a total movement recording system will allow rapid, accurate and relatively easy analysis of limb and body movements to be carried out.

References

GRIEVE, D.W.: A device called the Polgon for the measurement of the orientation of parts of the body relative to a fixed external axis. J. Physiol, Lond. *201* (1969).

REED, D.J. and REYNOLDS, P.J.: A joint angle detector. J. appl. Physiol. *27*(5): 745–748 (1969).

Author's address: Mr. D.L. MITCHELSON, Institute for Consumer Ergonomics, University of Technology, *Loughborough, Leicestershire* (England)

Medicine and Sport, vol. 8: Biomechanics III, pp. 185–188 (Karger, Basel 1973)

A Three-Dimensional Method for Measuring Small Articular Movements

P. C. ANTHONIS

Laboratory of Human Anatomy and Embryology, Université Libre de Bruxelles

Analysis of movement such as we conceive it classically is an analysis on the 2-dimensional plane and not in 3-dimensional space. For investigating any movement in space, we may select 2 planes of reference, generally placed perpendicularly, and in each of these we shall observe a projection of the true movement, i.e. a component of this actual movement in accordance with this plane.

Experimental Device

For the purpose of studying the movements of cervical vertebrae we took them from a fresh corpse and preserved them in 45% alcohol during the few days of experimenting. Vertebrae pairs were dissected so as to preserve only the intervertebral discs, the joint capsules and the ligaments.

Each pair of vertebrae was fittet on a wooden block in such a manner that the lower vertebra was fixed and the upper vertebra mobile on the first one. A firm fixation was ensured by partially bedding the lower vertebra in plaster. For the purpose of studying – in space – the movements of the mobile vertebra on the fixed one, we implanted into the body of the upper vertebra an iron wire, white-coated 'T'. The 'block-vertebra a–T' unit was fitted on a horizontal plate revolving around a strictly vertical shaft. This plate carried a tray in which the base was attached by means of 4 long screws. The apparatus body was fitted with a pitch circle and could be blocked in any position. It was fitted with a nonius and micrometric setting which allowed for orientating in the direction chosen with a 5-arc minute accuracy. Behind this assembly was fitted a black screen with a pattern of squares, the lines of which were horizontal and vertical.

Data Recording Principle

Definition in space of any position of the upper vertebra is obtained by a face-view photograph of the 'T' with a subsequent photograph after a 90° rotation of the element. Both photographs are then projected, one next to the other, on square-ruled paper, while taking care that the squares of the screen are accurately aligned and also with respect to those of the paper.

In this manner we obtain a working drawing that only has to be solved in order to know the angle of bending-stretching, of lateral inclination and of rotation.

Statistical Interpretation of Results

The 10 actual values of movement components obtained after having deducted rest values from values observed, were grouped in a chart in order to compute the average, the variance, the guage margin, the average standard error, the variation coefficient in percent and the correlation between the components.

The components of movements with 2 or 3 components were then compared with pure bending-stretching, left or right inclination and left or right rotation movements by means of the Student-Fisher t-test for small samples. Each pair of vertebrae requires 3 types of diagrams in order to show the 3-component reciprocal influence, two by two: bending-stretching and inclinations; (2) bending-stretching and rotations, and (3)inclinations and rotations. Differences of averages are represented by arrows parallel to abscisses and to ordinates. The asterisks indicate the degree of significance of eventual correlations within the group of measures or that of the differences of averages.

In reality, we used 6 diagrams for a single pair of vertebrae for the sake of greater clarity: the first 3 are applied to movements with 2 components, the other 3 to movements with 3 components. In this latter case, we obtain 2 groups of points for each combination of movements.

Results

A. Bending, Stretching Inclinations (2 Components)

Stretching components are smaller than pure stretchings. Bending components are, in general, not significantly different from pure bendings. Components are smaller than pure inclinations, in stretching-bending; they are not different in cases of bending-inclination.

B. Bending, Stretching Rotations (2 Components)

Stretching components are smaller than pure stretchings. Bending components are not or only slightly different from pure bendings. Rotation components are not different from pure rotations in stretching-rotations; they are different, but in various directions, in cases of bending-rotation. Pure rotations and corresponding components are smaller at the side of distorted posterior joints than at the healthy side.

C. Inclinations and Rotations (2 Components)

Inclination and rotation components are increased when their direction is the same, they are decreased when they have opposite directions. Inclinations and corresponding components are affected very little or not at all by articular distortions, while rotations and corresponding components are decreased on the affected side.

D. Bending, Stretching Inclinations (3 Components)

Bending and stretching components behave as shown in section A. Inclination components are different, depending upon whether the rotation components show the same direction or an opposite one.

E. Bending, Stretching Rotations (3 Components)

Bending and stretching components are smaller than pure corresponding movements. Rotation movements are different depending upon whether the inclination component is of the same or opposite direction (cf. sections C and D above).

F. Inclinations and Rotations (3 Components)

Inclination and rotation components behave as in section C above. It does not seem possible to deduce from the observations a clear and systematic influence of bending or stretching components on combined inclination and rotation components.

By this method we also investigated the movements of the carpian scaphoid in the movements of radial and cubital inclination of the wrist. The metacarpals of the wrists used were nailed to a round piece of wood corresponding with the hollow of the hand, itself attached to a supporting board. Then the radius and the ulna were severed at the junction of the upper third and the lower two-thirds were nailed to a wood block which, while maintaining them in a secure position, allowed for imprinting on the wrist strict radial or cubital inclination movements by sliding of the supporting board. The witness 'T' was implanted into the scaphoid and the whole was attached to the revolving plate.

The analysis shows the following results: (1) a bending movement is significantly connected with the cubital inclination of the scaphoid during the radial inclination of the wrist; (2) a stretching movement is significantly connected with the radial inclination of the scaphoid during the cubital inclination of the wrist, as well as a slight supination movement, and (3) a significant correlation exists in scaphoid movements between the bending-stretching components and the radial or cubital inclination.

Conclusions

This method allows for significantly measuring at the 0.001 threshold angle differences of a quarter degree and sometimes less (0.22°), in the 3 dimensions of space; it furthermore allows for: (1) defining the accurate position of an object in space as compared to 2 planes of reference, and (2) showing the influence of one movement component on another, within a movement taking place in space at 3 dimensions.

Summary

A method for analysing small articular movements in space has been elaborated and applied to cervical vertebrae as well as to the movements of the carpian scaphoid. This method consists in solving the working drawing, the two views of which are face and side photographs of a witness 'T' implanted into the body of the upper vertebra moving on the lower one, or into the scaphoid. It allows, with great accuracy, for defining in space the position of an articulation and for studying the influence of various movements or movement components, one in relation to another.

Author's address: Dr. P. C. ANTHONIS, Laboratory of Human Anatomy and Embryology, Université Libre de Bruxelles, rue aux Laines, 97, *B-1000 Bruxelles* (Belgium)

Medicine and Sport, vol. 8: Biomechanics III, pp. 189–195 (Karger, Basel 1973)

Evaluation of Athletic Fitness
in Weight-Lifters through Biomechanical,
Bioelectrical and Bioacoustical Data

S. Cerquiglini, F. Figura, M. Marchetti and A. Salleo

Institute of Human Physiology, Rome

It is commonly known that the athletic fitness of the weight-lifter depends chiefly on the degree of his skill as well as on the strength of muscles chiefly involved in performing the lifting. It was our intention to acquire knowledge about both these facets of athletic fitness, which could be directly used by coaches and athletes to improve training.

To evaluate the skill, a biomechanical study has been carried out concerning the pattern of movements in the kinds of lifting ruled for the Olympic Games, i.e. press, snatch and jerk. The following physical quantities and geometrical data have been taken into account: velocity of upward motion, force applied and trajectory of the barbell, as well as the angles formed by the skeletal segments at the joints. Instead of current kinematographic techniques, stroboscopic photokymography has been preferred, since through the latter more accurate data are more easily achieved. Muscle strength has been measured through isometric dynamometry. With the aim of selecting the intrinsic aspects of muscle function which could be monitored simultaneously with the performance of the exercise, to the trainers' and athletes' advantage, a physiological study has also been made of the EMG signal and the muscle sound, which have been recorded together with the muscle strength.

The muscle sound, neglected in the past, has been recently reconsidered by us in studies on the processes of muscular activation; we have proposed for such a recording method the name 'phonomyography' (PMG) [1].

Both the EMG and the PMG signals were submitted to frequency analysis which was carried out by means of a spectro-analyser; the instrument used by us plots the frequency and the amplitude of the harmonics which compose the signal in a tridimensional graphic, as function of time. Also the profile of the spectra has been obtained by means of a photoelectrical device designed by us.

On the whole, 11 subjects have been submitted to this study: 9 weight-lifting competitors of various levels of fitness, and 2 not used to physical exercises. Biomechanical measurements have been carried out on 8 of these subjects (all of them were weight-lifters, 4 being champions of Italy).

The three kinds of lifting, with loads corresponding to 80% of personal records of each athlete, were always carried out in the same sequence, i.e. press, jerk and snatch. Such a sequence was repeated for many days. The technique adopted for the strobokymographic recordings was similar to that used by other authors for analysing different physical exercises.

In short, such a procedure consists of lighting the subject by means of many flash-lamps synchronously activated at high frequencies and of shooting it on a film which is running slowly and continuously. The movement of the athlete is shown in a sequence of separate images which allows direct measurement of some geometrical quantities, as well as calculating, according to the method proposed by ZAJACZKOWSKA [2], the most interesting physical quantities such as velocity, acceleration and force. The maximum isometric strength of the lower limbs of all subjects has been measured in two typical lifting positions. The EMG and PMG recordings have been obtained on gastronemius (medial head) and quadriceps femoris (vastus lateralis). Muscle potentials were picked up by surface electrodes and the muscle sound by a microphone, screened for acoustic noise, and applied to the surface of the skin between the EMG electrodes. On the two inexperienced subjects and on two of the athletes that had been resting untrained until the beginning of the experiment, such measurements were repeated weekly for 2 months, during which all were submitted to an intense training program as is usual for weight-lifting.

The biomechanical study has given the following results. Each athlete is spontaneously able to repeat each exercise quite uniformly, even for a long period of time. The comparison of the various athletes, performing the same exercise, revealed that some of the quantities considered are very different, such as the force applied to the barbell and the angles at the joints. Also the trajectories of the barbell were found to be slightly different. On the other hand, the upward velocity of the barbell and, therefore, the acceleration transmitted to it, were found to be very similar, even in subjects of different weight divisions.

In figure 1 the lifting velocity during a snatch expressed by the best fitting curve of experimental data obtained from different athletes is plotted. Re-

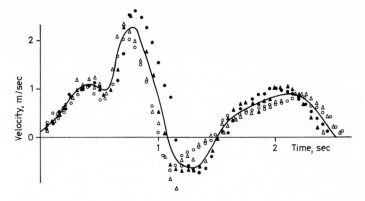

Fig. 1. Snatch. ○ = T.P., featherweight; △ = S.A., lightweight; ▲ = T.D., middle heavyweight; ● = V.R., heavyweight.

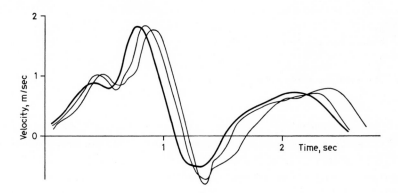

Fig. 2

garding the less skilful subjects, we can say that the more adroit they become, the closer they get to the champions from the point of view of lifting velocity.

Mistakes made during lifting are clearly revealed by an inspection of the velocity diagram. Such an effect is shown in figure 2: velocity curves of liftings during which the subjects themselves were aware that a mistake had been made, are compared to the curve of figure 1.

To summarise, the results of the physiological study are as follows:

1. Training causes changes in the maximum voluntary activation which are recognisable in EMG. Although such modifications, consisting essen-

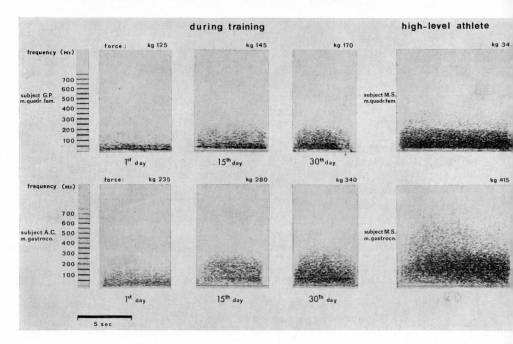

Fig. 3. Spectral analysis of EMG.

tially of higher voltage and frequency of the waves, are hardly distinguishable in the interference pattern of surface recordings, they are very evident in frequency analyses. In figure 3, three frequency spectra of EMG obtained at intervals of 15 days from 2 subjects submitted to intense training are shown. On the right are shown spectra of EMG obtained from champions. It is evident that during training, together with an increase in maximum isometric strength, the spectra are enriched by ever-increasing frequencies and, moreover, the amplitude of frequencies previously present is also raised. In subjects in excellent physical condition, such aspects of the EMG are so evident that the spectrum is absolutely distinctive compared to the untrained or simply less skilful subjects. A detailed examination of the profile of EMG spectra reveals that the modifications just described consist of a raised profile and, moreover, of an increase in the frequencies above 120 Hz as well as below 40 Hz (fig. 4).

2. Similar changes have been observed in muscle sound recordings. Also in this case the differences caused by training, which are hardly identifiable in the direct recordings, are clearly apparent in frequency analyses. An example

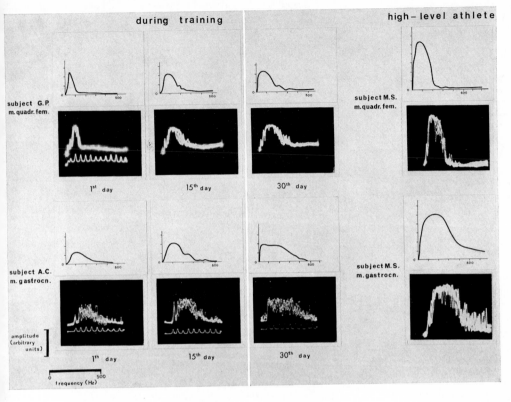

Fig. 4. Profiles of EMG spectra.

is shown in figure 5 referring to 2 subjects tested for PMG in the same way as for the EMG. The profile of the sound spectra (fig. 6) reveals that the PMG changes consist chiefly in a relative increase of higher frequencies (above 70 Hz).

The above results lead us to the following conclusions:

1. From the biomechanical analysis of lifting exercises performed by the best athletes, a model of motion has been reached. This is particularly meaningful as regards the acceleration transmitted to the barbell and, consequently, the velocity it acquires. In our opinion, conforming to such a model denotes good athletic skill. We suggest that an improvement of the training methods of junior athletes would result from a biomechanical exploration such as has been described in this paper. The most useful standard of judgement should result from the measurement of velocity which is, as we have noted, the

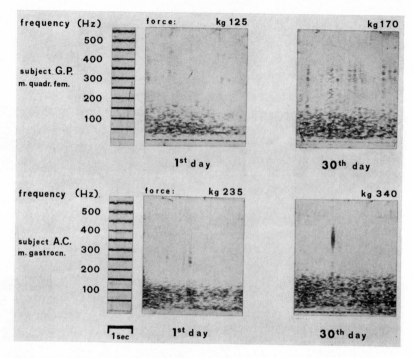

Fig.5. Spectral analysis of PMG during training.

quantity least dependent on body parameters as well as the more dependent on the style acquired. Such evaluation is equivalent to a comparison with the performance of champions.

2. The EMG as well as the PMG contains very relevant information about the functional improvement of muscles. As such information emerges from frequency analyses of the signals they could be easily picked up by the human ear, which has a high resolution power in the range of frequencies considered. Therefore, we suggest that the muscle potentials, transduced into acoustic signals, or the muscle sound *per se*, duly amplified, could be used to obtain a direct acoustic monitoring which should be presented to trainers and athletes during the course of lifting or during the basic exercises for this sport. Supplementing the proprioceptive information with the acoustic one, which is so meaningful as regards the effect of training on muscle, may represent a useful method to achieve a satisfactory athletic fitness more promptly.

Finally, we would emphasise the importance, from the general point of view, of being able to quantify the effects of training on muscles. We find this

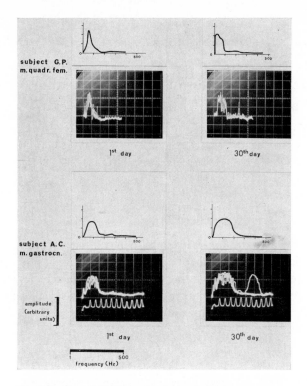

Fig.6. Profiles of PMG spectra during training.

following explanation of the frequency increase in EMG and PMG particularly interesting: it could depend on an increase of conduction velocity of action potentials in the case of EMG, and of activation velocity in the case of PMG. At present, the number of variables involved prevents us from drawing definite conclusions.

References

1   CERQUIGLINI, S.; MARCHETTI, M. e VENTURINI, R.: Spettrofonomiografia di muscoli scheletrici umani. Arch. Fisiol. *66:* 173–182 (1968).
2   ZAJACZKOWSKA, A.: Constant velocity in lifting as a criterion of muscular skill. Ergonomics *5:* 337–356 (1962).

Authors' address: Dr. SERGIO CERQUIGLINI, Dr. FRANCESCO FIGURA, Dr. MARCO MARCHETTI and Dr. ALBERTO SALLEO, Istituto di Fisiologia Umana, Università degli Studi, *I-00100 Rome* (Italy)

Medicine and Sport, vol. 8: Biomechanics III, pp. 196–200 (Karger, Basel 1973)

Possible Methods of Determining the Individual Motion Characteristics of Man

T. BOBER

Biomechanics Laboratory, Higher School of Physical Education, Breslau

The purpose of our investigations is to establish the existence of individual motion characteristics of man, specific for a given person, as well as to work out the most effective methods of measuring them. These characteristics were searched for among the following components of the motion structure: the spatial form, the rhythm, and the change of the reaction force [1, 2].

Method

The spatial analysis of the motion was carried out by means of cyclophotograms of direct shots in basketball. The rhythm was investigated by the acoustic recording of the high-jump run. The change of the reaction force was studied from oscillograms of the following motions on the tensometric platform: (1) vertical standing jump, and (2) knee-bending.

Material

The direct shots in a series of 10 were executed by juniors with 5-year practice and by 1 member of the national team. The running start for the high-jump was performed by 6 students of physical education and by 1 competitor of the master class (height 209 cm). The jumping bar was placed at about 75% of the subject's maximum jumping height. The test on the force platform was carried out by 3- to 7-person groups of students and skiers, each person performing 4–6 tests.

Results

The first investigations of use of reaction force recordings in identifying individual motion characteristics gave positive results [1]. By comparing a

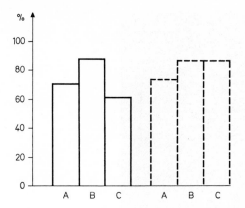

Fig. 1. Percentage of identified reaction force records of the vertical jump (—) and knee-bending (- - - -). The symbols A, B and C denote 3 different identifiers.

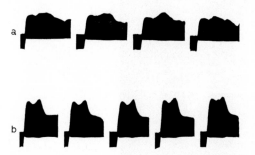

Fig. 2. Vertical component of the reaction force in the imitated ski-jump take-off. a ski-jumper's tests, b downhill racer's tests.

particular oscillogram of a particular subject with a series of coded oscillograms, we were able to identify the subject in 62–88% of cases from tests of vertical standing jumps, and in 75–88% of cases from knee-bending tests (fig. 1).

In this work, attention was also paid to the dependence of the motion characteristics on the sport specialization. For example, when skiers imitated the ski-jump take-off, it was easy to distinguish a ski-jumper from a downhill racer, as well as downhill racers among themselves. However, the reaction force recordings of the latter were less stable than those of the ski-jumper who specializes in this type of motion (fig. 2).

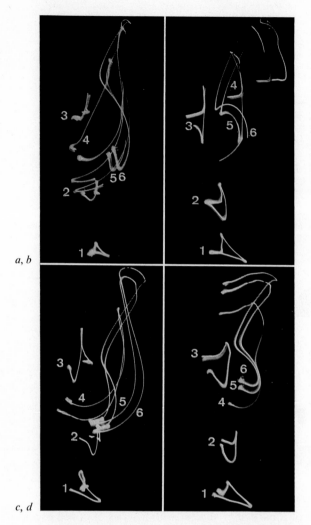

a, b

c, d

Fig. 3. Cyclophotogram samples of direct shots of 4 different subjects.

The next method of determining the individual human motion characteristics was the use of cyclophotograms of direct shots in basketball. The
light bulbs were installed at the following body sites: articulatio genus (1),
articulatio coxae (2), articulatio humero-scapularis (3), articulatio cubiti (4),
articulatio radio-carpea (5), and dorsum manus (6) (fig. 3). As illustrated in
figure 3, cyclophotograms of different persons differ essentially. On the other

Table I. Stability of individual motion characteristics

Kind of test	Examiners	Total number of tests	Number of positive identifications		
			first day	after 24 h	after 3 months
Knee-	A	35 =	32	30	32
bending	B	7 subjects	34	29	30
	C	× 5 tests	33	30	30
Vertical	A	30 =	22	23	20
standing	B	6 subjects	22	24	21
jump	C	× 5 tests	23	23	20

hand, succesive cyclophotograms of the same person are so alike that 4 examiners identified each subject faultlessly. It seems that, in this case, a quite correct identification is possible by limiting the spotlights to the following 3 sites: articulatio genus, articulatio coxae, articulatio humero-scapularis. It may be noted that the smoothest traces of the upper limb are those in oscillogram C in figure 3, which belong to the competitor with the highest hitting rate (8 scores from 10 shots).

While studying the rhythm of the high-jump run we confined ourselves to measuring the timing of the few final steps, without fully recording the acoustic effects of the run, i.e. the frequency and/or intensity. This is probably the reason why this method proved less effective than those described above. However, by comparing the timing of the two final steps we were able to distinguish a trained jumper from untrained subjects.

The above results have already been supplemented with preliminary investigations of the stability of the thus-established motion characteristics. By connecting the tensometric platform to an oscillograph, we examined repeatedly the vertical standing jump and the knee-bending at 24-hour and 3-month intervals. The results show fairly good stability and are given in table I.

At the time of the above experiment, the subjects had not attended any specialized training sessions.

Conclusions

Our results show that it is possible to identify a person by certain characteristic features of his motion structure. The identification certainly amounts to 70–90% when based on the reaction force change in simple motions, and is even higher for spatial motion characteristics. Rhythm analysis, when confined to measuring only the step-timing, appears to be a poor identification method. The reaction force change is a quite stable motion characteristic, as proved by experiments repeated after 3 months.

References

1   BOBER, T. i DOMITER, B.: Przyczynek do badań nad indywidualnymi cechami ruchu człowieka. Wychow. Fiz. Sport (1): 69–74 (1970).
2   FIDELUS, K. i GRADOWSKA, T.: Porównawcza ocena danych o wyskoku dosiężnym. Mat. Szkolen. PKOL 23 (5): 8–9 (1965).

Author's Address: Dr. TADEUSZ BOBER, Biomechanics Laboratory, Higher School of Physical Education, Breslau (Poland)

Medicine and Sport, vol. 8: Biomechanics III, pp. 201–208 (Karger, Basel 1973)

The Usefulness and Limitations of the Technique of Photo-Elastography in the Static and Dynamic Study of the Skeleton[1]

L. Perugia and S. Bonaccorsi

Institute of Orthopaedics and Traumatology, University of Rome, Rome

In the field of the anatomy and physiology of the skeleton there is no doubt that our knowledge of the qualitative and quantitative distribution of the mechanical forces is of particular interest. These forces act on bone segments subject to physiological stresses (such as the force of gravity, muscular tension and capsular and ligamental resistence) and are arranged in the positions allowed by joint movements under normal conditions.

It is equally understandable that in the field of pathology and, therefore, that of the therapies to be adopted, great value is placed on a rapid and complete knowledge of the modifications that take place in the distribution of static and dynamic forces in areas of the skeleton that are deformed, or parts acting abnormally on each other or subject to atypical stresses, whether in direction, intensity or point of application. As a consequence, the success of orthopaedic surgery, whether morphological or on axial orientation, can be judged by the extent to which the static and dynamic physiological condition, previously impaired, has returned to normal.

Photo-elastography has been shown to be a technique capable of providing an immediate and complete chart of the distribution of mechanical stresses on the structure of a given area. It is, as is well-known, a technique that permits analysis of the degree of density of lines of light in homogeneous transparent resin models (made of Perspex, Araldit or similar materials), either 2- or 3-dimensional, when such models are traversed by a beam of polarized light as they are subjected to varying degrees of mechanical pressure applied by suitable instruments (such as presses, clamps, etc.).

1   Paper presented at the 3rd International Seminar on Biomechanics, Rome, 27th September to 1st October 1971.

The ease with which models can be made in a form that reproduces fairly accurately the desired cutting plane (frontal, sagittal, transversal or oblique) of the bone segment under examination, generally leads to a preference for 2-dimensional photo-elastography. If particularly complex segments (the bones of the ileum, spine, etc.) are involved, however, it is necessary to build 3-dimensional models. This may incur greater difficulty in preparation, but it produces a closer correspondence to the anatomical reality.

This technique is especially useful because it allows us to examine segments of different forms, variously disposed in space, depending on the principal examination to be carried out. It also permits us to trace, by using suitable stress instruments, the direction, intensity and area of application of the forces, and makes it easier to compare immediately the various photo-elastographs that have been taken and even, sometimes, compare these with X-rays of the analyzed structures.

This becomes all the more true when we remember that the orientation of bone laminae in the various segments of the skeleton follows the orientation of the lines of force by which they are traversed.

Of course, photo-elastography is of maximum utility in that it offers the possibility of carrying out comparative analyses of normal and abnormal forms, or of models disposed spatially to correspond to possible joint movements or, again, of models that have been appropriately modified (which happens as the result of different surgical techniques).

It should, however, be made clear that photo-elastography can be of considerable use, but always relatively, never absolutely.

The properties of bone structures, with their various physical and chemical components are, in fact, always different from those of the materials subjected to photo-elastography.

The Institute of Orthopaedics and Traumatology of the University of Rome has carried out studies on a number of normal and pathological conditions in various osteo-articular areas (knee, hip, sacroilium, pelvis, spine, etc.) using the 2-dimensional and sometimes the 3-dimensional photo-elastographic techniques. This research has proved extremely useful both in casting light on aetiopathogenetic problems and in helping to determine the orientation of surgery on certain important orthopaedic and traumatological conditions. For the most part, these results have been confirmed by parallel research in the clinical and radiological fields, which were deliberately carried out in conjunction with purely experimental work.

On the knee, for example (fig. 1), studies have been made on the distribution of the load on the interior structure and an analysis made of the distri-

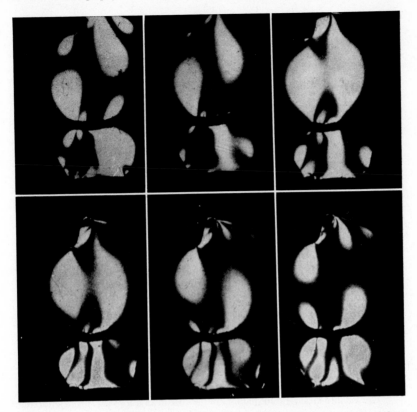

Fig. 1. Photo-elastographic pictures of the knee in maximum extension. From left to right and from top to bottom: rotations of polarized plane of 15° in 15°.

bution of weight-bearing forces during the flexion-extension movement over a range corresponding to the first 30° (beginning from maximum extension), since this range corresponds to the amount of movement generally made by a person in standing position or during physiological deambulation. The result has been a specification of the areas most frequently involved. In order to establish the division of stresses according to the principal of biological economy, comparison was then made with trabecular areas that can be observed in sagittal anatomical section, producing exceptional similarity of distribution.

The static function of the knee (fig. 2), under conditions of normality, valgismus and varismus was emphasized here, too, the presence of areas of high pressure which corresponded anatomically and radiologically, to analogous areas of trabecular accumulation.

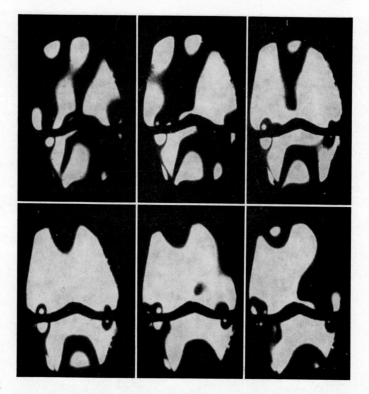

Fig. 2. Photo-elastographic pictures of the knee in anterior view. From left to right and from top to bottom: rotations of polarized plane of 15° in 15°.

As regards the hip, the 2-dimensional technique was used for experiments on certain kinds of traumatic lesion of the acetabulum. Concentrating on the most frequently used pathogenetic mechanism, i.e. the force transmitted from the knee along the femoral diaphysis and epiphysis to the pelvic segment with the hip in flexion, suitable stresses were applied to 2 models reproducing the morphology of the pelvic and femoral segment as shown in transversal section for a seated subject with a variation between them of an angle of 30° (open anteriorly). It was possible, from the different data, to confirm that when the abduction-adduction angle of the femur and the flexion of the hip at the moment of application of the traumatic force is modified, the posterior part of the acetabulum is subjected to different stresses. Considerable influence is also exercised by the degree of intra-rotation of the proximal epiphysis of the femur at the moment of trauma. The distribution of forces inside the model of

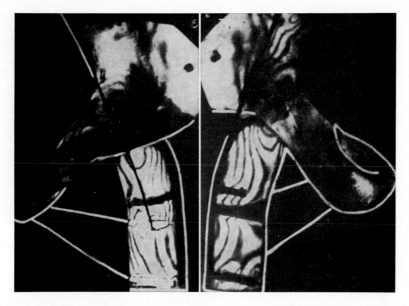

Fig. 3. Photo-elastography on 3-dimensional models of the hip with subluxation after a Salter pelvic osteotomy.

the pelvis is further influenced to an important extent by an increase of tension at the level of the horizontal branch of the pubic bone. This latter factor appears to contribute towards concentration of the traumatic shock towards the acetabulum socket.

Given the complex nature of the structure of the pelvis, 3-dimensional models were used for photo-elastography for studing the pathogenetic mechanism of fractures of the acetabulum, and straingauges were used to take measurements on analogous models. This double experiment served for a quantitative and qualitative demonstration of the mechanical forces induced on the models by the appropriate compression apparatus. When force was applied along the vertical axis of the pelvis, a remarkable amount of pressure was observed at the level of the posterior wall of the acetabulum and on the ileo-pubic branch. Strain was observed on the lower part of the pubic symphysis and on the whole internal curve of the obturating foramen. Applied to the clinical field, these discoveries may explain certain fractures of the ileo-pubic and ileo-ischiatic columns, given the concentration and, therefore, interference of their actions at the level of the base of the acetabulum. By varying the position of the extremity of the femur it was possible to visualize,

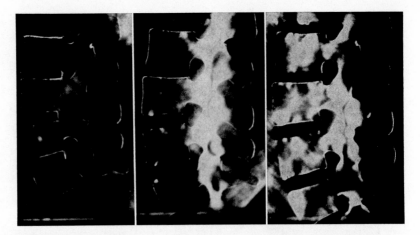

Fig.4. Photo-elastographic picture of the lumbar spine in lateral view. Orthomorphic position.

in particular, the evident pressure on the anterior part of the acetabulum when the hips are in abduction and extra-rotation.

Still in the field of traumatology, 2-dimensional photo-elastography was used for an analysis of the biomechanics of traumatic stresses acting on the pelvic girdle as the result of forces applied in sagittal, trasversal and longitudinal directions by moving the position of the hip. Further studies were made, also using the 3-dimensional technique, on the mechanism of lesions of the sacrum and ileum. In this sector, too, the importance was revealed of the moment of compression causing the lesion, and of the integrity of the pubic symphysis and the robust posterior ligamental formations in the ileum and sacrum when the pressure was exerted sagittally. When forces were applied longitudinally, it was further observed that there is a particular concentration of pressure at the level of the ischiatic notch and the lower posterior iliac spine.

In the field of experimental examination of surgery leading to morphological bone changes, redistribution of weight-bearing was studied and variations in muscular stress on the pelvis and proximal extremity of the femur in a case of subluxant dysplasia of the hip after a Salter's osteotomy. Photo-elastography on 3-dimensional models revealed that though, when subluxation is present, the weight of the sacrum and ileum is distributed almost exclusively towards the exterior, the normal model shows the load equally distributed between exterior and interior. After similar morphological changes to those encountered as a result of Salter's osteotomy had been applied to the

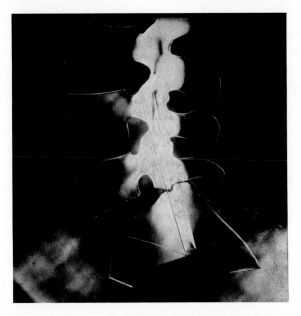

Fig. 5. In the model of a spine with spondylolisis of 5 lumbar vertebra the load is distributed almost exclusively on the juxtaposed transplant.

model, reproducing the state of subluxation, it was observed that the load which had been exerted towards the exterior returned towards the interior (fig. 3). From this it may be deduced that the pressures exerted on the ischiatic notch in a load-bearing area would, in time, lead to certain morphological changes in the pelvis itself and the proximal extremity of the femur.

As far as the spine is concerned, an analysis was made of changes in the distribution of the load-bearing on the lumbar spine, with respect to the orthomorphic position, in flexion and extension movements (fig. 4). The 2-dimensional technique revealed that the area of the isthmus was constantly affected by static and dynamic pressures. When compared with clinical data on lesions of the isthmus in gymnasts and weight-lifters – who subject their lumbar spines to repeated forced states of hyperextension – these results suggest some interesting theories about the much-debated pathogenetic mechanism of spondylolisis.

Equally interesting was the experimental confirmation of thickening in a model of a transplant juxtaposed to correspond to the model of a spine with spondylolisis of a lumbar vertebra (fig. 5).

From the findings reported above it is evident that photo-elastography in both 2- and 3-dimensional models is a useful technique for clarifying biomechanical problems, not only in the field of the anatomy and physiology of the osteo-articular apparatus, but also in the search for corrective surgery that will provide solutions for a number of orthopaedic and traumatological conditions.

References

BONACCORSI, S. e ROMANO, M.: Studio sperimentale sui meccanismi lesivi interessanti le sacro-iliache. Atti del 52nd Congr. SIOT, Rome 1967.

PERUGIA, L.; BONACCORSI, S. e ROMANO, M.: Studio sulla struttura interna del ginocchio (analisi fotoelastografica nel ginocchio normale, valgo, varo). Ortop. Traum. App. Mot. *34:* 121–131 (1966a).

PERUGIA, L.; BONACCORSI, S. e ROMANO, M.: Modifiche biomeccaniche indotte dall'osteotomia innominata tipo Salter (sperimentazione di fotoelastografia tridimensionale). Ortop. Traum. App. Mot. *34:* 487–498 (1966b).

PERUGIA, L.; BONACCORSI, S. e ROMANO, M.: Studio sperimentale sul meccanismo delle fratture acetabolari. Atti del 52nd Congr. SIOT, Rome 1967.

PERUGIA, L.; ROMANO, M. e BONACCORSI, S.: Influenza della ripartizione del carico sulla struttura interna del ginocchio (studio fotoelastografico). Ortop. Traum. App. Mot. *34:* 1–14 (1966a).

PERUGIA, L.; ROMANO, M. e BONACCORSI, S.: Rilievi sperimentali su alcuni tipi di lesioni traumatiche cotiloidee. Ortop. Traum. App. Mot. *34:* 145–160 (1966b).

ROMANO, M. e BONACCORSI, S.: Analisi biomeccanica delle sollecitazioni traumatiche agenti sul cingolo pelvico. Atti del 52nd Congr. SIOT, Rome 1967.

Author's address: Prof. L. PERUGIA, Institute of Orthopaedics and Traumatology, University of Rome, *Rome* (Italy)

II. Fundamental Research

Medicine and Sport, vol. 8: Biomechanics III, pp. 210–217 (Karger, Basel 1973)

Relationships between Selected Biomechanical Parameters of Static and Dynamic Muscle Performance

J. P. STOTHART

Faculty of Physical Education, University of Western Ontario, London

The study of static muscular performance characteristics is of concern for two specific reasons. Firstly, static performance can simply be considered as a special case of dynamic performance in which the force is maximal and velocity is zero. Secondly, the measurement of static performance, by its nature, has been and is much simpler than the measurement of dynamic performance. Thus, if meaningful relationships can be established between characteristics of both kinds of performance, then prediction of certain aspects of dynamic performance may be possible from a knowledge of static performance.

The object of this investigation was to study the relationships between selected characteristics of static elbow flexion performance and the biomechanical aspects of dynamic elbow flexion performed under 3 different load conditions.

Of specific hypothetical concern were the relationships between static force and dynamic force, static force and velocity, and dynamic force and velocity as well as the effect of an inertia-relative loading system upon those relationships.

Methods

22 male university students (mean age, 26.7 years; mean weight, 80.8 kg; mean height, 181 cm) were the subjects for this study. Tests included the recording of static force-time curves, forearm volume measurement, and acceleration, velocity and displacement time curves. Both static and dynamic tests were conducted in the horizontal plane. The elbow was at an angle of 170° for the static test and the start of the dynamic test.

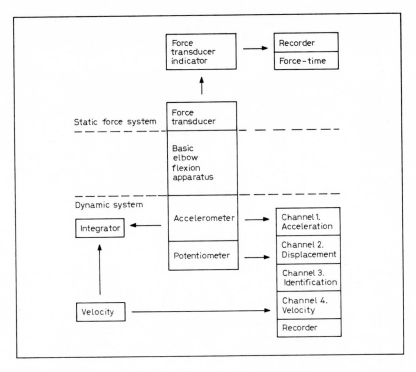

Fig. 1. Schema of the elbow flexion measurement system.

For static force-time measurement, an electronic force transducer was employed with the recording being made on a paper recorder (fig. 1). An accelerometer, potentiometer and integrator (fig. 1) were utilized to obtain the dynamic performance parameters acceleration, angular displacement and velocity curves, respectively. Forearm volume was measured using the water immersion method.

Analysis of static force-time curves yielded both force and time variables in addition to a maximum rate of force development variable, tangent to the force-time curve (fig. 2). The methods and techniques for the analysis of force-time curves were derived from SUKOP *et al.* [1968]. Forearm volume was utilized to compute the moments of inertia of the forearm. Acceleration, velocity and displacement curves were analyzed and basic data were extracted for simultaneous acceleration and velocity variables at 7 points through the movement (15, 30, 45, 60, 75, 90 and 105° from the start). From these data dynamic torque variables were derived.

Individual forearm moments of inertia were determined and applied to 3 load conditions. Condition A load was the combined moment of inertia of the forearm and movement apparatus. Condition B and C loads were adjusted with weights so that they were twice and 3 times the condition A load, respectively. All dynamic variables were determined for all 3 load conditions.

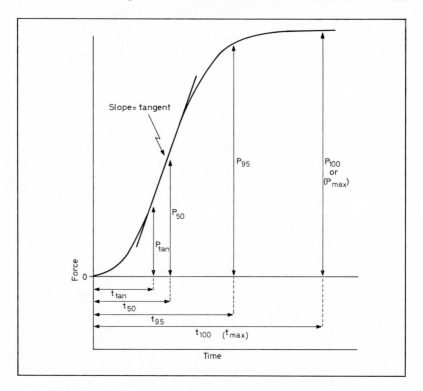

Fig. 2. Force-time curve parameters.

Results and Discussion

Moments of inertia (MI) of the subjects' forearms measured in this study are slightly higher than other researchers have found. For example, BOUISSET and PERTUZON [1968], using 11 subjects, reported a mean of 0.0599 kgm² and a range of 0.0430–0.0797 as compared with a mean of 0.0790 and a range of 0.0560–0.1120 found in the present study. This may be a factor in the results concerning the dynamic torque variables which are dependent upon MI.

The mean of maximum static torque developed was 49.4 Nm. The mean maximum rate of force development (144 kg/sec) was achieved on the average 0.052 sec after the beginning of force production. Mean time taken to reach maximum static torque was 0.415 sec. The present results are in close accord with results reported by other investigators [ROYCE, 1962; SUKOP, 1969].

Very low or nonsignificant correlations were found between static time

Table I. Means of maximum dynamic torque, acceleration and velocity variables

	Units	Condition A	Condition B	Condition C
Maximum dynamic torque	Nm	25.4	30.1	33.0
Maximum acceleration	rad/sec^2	155.9	92.5	67.6
Maximum velocity	rad/sec	17.5	14.3	12.4

variables and static force variables but a moderate correlation was evident between maximum static torque and the maximum rate of force development (tangent), r = 0.73 (critical r 0.05 = 0.42 for N = 22). Limited literature is available to compare with some aspects of intracorrelations among static variables. CLARKE's approach [1964] has involved somewhat different variables, and SUKOP [1969] is the only researcher who has used a comparable approach to static force-time curve measurements. SUKOP's high coefficients between maximum static torque and 'tangent' (r = 0.92) as well as between time to 50% of maximum static torque and time to 'tangent' (r = 0.83) are supported by the present results. Since, as WILKIE [1950] stated that the excitation of muscle begins with only a few motor units and builds up to the total excitation in approximately 40 msec, the time to 'tangent' (equal to 0.052 sec) which is the time during a static contraction taken to build up to the maximum rate of force development would appear to manifest the state of total excitation.

The maximum dynamic torque and velocity results (table I, fig. 3) are all consistent with the research literature [DERN et al., 1947]. Mean maximum dynamic torque was achieved at a very early position in the movement (12.2–19.4° from start) and increased (maximum acceleration decreased) with an increase in load as velocity decreased. As load increased from conditions A to C, mean maximum dynamic torque achieved proportions of 51.4, 60.9 and 66.8% of mean maximum static torque.

Although there were no significant correlations between static and dynamic time variables, a comparison of their relative sequence of events is revealing (fig. 4).

Maximum dynamic torque, maximum static torque and 'tangent' variables are substantially related to all velocity variables (table II). As hypothesized, high correlations were evident between maximum dynamic torque and velocity variables at some positions.

Little difference existed between correlations of maximum static torque – velocity and 'tangent' – velocity with those involving the 'tangent' variable

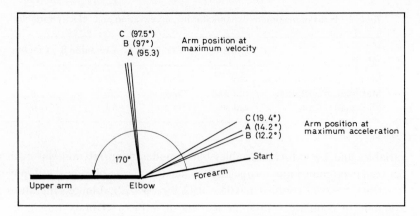

Fig. 3. Angular positions at maximum acceleration and maximum velocity.

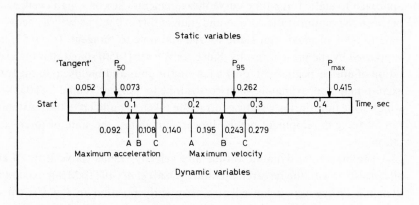

Fig. 4. Time sequence schema for static and dynamic performance.

being slightly higher in most cases. The relationships formed between maximum static torque and velocity variables are consistent with some research [LARSON and NELSON, 1969; STOTHART and NELSON, 1968] and not with others [HENRY and WHITLEY, 1960; CLARKE, 1964; SMITH, 1969]. A partial explanation of why velocity and static force should relate at least moderately is provided by WILKIE [1956] who said that the chemical reactions which produce muscular energy are themselves controlled by the tension in the muscle. That is, energy which is available in the muscle for ultimate conversion to kinetic energy is *dependent* upon the tension in the muscle. SMITH [1969] reasoned that the moderate to high relationships found by researchers between maxi-

Table II. Intercorrelations between velocity variables and maximum dynamic torque, maximum static torque and maximum rate of static force development

Velocity variables	Maximum dynamic torque in conditions:			Maximum static torque in conditions:			Maximum rate of static force development in conditions:		
	A	B	C	A	B	C	A	B	C
Maximum velocity	0.67	0.76	0.66	0.32	0.48	0.45	0.47	0.65	0.47
Time to maximum velocity	−0.73	−0.76	−0.79	−0.63	−0.61	−0.67	−0.71	−0.60	−0.75
Average velocity	0.79	0.83	0.81	0.58	0.56	0.59	0.68	0.63	0.73
Velocity at:									
15°	0.55	0.69	0.80	0.49	0.59	0.68	0.61	0.63	0.72
30°	0.67	0.83	0.83	0.47	0.72	0.69	0.61	0.79	0.77
45°	0.71	0.89	0.82	0.47	0.66	0.62	0.61	0.83	0.69
60°	0.75	0.87	0.78	0.48	0.64	0.59	0.63	0.78	0.66
75°	0.73	0.87	0.76	0.40	0.59	0.55	0.57	0.75	0.60
95°	0.69	0.80	0.69	0.35	0.51	0.48	0.50	0.70	0.51
105°	0.67	0.68	0.58	0.34	0.36	0.37	0.43	0.54	0.37

Critical correlation coefficient at 0.05 level of probability is ± 0.42.

Table III. Intercorrelations between dynamic torque variables, and maximum static torque and maximum rate of static force development

	Maximum static torque in conditions:			Maximum rate of static force development in conditions:		
	A	B	C	A	B	C
Maximum dynamic torque	0.73	0.71	0.76	0.74	0.75	0.82
Dynamic torque						
15°	0.70	0.75	0.73	0.67	0.75	0.78
30°	0.60	0.66	0.70	0.64	0.75	0.71
45°	0.47	0.59	0.58	0.49	0.65	0.61
60°	0.25	0.45	0.37	0.32	0.57	0.37
75°	−0.02	0.19	0.08	0.01	0.34	0.06
90°	−0.16	−0.13	−0.12	−0.22	0.04	−0.17
105°	−0.05	−0.20	−0.25	−0.21	−0.11	−0.26

mum strength and velocity variables was due to the load placed on the muscle by the apparatus used. Since the apparatus used in this study had an MI almost equal to the mean MI of the subjects' arms, SMITH's explanation appears tenable.

The effects of adding load to dynamic performance upon mean values were predictable. However, the lack of effect from dynamic loading upon relationships among the static and dynamic variables was of particular interest (table II, III). This lack of effect may be interpreted more specifically as a lack of interaction between loads moved and the variability of dynamic performance characteristics which is contrary to the hypothesized result. The adding of load to dynamic performance should increase relationships between static and dynamic variables by making the dynamic performance more like a static performance (i.e., by decreasing velocity).

Both maximum static torque and 'tangent' are moderately related to dynamic torque in earlier positions of the movement. The results concerning maximum static torque and dynamic torque are in accordance with the expressed hypotheses and the experimental results of other researchers [ASMUSSEN et al., 1965; STOTHART and NELSON, 1968]. That is to say, that since dynamic torque in earlier positions more closely approximates static torque, but diminishes near the end of the movement, higher relationships should exist in those earlier positions. 'Tangent', at the same time (as a variable made up of both force and time), relates similarly to dynamic torque for the same reasons.

Conclusions

With respect to the hypotheses held, these conclusions were deemed justified:

1. Static and dynamic force are moderately related in early phases of movement.

2. Velocity is highly related to dynamic torque, and moderately related to both static torque and maximum rate of force development.

3. Loading the muscle with multiples of the natural moment of inertia of the forearm does *not* increase relationships between static and dynamic performance but, in fact, tends to maintain those relationships.

References

ASMUSSEN, E.; HANSEN, O., and LAMMERT, O.: The relation between isometric and dynamic muscle strength in man. Communications from the Testing and Observation Institute of the Danish Nat. Ass. for Infantile Paralysis. Hellerup, Denmark, No. 20 (1965).

BOUISSET, S. and PERTUZON, E.: Experimental determination of the moment of inertia of limb segments. In: Biomechanics. Proc. 1st Int. Seminar on Biomechanics, Zurich 1967, pp. 106–109 (Karger, Basel 1968).

CLARKE, D. H.: The correlation between strength and the rate of tension development of a static muscular contraction. Int. Z. angew. Physiol. 20: 202–206 (1964).

DERN, R. J.; LEVENE, J. M., and BLAIR, H. A.: Forces exerted at different velocities in human arm movements. Amer. J. Physiol. 151: 415–437 (1947).

HENRY, F. M. and WHITLEY, J. D.: Relationships between individual differences in strength, speed and mass in an arm movement. Res. Quart. amer. Ass. Hlth phys. Educ. 31: 24–33 (1960).

LARSON, C. L. and NELSON, R. C.: An analysis of strength, speed and acceleration of elbow flexion. Arch. phys. Med. Rehabil. 50: 274–278 (1969).

ROYCE, J.: Force-time characteristics of the exertion and release of hand grip strength under normal and fatigued conditions. Res. Quart. amer. Ass. Hlth phys. Educ. 33: 444 (1962).

SMITH, L. E.: Specificity of individual differences of relationships between forearm strengths and speed of forearm flexion. Res. Quart. amer. Ass. Hlth phys. Educ. 40: 191–197 (1969).

STOTHART, J. P. and NELSON, R. C.: Interrelationships among selected biomechanics components in a simple flexion movement. Presented at Research Section of AAHPER Convention, St. Louis 1968.

SUKOP, J.: Personal communications (1969).

SUKOP, J.; REISENAUR, R., and TSCHERNOSTER, E.: Evaluation of curves of isometric muscular contraction. Unpublished paper (1968).

WILKIE, D. R.: The relation between force and velocity in human muscle. J. Physiol., Lond. 110: 249–280 (1950).

WILKIE, D. R.: The mechanical properties of muscle. Brit. Med. Bull. 12: 177–182 (1956).

Author's address: Dr. J.-P. STOTHART, Faculty of Physical Education, University of Western Ontario, *London* (Canada)

Medicine and Sport, vol. 8: Biomechanics III, pp. 218–223 (Karger, Basel 1973)

The Relationship between the Rate of Tension Development and the Strength of a Voluntary Isometric Muscular Contraction in Man

E. J. WILLEMS

University of Leuven, Institute of Physical Education, Leuven

An overview of the literature concerning the correlation between the maximum static strength and the rate of tension development of a voluntary muscular contraction reveals some differences between the opinions of several investigators [1–5].

The scope of the study here reported was to re-examine this problem with a technique that enables a rigorous isometric contraction of the hand gripping muscles and the simultaneous recording of the force-time curves on fast-moving paper. The hand dynamometer is shown in figure 1.

As pointed out in a previous study [6], maximum forces are obtained by setting the dynamometer grip width according to the proximal phalanx of the middle finger. During several consecutive years, we tested 5 groups consisting of 169 male and 85 female young adults. In general, the slope of the curves coincides rather well with those of previous investigators. A math-

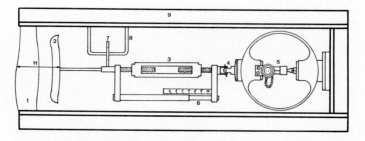

Fig. 1. Hand dynamometer. 1 = Fixed part of the grip; 2 = adjustable part of the grip; 3 = turn-buckle; 4 = toggle-joint; 5 = force transducer; 6 = scale, mm; 7,8 = ring and reference bar limiting the deviation to 2°; 9 = U-shaped frame, and H = hand grip width.

ematical description using a single exponential equation as proposed by
ZACIORSKII [7] seems to be unsatisfactory (fig. 2).

The cumulative exponential equation [2]:

$$P_t = A_2(1 - e^{-k_2 t}) + A_3(1 - e^{-k_3 t}) - A_1(1 - e^{-k_1 t}),$$

fits quite accurately the experimental values, except for the early phase of the
contraction.

This may best be demonstrated by computation of the gradient for t = 0.
For all parameters published by ROYCE [2] and CLARKE [1] and for all our
results, the gradient seemed to be negative, indicating a negative value of the
force by the onset of the contraction; this is, of course, impossible. A further
disadvantage of this method lies in the fact that this kind of curve fitting is
not applicable to individual results, due to the irregularities of these curves
(fig. 3).

To quantitate the rate of tension development of each subject, we propose
2 parameters:

$$I_{90} = \alpha = \arctan \frac{P_{90}}{T_{90}} \quad \text{and}$$

$$I_{50} = \beta = \arctan \frac{P_{50}}{T_{50}}, \quad \text{where}$$

P_{90} and P_{50} equal a proportion (90 and 50%) of the maximum static force;
T_{90} and T_{50} the time taken to achieve P_{90} and P_{50}. α and β are expressed in
degrees (fig. 4). The results of calculating these parameters for the 5 different
groups are summarised in table I.

These results are consistent with the general concept that the maximal
static strength of woman is about two-thirds that of men. In spite of the lower
maximal strength to be achieved we can see that the rate of tension development is much slower for female groups. The coefficients of correlation between these parameters and the maximal isometric strength (P_{max}) recorded
during the same experiment lies between 0.40 and 0.60 for the relation
$I_{90} - P_{max}$ and between 0.40 and 0.80 for the relation $I_{50} - P_{max}$.

In each of the 5 groups, the correlation between P_{max} and I_{50} has always
a higher value than those obtained between P_{max} and I_{90}, indicating that the
rate of tension development in achieving submaximal forces depends more
on the maximal available strength than the rate of tension development in
achieving maximal forces. After a short period (20 min) of unrelated warm-up
exercises of medium intensity no great differences were found in these correlation coefficients (N = 68 $\male$).

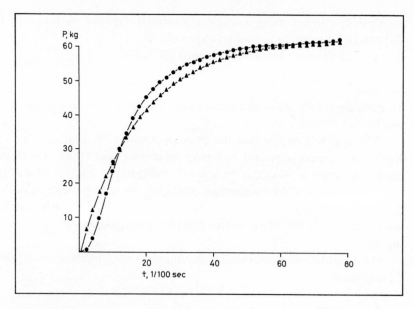

Fig. 2. Force-time curve of a maximal isometric contraction (N = 68). ● = experimental values; ▲ denote values calculated from $P_t = P_{max}(1 - e^{-kt})$.

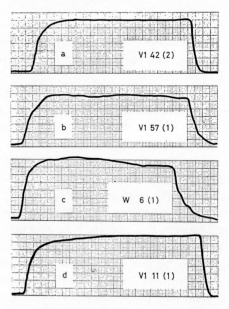

Fig. 3. Individual force-time curves of a maximal isometric contraction.

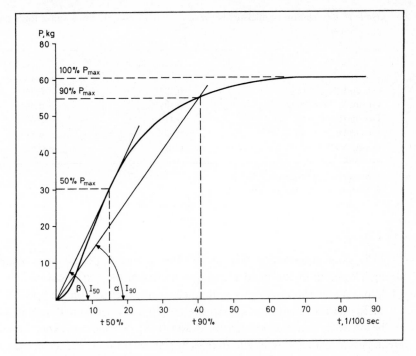

Fig. 4. Parameters to quantitate the rate of tension development in an isometric contraction.

Table I. Means and SD of mean of maximal forces, I_{90} and I_{50} in 5 different groups

	P_{max}	I_{90}	I_{50}	SD$\overline{X}$
Group I, N = 33, ♂	60.1	–	–	1.8
	–	54.1	–	1.9
	–	–	67.3	1.3
Group II, N = 68, ♂	62.3	–	–	1.6
	–	59.1	–	1.2
	–	–	68.2	1.2
Group III, N = 68, ♂	56.2	–	–	1.1
	–	60.9	–	0.9
	–	–	70.5	0.7
Group IV, N = 40, ♀	39.4	–	–	1.1
	–	46.6	–	1.7
	–	–	55.7	1.3
Group V, N = 45, ♀	36.2	–	–	1.0
	–	38.8	–	1.7
	–	–	49.4	1.7

Table II. Correlation coefficients between maximal isometric force and 2 force-time parameters

	$P_{max} - I_{90}$	$P_{max} - I_{50}$
Group I, N = 33, ♂	0.55[2]	0.76[2]
Group II, N = 68, ♂	0.38[1]	0.46[2]
Group III, N = 68, ♂	0.41[2]	0.69[2]
Group IV, N = 40, ♀	0.37[1]	0.81[2]
Group V, N = 45, ♀	0.61[2]	0.62[2]

1 Statistical significance, 5%.
2 Statistical significance, 1%.

The same observation was made after a 4-week period of isometric strength training of the hand gripping muscles on the test apparatus (N = 34 ♂), or on a similar apparatus without electronic registration facilities (N = 21 ♀). In the last-mentioned group the correlations between P_{max} and I_{90}, I_{50} becomes more differentiated (0.53 and 0.66). We cannot expect a very great gain in maximal strength after such a short training. Nevertheless, we noted a mean gain of 3.8 kg (significant 1%) for the male subjects and 2.2 kg (not significant) for the female subjects.

The influence of the training on the rate of tension development is rather obscure. By no means was the difference between I_{90} and I_{50} before and after training statistically significant. In the young men's group there was a slight increase for the I_{50} values, but a decrease in I_{90}. This may indicate an acceleration for the submaximal strength and a rather decrease in the rate of developing maximal forces. This was not confirmed in the young women's group where both parameters increased slightly.

References

1. CLARKE, D.H.: The correlation between strength and the rate of tension development of a static muscular contraction. Int. Z. angew. Physiol. *20:* 202–206 (1964).
2. ROYCE, J.: Force-time characteristics of the exertion and release of hand grip strength under normal and fatigued conditions. Res. Quart. amer. Ass. Hlth phys. Educ. *33:* 444–450 (1962).
3. SMIRNOV, I.I. et PODLIVAEV, B.A.: Relations entre les caractéristiques de force et de vitesse dans une contraction musculaire. Kinanthropologie *1:* 245–249 (1969).

4. Stothart, J.P.: A biomechanical analysis of static and dynamic muscular contraction. Unpublished master's thesis, Pennsylvania State University (1970).

5. Sukop, J. and Reisenauer, R.: The changes in muscular contraction and relaxation after the static load in 16-years boys and girls. Čs. Hyg. *13:* 458–466 (1968).

6. Willems, E. en Sauveniere, E.: Het verband tussen de kracht en de snelheid bij de isometrische spiersamentrekking. Werk. belg. Geneesk. Veren. Lich. Opl. Sport *17:* 43–52 (1964).

7. Zaciorskii, V. M.: Die körperlichen Eigenschaften des Sportlers (Sportverlag, East Berlin 1968).

Author's address: Dr. E. Willems, Institute of Physical Education, University of Leuven, Tervuurse vest 101, *B-3030 Heverlee* (Belgium)

Medicine and Sport, vol. 8: Biomechanics III, pp. 224–229 (Karger, Basel 1973)

Measurement of the Force-Velocity Relationship in Human Muscle under Concentric and Eccentric Contractions

P. V. KOMI

Kinesiology Laboratory, University of Jyväskylä, Jyväskylä

Since HILL [1965] made the suggestion that an attempt should be made to develop an instrument to record eccentric and concentric forces in human muscle, considerable interest has been devoted to the subject. ASMUSSEN *et al.* [1965] constructed an electrical dynamometer to measure isotonic forces of the arm-shoulder muscles. DOSS and KARPOVICH [1965] developed a device to obtain maximum isotonic forces of the forearm flexors with a manually operated dynamometer, in which the tester applied the pulling and resisting force. Later, SINGH and KARPOVICH [1966] designed and built an electrically operated dynamometer for the same purpose. Recently, we reported on a dynamometer [KOMI, 1969] in which the muscle contraction could be exerted at constant speeds both in concentric and eccentric work. The present report deals with our latest dynamometer which was designed for the measurement of the force-velocity relationship of the human forearm flexors and extensors.

Dynamometer

The dynamometer is capable of recording both the isotonic force (either eccentric or concentric) and a change in muscle length (elbow angle) with 8 different velocities of shortening and lengthening of the elbow flexors and extensors. Thus, to obtain the force-velocity relationship, a total of 16 different constant speeds could be selected along the velocity axis. The dynamometer was so constructed that the velocity of lengthening and shortening of the biceps brachii muscle stays constant throughout the movement range of approximately 120°. This corresponds to a 7-cm change in the length of the biceps muscle of an adult man.

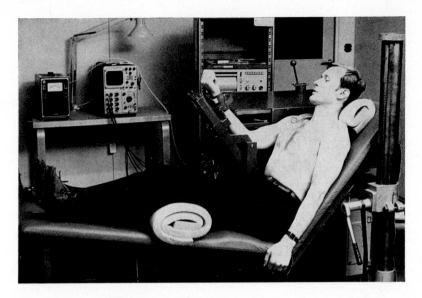

Fig. 1. General arrangement of the subject-dynamometer system.

Fig. 2. Different components of the dynamometer. A = motor; B = gear-box; C = spindle; D = magnetic clutch; E = photoelectric transducer; F = lever arm; G = wrist cuff.

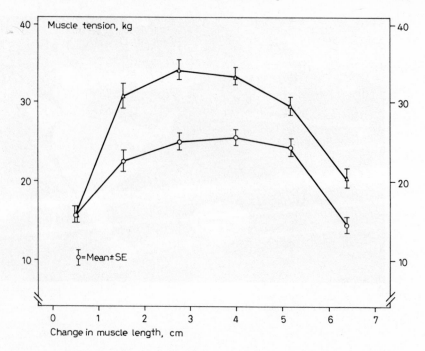

Fig. 3. An example of the length-tension relationship for the elbow flexor muscles in eccentric and concentric work. Velocity = 0.8 cm/sec. Δ = eccentric tension; o = concentric tension, n = 16.

The general arrangement of the dynamometer-subject system is shown in figure 1. The different components of the dynamometer are shown in figure 2. As can be seen, the dynamometer is a purely electromechanical system, which we feel is an improvement on our earlier used hydraulic system [KOMI, 1969]. Power for the dynamometer was provided by an electric motor (ASEA M 112). The motor was connected through a belt drive to a gear-box, which further actuated a steel spindle, through which a rotatory movement was converted to a linear motion. This linear movement in the spindle was connected to a muscle (biceps brachii) through a proper geometrical arrangement of the lever-arm system.

The speed range varied from 0.8 to 6.7 cm/sec when measured from the biceps muscle. Because the purpose was to keep each speed of contraction as constant as possible, magnetic clutches were installed along the spindle. Thus, the acceleration phase of the movement was minimal and the movement

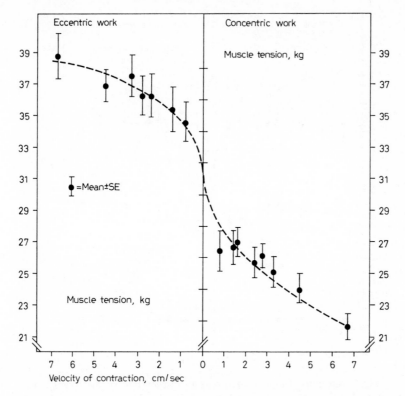

Fig. 4. Force-velocity relationship for the elbow flexor muscles. The values at each velocity represent the mean force for the mid-portion of the movement (the 4 innermost points in fig. 3).

could be stopped within 1 spindle revolution at the highest speed. One spindle revolution corresponds to 8 mm in linear distance, which equals a 2.35-mm change in the length of the biceps muscle. Lest the electromagnetic clutch mechanism should fail to operate properly, a special mechanical stopping mechanism was also installed in order to assure that the subject would not, in any event, break his arm.

The velocity of contraction was obtained with a photoelectric transducer which gave an impulse on the oscillograph paper at each spindle revolution. Strain gauges to record the force were installed on both sides of a special wrist cuff, with which the wrist could be fixed in any desired position between full supination and full pronation.

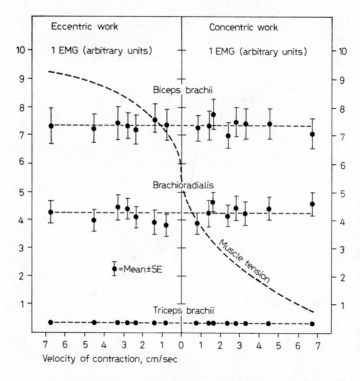

Fig.5. Integrated EMG (iEMG)-velocity relationship for the biceps brachii and brachioradialis muscles and their antagonist (triceps brachii).

Results

Figure 3 gives an example of the length-tension relationship for the elbow flexors both in eccentric and concentric contraction. The form of the curves is in good agreement with our earlier records [Komi, 1969]. If the velocity of contraction is increased, then the maximum eccentric tension at each muscle length increases and the concentric force decreases. This is shown in figure 4, which gives the force-velocity relationship for the elbow flexor muscles. The curve, although slightly smoothed, shows that it follows closely the classical force-velocity relationship obtained with isolated muscle [e.g., Levin and Wyman, 1928]. The result is also in a fairly good agreement with that of Asmussen *et al.* [1965], although Asmussen used different muscles. Thus, the maximum tension of the forearm flexors is always greater in eccentric than in concentric work. The result also confirms that the faster the muscle contracts the greater becomes the difference in maximum tension between eccentric and concentric work.

When designing the dynamometer, careful consideration was given to its use to record faithfully electromyographic activity during both contraction types and different contraction velocities. When the maximum integrated EMG activity was measured from the same subjects using the entire speed range, the EMG-velocity relationship looked as is shown in figure 5. This preliminary result suggests a possibility that while the maximum tension is highly dependent upon type and velocity of contraction, the maximum integrated EMG measured from the same muscle remains the same, despite the type of contraction or the velocity used. When interpreting this result one should keep in mind, however, that it may not be applicable to all situations of concentric and eccentric work, because the velocity range of the dynamometer is after all only a portion of the entire physiologically possible range.

References

ASMUSSEN, E.; HANSEN, O., and LAMMERT, O.: The relation between isometric and dynamic muscle strength in man. Communications from the Testing and Observation Institute of the Danish National Association for Infantile Paralysis, No. 20 (1965).

DOSS, W.S. and KARPOVICH, P.V.: A comparison of concentric, eccentric, and isometric strength of elbow flexors. J. appl. Physiol. 20(2): 351–353 (1965).

HILL, A.V.: Trails and trials in physiology, p. 357 (Arnold, London 1965).

KOMI, P.V.: Effect of eccentric and concentric muscle conditioning on tension and electrical activity of human muscle. Ph. D. thesis, Penn State University (1969).

LEVIN, A. and WYMAN, J.: The viscous elastic properties of muscle. Proc. roy. Soc. B 101: 218–243 (1928).

SINGH, M. and KARPOVICH, P.V.: Isotonic and isometric forces of forearm flexors and extensors. J. appl. Physiol. 21(4): 1435–1437 (1966).

Author's address: Dr. PAAVO V. KOMI, Kinesiology Laboratory, University of Jyväskylä, *Jyväskylä* (Finland)

Medicine and Sport, vol. 8: Biomechanics III, pp. 230–234 (Karger, Basel 1973)

Instantaneous Force-Velocity Relationship in Human Muscle

E. PERTUZON and S. BOUISSET

Laboratoire de Physiologie Générale, Université des Sciences et Techniques, Lille

The conditions under which natural movements are performed, appear to be very different from the isotonic conditions usually chosen for studying the force-velocity relationship in isolated muscles [FENN and MARSH, 1935; HILL, 1938] and, at least to a certain extent, in human muscles *in situ* [RALSTON *et al.*, 1949; WILKIE, 1950].

In most natural movements performed against constant inertia, whether or not in the plane of gravity, force and speed vary instantaneously. Moreover, even in the latter case, the presence of a joint transforms rectilinear movement of the muscle into circular movement of the limb segment. From this, the moment of inertia of the limb segment must be converted against muscle into a variable mass which is a complex function of the joint angle.

Therefore, it appeared interesting to estimate from voluntary movements against inertia a relationship between the *instantaneous* velocity of muscle shortening and the corresponding force. Since the activation level must be constant, it was advisable to choose the one voluntary movement likely to agree with this criterion, i.e. the maximal movement.

Technique and Procedure

Each subject was seated and asked to carry out voluntary flexions of the right elbow. His arm was held in a horizontal position and the forearm was fitted into a splint which had previously been moulded from the forearm itself. This splint was connected to an iron lever. The axis of rotation of this lever was aligned with that of the elbow. Five inertia levels (from 0.1 to 1.0 m². kg) were added to the inertia of the mechanical device (0.118 m². kg). The moment of inertia (I) of the forearm and the hand of each subject had been previously measured [BOUISSET and PERTUZON, 1968]. The angle of the elbow joint (Θ) was calculated using the fully extended arm as a base, and was detected by a goniometer whose axis was driven by the axis of the iron lever. The goniometrical signal was instantaneously differentiated in order to obtain the angular velocity (Θ') of the forearm. The lever was also provided with a tangential accelerometer so as to measure the angular acceleration (Θ''). This tangential accelerometer was situated at a given distance from the rotation axis of the lever.

On the basis of certain known anthropometric constants, it is possible to calculate the instantaneous velocity of biceps shortening (fig. 1), with an on-line analog computer [PER-

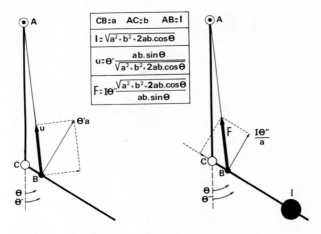

Fig.1. Calculation of the velocity of shortening of biceps and of the corresponding force. Explanations in text.

TUZON and BOUISSET, 1971], and to transform the torque $(I\Theta'')$ into a force (F) with respect to the biceps brachii. In this case, the biceps is considered as the chief flexor of the forearm [WACHHOLDER and ALTENBURGER, 1926]. Global (and integrated) surface EMG of the biceps brachii, the brachioradialis and the triceps were simultaneously recorded.

Five subjects carried out 2 experimental tests each. The order in presenting the added inertia conditions was reversed in the 2 tests. The movement was started at an angle of 30° and stopped at an angle of 120° by means of a safety buffer so as to keep the subjects free of any apprehension. The subjects were asked to perform the movements and to develop a maximal effort as fast as possible, continuing this effort until the buffer was struck. Thus, it was possible to obtain a maximum level of activation during nearly the whole movement. Among the 5 movements performed in each condition of inertia, only those complying with these criteria and during which the triceps was silent, were taken into account. In order to calculate the force-velocity relationship, we chose from these movements the one which presented the highest value of the muscular velocity of shortening.

Results and Discussion

Figure 2 represents the record of a flexion movement; any instantaneous value of biceps velocity of shortening can be associated with a corresponding value of force. Thus, for each movement, at least 10 pairs of values are available. A diagram was drawn for every subject and the points were plotted for each movement performed against 6 inertia values (fig. 3). The shape of the relation is comparable to the one established by ASMUSSEN *et al.* [1965] where

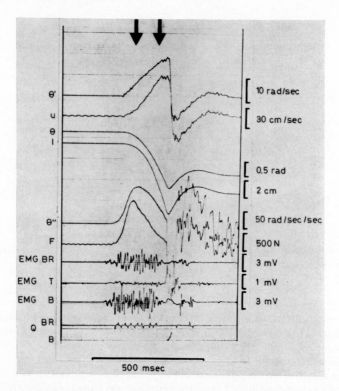

Fig. 2. Original record of a voluntary movement of elbow flexion (inertia: 0.396 m². kg). The part of the movement used for calculating the instantaneous force-velocity relationship is that which lies between the 2 arrows.

constant velocity movements were performed under conditions closely related to isotonic activity.

As can be seen from figure 2, maximal force appears after a certain delay which is related to the conditions for performing the movement. For each subject, both the time of rising force and the EMG duration increase proportionally with inertia.

This phenomenon, considered by ASMUSSEN *et al.* [1965] as the 'inability of the subject to immediately mobilize full strength at the beginning of the movement' could be related to the regulation of the recruitment of motor units. For these reasons, the instantaneous force-velocity relation was calculated only during the part of the movement where the amplitude of the flexor electromyograms was constant.

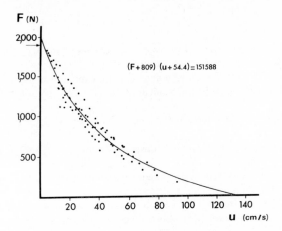

Fig. 3. Instantaneous force-velocity relationship of 1 subject. The points correspond to 6 movements performed against various inertias. The arrow points to the maximal isometric force calculated from static pulls.

All experimental points were subjected to numerical analysis, by the least squares method, so as to fit the relationship to the general equation of a hyperbola, $(F + a) (u + b) = c$, the parameters of which are a, b, and c. The values through extrapolation from $F = 0$ and $u = 0$ determine the maximal velocity of biceps shortening (u_0) and the maximal isometrical force (F_0). This value of F_0 agrees with the maximal isometrical force previously measured for the initial angle of the movement [PERTUZON, 1970]: it can be seen in figure 3 that it differs only slightly from the one that was measured directly. The value of u_0 cannot be measured directly through our experimental conditions [PERTUZON and BOUISSET, 1971]. However, it can be compared, although it is greater, to the one determined by RALSTON *et al.* [1949] in their study of disconnected muscles (92.5 cm/sec). The ratios of $a : F_0$ (from 0.21 to 0.49) are greater than those established by HILL [1938] for isolated muscles, but comparable to those calculated by WILKIE [1950].

In conclusion, the *instantaneous* relationship between force and velocity does not seem to be very different from the relationships already established between the *maximal* values of force and velocity. With regard to muscles, the interpretation of this relation would appear to be premature, one of the reasons being that it has not yet been ascertained whether the instantaneous force is in a constant ratio to the force actually developed by the biceps or not.

References

ASMUSSEN, E.; HANSEN, O., and LAMMERT, O.: The relation between isometric and dynamic muscle strength in man. Dan. nat. Ass. infant. Paralysis *20*: 3–11 (1965).

BOUISSET, S. and PERTUZON, E.: Experimental determination of the moment of inertia of limb segments. Biomechanics, vol. I. Ist Int. Seminar, Zurich 1967, pp. 106–109 (Karger, Basel 1968).

FENN, W. O. and MARSH, B. S.: Muscular force at different speed of shortening. J. Physiol., Lond. *85*: 277–297 (1935).

HILL, A. V.: The heat of shortening and the dynamic constants of muscle. Proc. roy. Soc. B. *126*: 126–195 (1938).

PERTUZON, E.: Relation force-longueur du muscle *in situ* en contraction maximale. J. Physiol., Paris *62*: suppl. 2, p. 303 (1970).

PERTUZON, E. and BOUISSET, S.: Maximal velocity of movement and maximal velocity of muscle shortening. Biomechanics, vol. II. 2nd Int. Seminar, Eindhoven 1969. Medicine and sport, vol. 6, pp. 170–173 (Karger, Basel 1971).

RALSTON, H. J.; POLISSAR, M. J.; INMAN, V. T.; CLOSE, J. R., and FEINSTEIN, B.: Dynamic features of human isolated voluntary muscle in isometric and free contractions. J. appl. Physiol. *1*: 526–533 (1949).

WACHHOLDER, K. und ALTENBURGER, H.: Beiträge zur Physiologie der willkürlichen Bewegung. VIII. Ueber die Beziehungen verschiedener synergisch arbeitender Muskelteile und Muskeln bei willkürlichen Bewegungen. Pflüger's Arch. ges. Physiol. *214*: 642–661 (1926).

WILKIE, D. R.: The relation between force and velocity in human muscle. J. Physiol., Lond. *110*: 249–280 (1950)

Authors' address: Dr. E. PERTUZON and Dr. S. BOUISSET, Laboratoire de Physiologie Générale, B. P. 36, 59-Villeneuve D'Ascq, *Lille* (France)

Medicine and Sport, vol. 8: Biomechanics III, pp. 235–238 (Karger, Basel 1973)

The Correlation between Static Muscular Force and Speed of Movement

G. Bergmaier and P. Neukomm

Swiss Federal Institute of Technology, Department of Physical Education, Zurich

Introduction

In sport, the impulse that is imparted to a certain body segment or mass is of a decisive importance. The velocity is, thereby, of great significance, as is evident from the physical terminology for an impulse: $p = m \cdot v$.

For years researchers have studied, therefore, the possibility of a relationship between the static muscular force and the speed of movement of a muscle. The results of Henry, Clarke, Smith, etc., who could not find any correlation between both values, are opposed to the latest results of Nelson and Jordan [1969], who determined correlation coefficients of 0.74–0.79. (A direct linear correlation is to be found at 1.0.) In view of the different results, a reconsideration of this question seemed appropriate.

A horizontal adductive movement of the arms was chosen for the present experiment.

Experiment

The subjects, 22 sportsmen of whom 8 were national champions, were chosen from 4 different groups of sports for comparison: gymnasts with relatively slow movements and great force input, boxers and swimmers with quicker movements, and volleyball players with very speedy movements of the arms.

The average age was 21.8 years, of the swimmers (as the youngest) 18 years and of the gymnasts (as the oldest) 24 years. The average arm length measured from the shoulder joint to the outer end of the middle hand bone amounted to 64.5 cm.

Force and velocity were measured in the radius of inertia of the subject's right arm. Thus, the body fitness of the subjects was taken into account. This seemed to have been neglected in tests made by some researchers. The average radius of inertia measured from the acromion amounted to 38,7 cm. The test apparatus was constructed with a racing-car seat and with individually adjustable measuring instruments. The total weight of the measuring device consisting of goniometer and accelerometer, attached in the radius of inertia, amounted to 110 g and added relatively little additional mass to the arm.

Static Force Measurement

The static force was determined in a horizontal position of outstretched arms at 0° and by an electric pressure and pull measuring device at the level of the inertia radius. The angle was measured with a potentiometer which was attached to a metal frame above the right shoulder joint. The angle was determined with 0° for the sideward, and with 90° for the forward out-stretched horizontal position of the arms (fig. 1).

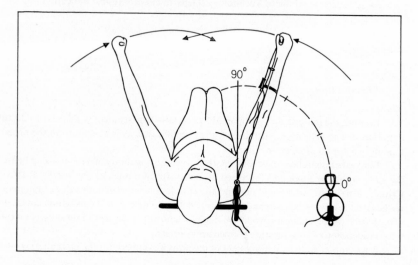

Fig. 1

Velocity Measurement

As the usual method of recording average speeds by means of stroboscopy and the measurement of angles seemed to be lacking in precision, we decided to try a new electronic method.

In electronic recording of measurement data, the velocity values must exactly correspond with values of other data (as, for instance, trajectory, acceleration, eventually EMG and force). Generally, two methods are used for determination of speed. The popular method of *differentiation* uses a potentiometer, built into a goniometer, as a recorder of measurement values. With this apparatus any movement can be measured. However, the quality of the recorded velocity is generally unsatisfactorily, especially in the domain of small speeds. Every deviation in the registration of angles is differentiated and thus causes erroneous signals in the recording. Practically, this means that all error signals will have to be subdued by deep bass filters. This procedure causes a delay in the registration of speed of 0.01–0.1 sec. In this study we preferred to use the *integration* method, and an accelerometer to record the measurement values. Thus, only those movements can be measured during which the acceleration of the earth influences the accelerometer at the time of measurement with a constant error, which can be compensated afterwards. If this influence is not heeded, the registration indicator of the velocity will start to drift, especially in the lower areas of speed. On the other hand, the quality of the registration is excellent and does not show any delay with regard to the other simultaneously measured values.

The *AVP integrator*, developed in our laboratory by Mr. NEUKOMM, is used for measuring the simultaneous acceleration (a), velocity (v) and position (p). Furthermore, this apparatus measures the change of impulse as an integral of force as well as the speed as an integral of acceleration. In horizontal guided movements, measurement values can be recorded with a precision of 1%. Data on reaction times and signal values – recorded on ultraviolet-sensitive paper strips – are immediately provided by means of an automatic calibration gauge and an audible starting signal.

Results

The following schedule shows the results of the statistically evaluated measurement values:

Table I

	n	Static force, x, kp	Velocity, y, m/sec	r_{xy}
Gymnasts	5	34.23 ± 4.8	4.38 ± 0.42	0.36
Boxers	6	35.82 ± 3.3	4.61 ± 0.24	0.24
Swimmers	6	22.24 ± 1.8	4.64 ± 0.28	0.48
Volleyball players	5	31.90 ± 1.6	4.76 ± 0.26	0.29
Total	22	30.30 ± 1.8	4.61 ± 0.14	0.06

None of the 4 groups showed a close correlation of static force and speed. Swimmers achieved a somewhat higher coefficient, contrary to those of the total amount of test subjects. In this case, an absolute independency of both values was affirmed.

Conclusion

The above experiment has confirmed the results of the other researchers, with the exception of NELSON and JORDAN [1969]. These authors registered velocities of 3.5 m/sec for the same movement, about 1 m/sec less than registered by us. As a correlation is lacking between velocity and the static force, it can be assumed that the velocity cannot be increased by augmentation of the static force of well trained sportsmen. It might have been anticipated, that gymnasts would obtain a greater static force because of their great input of forces during training, contrary to volleyball players who would obtain increased speed during their dynamic training, whereas the boxers and swimmers would be situated in between.

However, the Studen t-test, comparing the average velocities, shows that the deviations between the 4 groups of subjects are limited and incidental.

Former experiments in our laboratory have affirmed that training aimed at the static force increase causes an increase of the dynamic force. Obviously an increase of the velocity of movement is not included. This generalization could be valid for all movements, whereby body parts are accelerated without additional mass to the system.

It now remains to clarify how much influence a training aimed at the increase of static muscular force can have on the factors causing velocities, and in what respect these can retard or even limit the speed of a movement.

References

LARSON, C. and NELSON, R.: An analysis of strength, speed, and acceleration of elbow flexion. Arch. phys. Med. *50* (1969).

NELSON, R. and JORDAN, B.: Relationship between arm strength and speed in the horizontal adductive arm movement. Amer. corr. Ther. J. (1969).

SMITH, L. E.: Specificity of individual differences of relationship between forearm 'strengths' and speed of forearm flexion. Res. quart. *40*: 1 (1969).

Author's address: Dr. GUIDO BERGMAIER, Eidgenössische Technische Hochschule, Turnen und Sport, Leonhardstr. 33, *CH-8006 Zürich* (Switzerland)

Medicine and Sport, vol. 8: Biomechanics III, pp. 239–242 (Karger, Basel 1973)

Muscle Coordination in Simple Movements

J. VREDENBREGT and G. RAU

Instituut voor Perceptie Onderzoek, Eindhoven

Voluntary movements of a limb are generated by contracting muscles of which the patterns of activity are coordinated in such a way that a desired movement is performed according to the preset aim. With respect to this process one of the main questions is: 'Are there any fundamental rules in the pattern of successive muscle activities during cooperation of the various muscles.'

The aim of this paper is to show some methods and techniques which may be helpful in studying this question. Two different approaches are used in this study, namely analysis and synthesis of the movement patterns. In the analysis studies we used movement recording techniques as well as surface EMG. The EMG signals offer the possibility of observing the activity of the different muscles involved in relation to the movement pattern, and especially the correlation between successive periods in time during which the muscles are active.

In figure 1 an example is shown of a voluntary circular hand movement, together with the detected EMG signals. The arrows indicate the directions of the movements. In this experiment the movement was recorded by a normal pencil held in the hand. The movement may be broken down into mutually perpendicular directions. The main EMG activity was picked up from 4 muscle groups involved in the movement. During ventral flexion the main activity was found near the flexor digitorum superfacialis and flexor carpi ulnaris, and during dorsal flexion near the extensor digitorum and extensor carpi radialis brevis and longus.

The main activity during ulnar abduction and radial abduction was found, respectively, near the flexor carpi ulnaris and the extensor digiti minimi.

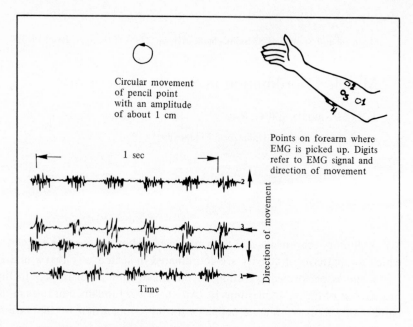

Circular movement
of pencil point
with an amplitude
of about 1 cm

Points on forearm where
EMG is picked up. Digits
refer to EMG signal and
direction of movement

Fig. 1. EMG activity of forearm muscles during control of pencil movement.

It may be noticed: (1) that a clear successive pattern of muscle activity is correlated with the direction of the movement. The various periods of activity and non-activity appear to be regular and well-defined in time, and (2) there is no or hardly any overlapping of muscle activities of the muscles acting in the opposite direction. This may be an indication that in this movement the inertia of the moving system plays a non-neglectable role.

The EMG detectors used consist of a bipolar surface electrode system of the suction cup type together with the integrated amplifiers in between. The electrodes are fixed to the skin by vacuum, assuring very good contact between electrodes and the skin and reducing movement artefacts.

It is of great importance to know the influence of changes in the time patterns of the muscle activities upon the resulting movement pattern.

In voluntary movements it is difficult to induce well-defined changes in this time pattern. This difficulty can be overcome by electrical stimulation of the muscles in comparable time patterns of activity. In this way we can synthetise movements generated by the original system with the same mechanical properties. However, this way of stimulation differs from the physiological one. This technique has the advantage of defining precisely, in time, the

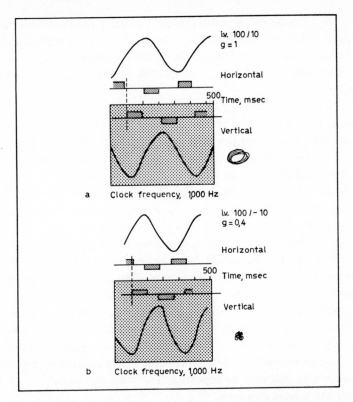

Fig. 2. Electrical stimulation of forearm muscles. *a* Refers to a stimulation programme in which the period of stimulation is 100 msec, followed by a period of non-stimulation of 10 msec. The result of this stimulation programme is a smoothed circular movement of the pencil. *b* Shows the results of a stimulation programme in which the successive period of stimulation has an overlap of 10 msec. The result is an uncontrolled and somewhat spastic movement pattern.

periods of muscle activity and within certain limits the amount of muscle force. The influence upon the movement pattern of one single change in the stimulation programme can be studied as well.

The apparatus used consists of an electrical muscle stimulator, generating continuously electric impulses of 0.5-msec duration at a repetition frequency of 50 Hz. In order to stimulate the adequate muscles at the right point of time, relays were used, of which the switch time was 1 msec. They are switched on and off by a preset cascade counter, which is driven by a frequency generator. In this way any interval ratio between periods of stimulation and no stimulation can be obtained for every muscle. Changing the clock frequency of the frequency generator gives a multiplication of the total stimulation programme, making it longer or shorter. To obtain the same amplitude in all directions of the movement, adequate stimulation amplitudes are adjusted by separate potentiometers.

Figure 2a shows an example of a stimulation programme as well as the movements in both directions. Moreover, the trace of the complete movement, recorded by a ballpoint-pen on a sheet of paper, is shown. The upper part of the stimulation programme, indicated by 'horizontal', refers to the ulnar and radial abductions of the hand, the lower part, indicated by 'vertical' to the dorsal and ventral flexions. The stimulation period is a 100-msec, followed by a 10-msec period of no stimulation. From the trace a smooth movement can be observed. The movement is well reproducable.

In comparing the diplacement time diagram with the period of no stimulation we concluded that the inertia plays an important role in smoothing the movement. By changing the stimulation programme in such a way that an overlap of 10 msec between the successive stimulation periods was obtained, the well coordinated movement disappeared completely. Figure 2b gives the stimulation pattern with a 10-msec overlap as well as the displacement-time diagram for one period. From the trace it is shown that the movement is not well coordinated and that the amplitude has greatly diminished. The reproducability has also disappeared. The total movement shows an uncontrolled and spastic character.

From these examples it may be seen that a certain relation between the period of muscle activity and non-activity may exist, in which a particular ratio between the two periods leads to a smoothed circular movement. In conclusion, we can say that the methods presented seem to be very helpful in studying muscle coordination for various forms of movement.

Author's address: J. VREDENBREGT, Instituut voor Perceptie Onderzoek, Insulinde-laan 2, *Eindhoven* (Holland)

Medicine and Sport, vol. 8: Biomechanics III, pp. 243–248 (Karger, Basel 1973)

The Mechanical Behaviour of the Passive Arm

K. L. Boon, A. L. Hof and W. Wallinga-de Jonge

Twente University of Technology, Enschede

Introduction

This article describes the mechanical impedance (i.e., the mechanical resistance that is exerted in response to passive motion) of the forearms of healthy subjects. The mechanical impedance is determined by moving the arm of the subject sinusoidally and measuring the force it applies.

In the medical world, the mechanical impedance is known as 'rigidity'. Many authors [1, 7, 13] have investigated the clinical aspects of rigidity. These clinical investigations have the disadvantage that they are little systematic; therefore, the results can hardly be used for a specifically system-analytical approach as proposed by several authors [2, 4, 6].

Research not just concerned with the clinical aspects of the mechanical impedance is done by RACK [8, 9], ROBERTS [10, 11] and VREDENBREGT and KOSTER [12].

From this, one can conclude that the mechanical impedance is not a poorly studied object in 'biomechanics'. Indeed it is, but usually the mechanical impedance is determined at one (mostly very low) frequency. This article will reveal some of the characteristics of the mechanical impedance of human muscle as function of frequency and amplitude of sinusoidal movement.

In this study, mainly the results of measurements on a passive arm are shown. The measuring apparatus is extensively described by BOON [3]. The subject is seated on an adjustable chair, his arm is fixed in a horizontal position. The angle between upper arm and forearm is 90° in the mid-position. The elbow lies in a small cup, and the wrist is tied up in a clamp which is mounted on 2 springs with strain gauges, that measure the reaction force of the forearm upon moving it.

The experiments are done with 5 subjects, aged between 20 and 26 years (1 female).

Results

Figure 1 shows some results of an experiment in which the force is measured as a function of the forearm position. The amplitude is 0.06 rad, frequency changes from 2.5 to 9.2 rad/sec. From this figure we see that, even at the lowest frequency, a hysteresis appears; this means that on moving, energy is dissipated in the arm. In figure 2 we see force-position curves at different amplitudes at a low frequency (1.5 rad/sec).

From the hysteresis curve some typical characteristics can be determined: $\hat{M}$ = amplitude of the force (or momentum when indicated); W = energy, dissipated during one period of movement. The energy is determined by the area of the hysteresis curve; $\hat{\Phi}$ = amplitude of position; h = the slope of the curve; a = vertical distance between the 2 lines with slope h.

What are the relationships between these characteristics?
Figure 1 and 2 suggest that the amplitude of the momentum $\hat{M}$ can be described by:

$$\hat{M} = \tfrac{1}{2}a + h\hat{\Phi}. \tag{1}$$

The first term, $\tfrac{1}{2}$ a, indicates the hysteresis, the second term indicates a proportional relationship between force and position. At low frequencies ($\omega < 3$ rad/sec), the h is constant (say, h = k). Physically, this means we can compare the last term, k $\hat{\Phi}$, with the action of a spring with a spring constant, k (amplitude between 0.05 and 0.5 rad, see fig. 2). At high frequencies ($\omega >$ 10 rad/sec) the system will be mainly determined by the inertia:

$$M(t) = J\frac{d^2\Phi}{dt^2}. \tag{2}$$

On moving sinusoidally,

$$\Phi(t) = \hat{\Phi} \sin \omega t \tag{3}$$

equation (2) reduces to:

$$M(t) = -\omega^2 J \Phi(t). \text{ So } h = -\omega^2 J. \tag{4}$$

It is interesting to see if the mechanical impedance of the passive arm can be described uniquely by a hysteresis ($\tfrac{1}{2}$ a), a spring (k) and an inertia (J). Therefore, the relationship between h and ω has been investigated. If the characteristic slope, h, depends only on the spring and the inertia this would mean:

$$h = \left(\frac{\partial M}{\partial \Phi}\right)_{\Phi = 0} = -\omega^2 J + k. \tag{5}$$

In figure 3 we see the experimentally found relation between $-h$ and ω^2. It seems reasonable to apply equation (5); in which there is no term pro-

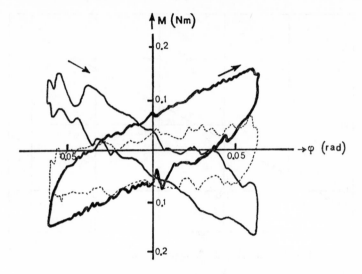

figure 1

Fig. 1. Momentum-position curves at different frequencies. ▬▬ ω = 2.5 rad/sec;
------ ω = 6.0 rad/sec; ▬▬▬ ω = 9.2 rad/sec. Subject K. L. B.

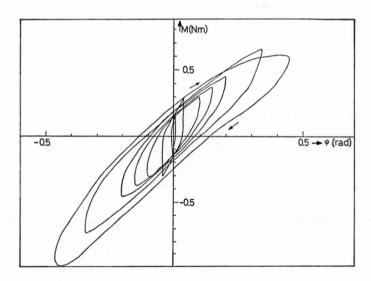

Fig. 2. Momentum-position curves at different amplitudes. Subject A. H. Frequency
1.5 rad/sec.

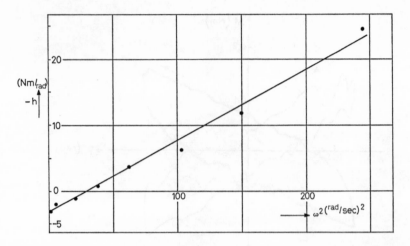

Fig. 3. The relation between — h and ω^2. Subject A. H.

portional to ω, which would indicate a viscous damping. A way to see if no viscous damping is actually present, is to investigate the relationship between W and ω. If a viscous damping occurs, it follows that:

$$W = W_{hysteresis} + W_{viscous},$$

and

$$W_{viscous} = \int_0^{2\pi} Md\Phi = \int_0^{2\pi} B\omega\cos\omega t\, dt = \omega B\hat{\Phi}\pi \ (B = \text{damping factor}). \tag{6}$$

In figure 4 the relation we found between W and ω is shown. Within certain limits, W seems to be independent of ω. The reproducibility of the experiments is satisfying. In 5 subjects, the spring constant, k, varies from 1.5 to 4.0 Nm/rad; the factor, a, varies from 0.1 to 0.3 Nm.

Discussion

The different phenomena that are found can be described with mechanical terms: a spring, an inertia and a dry friction. A dry friction can, of course, explain the hysteresis. However, it is not clear if these mechanical constants are purely mechanical or partially determined by neuronal feedback. To see if actually neuronal feedback exists some observations have been made. Measurements on a patient with a complete lesion of the plexus

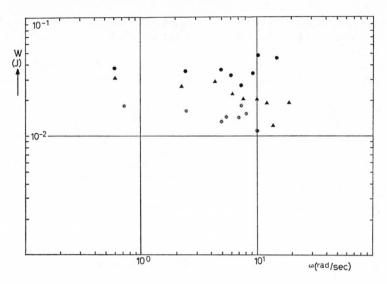

Fig. 4. The relation between W and ω. Subject: ● = K.L.B.; ▲ = A.H.; ✻ = H.E.

brachialis point out that hysteresis is strongly diminished. However, this can also be due to atrophy. With surface electrodes no EMG activity can be found during registrations on healthy subjects. This could indicate that neuronal feedback is not an important factor in the appearence of hysteresis.

The results have a good similarity with the measurements of ROBERTS [10] on the soleus muscle of the cat; however, force change after imposed length change is more predictable in our experiment. Maybe the hysteresis can be explained by means of the hypothesis of HILL [5] about a 'short range elastic component' in passive muscle.

Some orienting measurements have been done on subjects with voluntary contracted muscles. These experiments are less reproducible; it turned out, however, that the spring constant is a function of innervation. Furthermore, the viscous damping factor becomes apparent.

References

1 AQUILONIUS, S.M. and TISELIUS, P.: Measurements of rigidity tremor in Parkinson's disease. Acta neurol. scand. *45:* 327–334 (1969).
2 BOON, K.L.: The muscle spindle as a control element. Proc. 2nd Seminar on Biomechanics, Eindhoven 1969 (Karger, Basel 1971).

3 BOON, K. L.: Mechanische eigenschappen van ledematen, een systeem analytische studie. Nederlandse Orthopaedische Vereniging, Enschede 1970. Ned. T. Geneesk. *115:* 2133–2136 (1971).

4 GON, J.J. DENIER VAN DER; BOSMAN, F.; DIJKSTRA, S. en WIENEKE, G.H.: Enkele stuurkundige aspecten van de motorische functie bij de mens. Ned. T. Natuurk. *35:* 161–181 (1969).

5 HILL, D.K.: Tension due to interaction between the sliding filaments in resting striated muscle. The effect of stimulation. J. Physiol., Lond. *199:* 637–684 (1968).

6 MCRUER, D.T.; MAGDALENO, R.E., and MOORE, G.P.: A neuromuscular actuation system model. IEEE Trans. Man-Machine Syst. *9*(3): 61–71 (1968).

7 NASHOLD, B.S.: An electronic method for measuring and recording resistance to passive muscle stretch. J. Neurosurg. *24:* suppl., pp. 310–314 (1966).

8 RACK, P.M.H.: The reflex response to sinusoidal movement. Symp. on Control and Innervation of Skeletal Muscle, Dundee 1965 (Thomson, Dundee 1966).

9 RACK, P.M.H.: The behaviour of a mammalian muscle during sinusoidal stretching. J. Physiol., Lond. *183:* 1–14 (1966).

10 ROBERTS, T.D.M.: Rhythmic excitation of a stretch reflex, revealing (a) hysteresis and (b) a difference between the responses to pulling and to stretching. Quart. J. exp. Physiol. *48:* 328–345 (1963).

11 ROBERTS, T.D.M.: The nature of the controlled variable in the muscle servo. Symp. on Control and Innervation of Skeletal Muscle, Dundee 1965 (Thomson, Dundee 1966).

12 VREDENBREGT, J. and KOSTER, W.G.: A method for measuring the passive force-length relationship of the human biceps *in situ*. Report No. 83 (Institute for Perception Research, Eindhoven 1966).

13 WEBSTER, D.D.: Rigidity in extra pyramidal disease. J. Neurosurg. *24:* suppl., pp. 299–307 (1966).

Author's address: Ir. KASPER LODEWYK BOON, Technische Hogeschool Twente, *Enschede* (The Netherlands)

Medicine and Sport, vol. 8: Biomechanics III, pp. 249–260 (Karger, Basel 1973)

Biomechanical
and Neuromuscular Responses to Oscillating
and Transient Forces in Man and in the Cat

A. BERTHOZ

Laboratoires de Physiologie du Travail du CNRS et du CNAM, Paris

One of the most important problems in biomechanics is the evaluation of the transfer characteristics of the system including a limb, the corresponding group of skeletal muscles, and their neural control. These characteristics have been studied in several cases by using the methodology of system analysis, and it is well-known [7] that the human hand motion control system is efficient for movements in which frequency components are smaller than 3 Hz. Equivalent results have been obtained for eye pursuit capabilities in man or in animals.

Most of the work in this field has been done with voluntary movements. Another way of investigating motion control is to apply variable forces either to the whole body or to a particular limb. In the present work we shall summarise some work done with several collaborators concerning the response to oscillating or transient forces in man and in animals. The particular technique concerning the work has been described in previous papers and will only be dealt with very briefly. It will be shown that a general characteristic of human limb position control is that the system tends to oscillate around 3–5 Hz. Similar phenomena will be described in the cat. It will be demonstrated that the mass of the limb contributes to the existence of this oscillation. It will also be shown that within this frequency diameter, the capability of the neuromuscular system to generate an adequate stiffness is limited when a mass is present. A supraspinal contribution to these phenomena is probable, but has not been specifically investigated.

I. Muscular Activity and Movements during Oscillatory and Transient Forces in Man

It has been shown [13] that when submitted to sinusoidal oscillations, the human body behaves like a system of suspended masses. A resonant frequency has been found in the 3–5 Hz range. The amplification factor of this resonance ranges between 2 and 3. In other words, when a man is sitting on a vibrating stool the amplitude of his thorax may be up to 3 times the amplitude of his buttocks. The same order of magnitude has been found for head rotation on the cervical vertebrae. It was demonstrated by recording the EMG paravertebral activity that the tonic postural activity of these muscles was slowly modulated at frequencies lower than 2 Hz. Moreover, a rhythmical activity appeared simultaneously with the resonance [2].

A. Oscillation of the Head due to Low-Frequency Movements of the Body

We have further studied these phenomena on a particular sub-system of the human body: the one formed by the head and neck muscles.

Figure 1a shows the subject lying on a vibrating table. The body of the subject is firmly strapped to the table with only his head hanging free to move. A displacement transducer, a, measures the movement of the table Z_1, another displacement transducer, b, measures the *relative* displacement between the head and the table $Z_2 - Z_1 = \Delta L$. Tangential acceleration Z_2 at the summit of the head is measured by an accelerometer, c. Surface electrodes recorded the EMG from neck muscles and the movements of different points of the system were checked by photography of small lights. Figure 1b shows the variation of the relative displacement between the head and the oscillating table *versus* the frequency of sinusoidal oscillation for different peak-to-peak amplitudes. These preliminary results show that at a given frequency ΔL is somewhat proportional to the amplitude of input oscillation. For instance, at 4 Hz the mean values of ΔL are 5, 10 and 24 mm when the values of a are ± 2, ± 4 and ± 10 mm, respectively. Moreover it can also be noted that the increase of amplitude of ΔL between 1 and 4 Hz has a parabolic shape. This would suggest that within this frequency range the relative displacement is related to acceleration. Above 4 Hz the relative displacement remains constant in spite of a very high amplitude as frequencies increase with acceleration.

Figure 2 shows, for 1 subject, the EMG recordings made in the same conditions as in figure 1a. It can be seen that for very low frequencies the relative displacement, ΔL, during slow movements of the table, is small but oscillatory and has a frequency of 3–5 Hz. This observation had already been made by STARK during human hand tracking [8]. It suggests that the tendency to oscillate at these frequencies is indeed a basic characteristic of the limb position control system which STARK attributed to a sampling mechanism. When frequency increases are above 2 Hz, EMG activity tends to be organises in bursts.

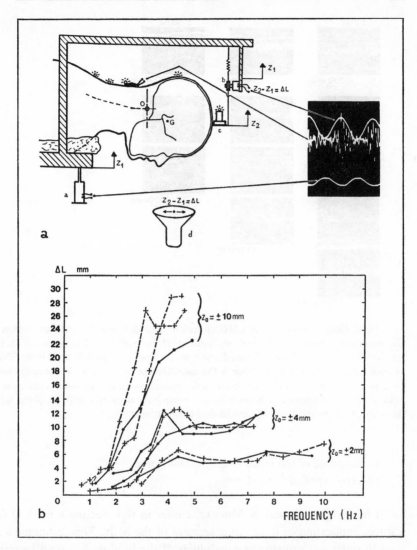

Fig. 1 a Experimental set-up for the study of oscillating forces applied to the head. The subject is strapped on to a vertically oscillating table. The head is free to move. The subject is asked to keep his head still while the table is oscillating. The position of his head is measured as described in the text and appears on an oscilloscope (d) allowing the subject to regulate this position very precisely. *b* Relative displacement ΔL between head and vibrating table (or between head and body) in the conditions of figure 1 for 2 subjects a ($\bullet$—$\bullet$) and b ($+$ $-$ $-$ $+$). The frequency of sinusoidal oscillation varied between 0 and 10 Hz and the amplitude Z_0 was ± 2, ± 4, ± 10 mm [BERTHOZ, GUERIN, and BANDET, unpublished information].

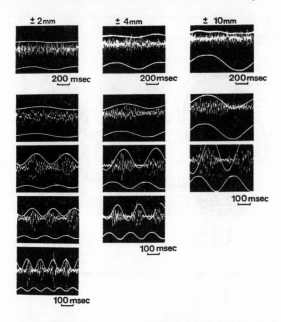

Fig. 2. Head movements and EMG of neck muscles during a sinusoidal oscillation in the conditions of figure 1a. Effect of the amplitude of the oscillation. The upper trace is the relative displacement ΔL between head and table (or body), the middle trace is the EMG of neck muscles and the lower trace is the displacement of the table. Three amplitudes (± 2 mm, ± 4 mm, ± 10 mm) have been used. Calibration for displacement is the same in both displacement traces and is given by the numerical values of amplitude of the table [BERTHOZ, GUERIN and BANDET, unpublished information].

B. Response of the Head to Sinusoidal Forces Directly Applied to the Skull

I have reported, with S. METRAL, earlier in this meeting a method for a direct application of forces to a segment of the body. This technique was used in order to evaluate the capabilities of the head-neck system to resist applied forces. As yet, very little data are available on such responses. This despite the relevance of such studies for the knowledge of head-neck dynamics which is essential to many biomechanical problems such as whiplash studies, vibration protection helmet design or rehabilitation.

Figure 3 shows an example of the experimental set-up. Tangential forces were applied directly to the summit of the skull in human subjects. As a first approach we measured the response to sinusoidal forces with frequencies ranging from 0 to 10 Hz. Mean value of the force was 500 g to 3 kg and sinusoidal amplitude was 200 g to 1 kg.

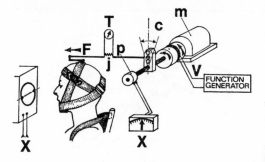

Fig. 3. Experimental set-up for the study of the head-neck system by application of forces. The subject is strapped to chair with only the head free to move. A special helmet allows the fixation of an articulated rod at the summit of the skull. This aluminium rod connects the head to the lever of the torque-producing device. The device consists of a motor (m) turning at constant speed which drives, through the effect of an electromagnetic clutch (c) the lever. This technique is similar to that described by BERTHOZ and METRAL [3]. A potentiometer (p) measures the angular displacement (X) which is a good approximation of head movement. The force is applied by a function generator through a voltage, delivering a voltage (V) to the clutch and a strain gauge (T) measures this force at a point in between the clutch and the head. The displacement (X) is sent to an oscilloscope which can be viewed by the subject; this allows him to see the position of his head.

The subject was instructed to maintain the position of his head as fixed as possible. To help him in his task, an oscilloscope displayed in front of him gave a precise measure of the exact position of his head. A displacement transducer placed on the torque-producing machine measured the tangential displacement of the head, while a strain gauge measured the force very near to the point of application. Only an example of the results will be given here to illustrate the basic properties of the response. In figure 4 are described the characteristics of the movements of the head. Former analyses were performed on output displacements on the summit of the skull and input sinusoidal forces with the methods described elsewere [4]. A Bode plot giving the gain and the phase between the input force and output displacement has been constructed (see legend of fig. 4). This gives a measure of the compliance of the system. Because rotation of the head was of small amplitude, the actual angular displacement was accurately measured by the tangential linear displacement of the summit of the skull. Harmonic distortion between input and output was low enough to allow computation of gain and phase on the bases of the fundamental of the input and output.

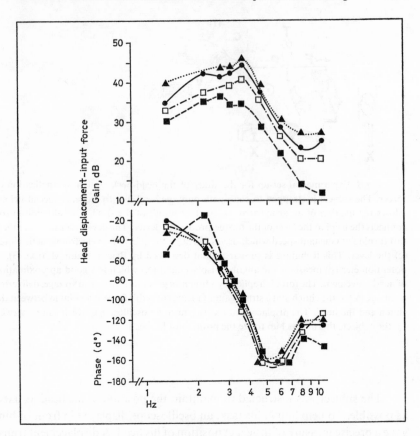

Fig. 4. Head displacement *versus* input sinusoidal force for a human subject. Tangential sinusoidal force was applied to a seated subject at the summit of the skull. The input force was of constant mean value (1 kg) and its sinusoidal amplitude was successively ± 200 (■), ± 400 (□), ± 600 (●) and ± 800 g (▲) g. The subject was instructed to resist the applied force. The amplitude ratio and phase relation between output displacement of the head and input force was plotted (compliance of the system). In this case, input force is measured directly at the control unit of the torque-producing machine. Amplitude ratio is expressed as gain in decibels computed as

$$20 \log_{10} \frac{\sqrt{a_1^2 + b_1^2}}{\sqrt{a_2^2 + b_2^2}} \; ; \; a_1 \text{ and } b_1 \text{ are the Fourier coefficients of}$$

the fundamental of output head displacement and a_2, b_2 the same coefficients for input sinusoidal force.

1. For frequencies lower than 2 Hz it is observed, as previously in the forearm, that the subject can well compensate the applied force by active contraction of his neck muscles. However, for these frequencies an additional oscillation of the head was seen to be superimposed to the applied frequency. This superimposed oscillation was at around 3–5 Hz. It is very similar to what has been observed on the vibrating table.

2. Between 3 and 5 Hz a resonance occurs. We were able to demonstrate by combined radio-cinematography and photography of small lights on the head that the centre of rotation for this case is in the proximity of the lower cervical vertebrae.

3. Above 5 Hz, the displacement decreases very quickly and the head rotation is then around a centre located in the proximity of the first cervical vertebrae. The atypical shape of the phase curve in the Bode plot is probably due to this change of rotation centre, and we suggest that different groups of muscles may be activited at these frequencies (deep muscles of the neck).

We concluded, from these preliminary experiments that the head-neck system behaves in a very similar manner to that described by BERTHOZ and METRAL [3] for the forearm. Detailed analysis of biomechanical and muscular responses of this system to transient and oscillatory forces needs to be continued.

C. Transients

Application of transient forces (force ramps) has shown that when submitted to *transient* forces, the displacement of the head is linked to the *speed of onset* of the force, and to the loading-unloading sequence, as was demonstrated in the forearm [3]. This result, we think, is of great importance in understanding the mechanisms of neck injuries occurring during small accelerations (and thus forces), but with a particular time course and sequence of loading-unloading [see discussion in reference 3].

II. *Behaviour of the Neuromuscular System Studied in the Cat*

In order to establish the fundamental mechanisms underlying the response in man to these forces, we designed some experiments to study the same phenomena in the cat.

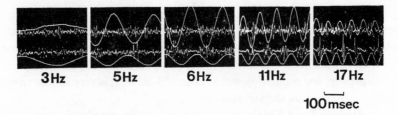

3 Hz   5 Hz   6 Hz   11 Hz   17 Hz

100 msec

Fig.5. Muscular activity and movements during whole body oscillations in a cat chronically implanted with EMG electrodes. The upper trace is the vertical acceleration of the head. The bottom trace is the displacement of the vibrating table (± 2 mm). EMG of neck muscles (top EMG) and front limb extensor (bottom EMG) have been recorded [BERTHOZ and PERRET, unpublished information].

A. Whole-Body Oscillation in the Cat

In experiments made with J. C. PERRET, EMG electrodes were chronically implanted in neck muscles and extensor muscles of cats. Figure 5 gives an example of the recordings: the lower trace is the sinusoidal displacement of the vibrating table, the trace just above is the EMG of the front limb triceps, then an EMG of the neck muscles; the upper trace is the vertical acceleration of the head recorded by an accelerometer on the summit of the head of the cat. The cat was standing freely on all four legs during oscillations of the table. The recordings have been made at different frequencies (3, 5, 6, 11 and 17 Hz) for ± 2 mn amplitude of the table. It can be seen that the acceleration recorded at the level of the head of the cat is maximum at 6 Hz, and if transmissibility is plotted in terms of displacements one sees that the 'resonant' frequency is around 5 Hz. There is then a great similarity between the overall behaviour in man and cat, a finding which is further confirmed by EMG recordings. These observations allowed us to study this problem in the cat.

B. Dynamic Characteristics of the Stretch Reflex

In the final motor response, the characteristics of the stretch reflex occur at different levels. We have submitted isolated extensor muscle (gastrocnemius) in the decerebrate cat to sinusoidal forces and the results demonstrate, by plotting the ratio between output displacement to input force as was demonstrated before by TERZUOLO *et al.* [11], that the overall transfer function of the stretch reflex in these conditions, is constant between 0 and 10 Hz (fig. 6). In

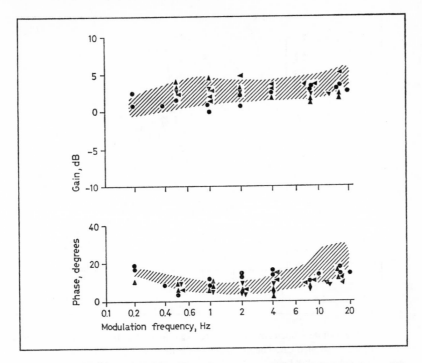

Fig. 6. Length-tension relationship of the gastrocnemius muscle with reflex intact. The data shown were obtained during sinusoidal modulation of either the gastrocnemius muscle length or tension in 10 decerebrate cats. The data were analysed as length (input) tension (output) ratios making the indicated gains a measure of the muscle stiffness (an increase of 6 dB means a doubling of the muscle stiffness). The gains were normalised to emphasise the relative change in muscle stiffness with modulation frequency. The discrete data points describe the relationship measured in 5 experiments using a force input device. The shaded areas encompass the data obtained in 5 additional experiments using a low-compliance, length input device [11]. The range of mean muscle tensions tested was 600–1,000 g weight with force inputs and 400–1,800 g weight with length inputs. The peak-to-peak input amplitudes ranged from 400 to 800 g weight using force inputs and 400–2,700 g using length inputs. This figure shows that the stretch reflex has constant gain characteristics over the frequency range studied and that, in the absence of mass, the displacement lags the input force by about 20° (360° is the lenth of a sinusoidal cycle) [4].

Copyright J. Neurophysiology (USA)

other words, the displacement resulting from the application of a sinusoidal force remains constant when frequency varies and the phase lag of displacement, with respect to tension, also remains constant. In the same work it was shown that by cutting the dorsal roots, i.e. by interrupting the reflex, the

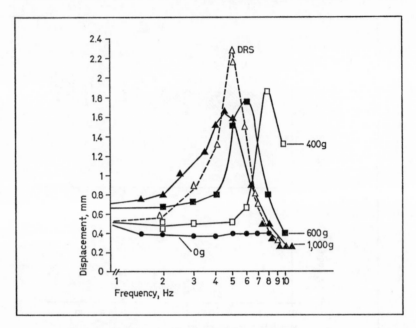

Fig.7. Absolute displacement (peak-to-peak in millimetres) of the gastrocnemius muscle of a decorticated cat. The extremity of the muscle was attached to the same force-producing device used in figure 1a. A constant amplitude (500 ± 300 g sinusoidal force was applied to the tendon of the muscle. When no weight was present at the extremity of the pulling device (0 g curve) the displacement remained constant for all frequencies of applied force. When weights of different values were fixed to the rod (400, 600 and 1,000 g) a resonance occurs. With a weight of 1,000 the dorsal roots of the spinal cord at L_6, L_7, S_1 levels were sectioned (DRS). The amplitude of the resonance increased (----), revealing the fact that when the reflex is present its action is to decrease the oscillation [BERTHOZ and PERRET unpublished information].

'stiffness' increases with increasing frequency, thus verifying that the reflex has the important ability to allow a dynamic compensation of rapid changes in applied force. But as no 'resonant' peak is observed under these conditions (when no mass is attached to the limb) one can assume that the presence of a mass is a condition for the resonance to occur. This has been verified in decorticated cats. An extensor (gastrocnemius) of a hind limb was subjected, as before, to sinusoidal forces and weights of different values were fixed to the force input device. It was clearly shown (fig. 7) that without any weight the displacement is constant with varying frequency, whereas a resonance peak appears when weights are added. When a weight of 1,000 g is added (which is within physiological range of normal weight on a limb in the cat) a definite peak appears at around 5 Hz. Cutting the dorsal root increased the amplitude of the oscillation.

It can, then, be concluded that this general characteristic of human or animal limb to oscillate at around 3–5 Hz in response to externally applied dynamic forces is not solely due to an oscillation inherent of the loop characteristics of the stretch reflex itself, but rather to a combination of the effect of the mass of the limb and the stretch reflex including the low-pass filter characteristics of the striated muscle. The supraspinal influences to reduce or increase this phenomenon have to be taken into account, as was shown in the case of parkinsonian patients [METRAL and SCHERRER unpublished observations]. Though we cannot present any evidence on this point, studies should be orientated towards long-loop reflexes, as suggested by ECCLES [5] involving the cerebellum or even other central structures whose activity, though not involved in a 'dynamic loop' could, by their ability to have a rhythmical discharge, contribute to these phenomena. We then feel that one should yet be very careful in interpreting the above data in favour of a purely peripherical (mass-induced) or central (long loops, or set rhythms) hypothesis.

Conclusion

It can be seen from the data presented here that the description of dynamic characteristics of the human muscular control in response to applied forces is still very incomplete. But recent developments in techniques and concepts allow us to have a more detailed view today of the respective rôle of peripheral and central mechanisms. The analysis of the resonance observed in the 3 to 5-Hz frequency range can throw some light onto some of the basic limitations of overall motor performance, though we cannot pretend here to have done more than make the first steps toward a new quantitative description of movement.

References

1   BERTHOZ, A.: Mise au point et validation sur le terrain d'un critère de tolérance aux vibrations de basse fréquence. Rapport CECA n° 6242/22/025 (CNAM, Paris 1970).
2   BERTHOZ, A. and WISNER, A.: Striated muscles activity and biomechanical effects in man submitted to low frequency vibrations. Proc. 1st Int. Congr. of Electromyography and Kinesiology. Electromyography 8: suppl. 1, pp. 101–109 (1968).
3   BERTHOZ, A. and METRAL, S.: Behavior of a muscular group subjected to a sinusoidal and trapezoidal variation of force. J. Physiol. 29 (3): 378–384 (1970).
4   BERTHOZ, A.; ROBERTS, W.J., and ROSENTHAL, N.P.: Dynamic characteristics of stretch reflex using force inputs. J. Neurophysiol. 34: 612–620 (1971).

5   ECCLES, J.: The way in which the cerebellum processes information from muscle; in
    YAHR and PURPURA Neurophysiological basis of normal and abnormal motor activi-
    ties, pp. 379–414 (Raven Press, Hewlett 1967).
6   EWING, C.L.; THOMAS, D.J.; BEELER, G.W.; PATRICK, L.M., and GILLIS, D.B.:
    Dynamic response of the head and neck of the living human to $-G_x$ impact accel-
    eration. Proc. 12th Stapp Car Crash Conf. (SAE, New York 1968).
7   MERTZ, H.J. and PATRICK, L.M.: Investigation of the kinematics and kinetics of
    whiplash. Proc. 11th Stapp Car Crash Conf. (SAE, New York 1967).
8   STARK, J.: Neurological control system (Plenum, New York 1968).
9   TARRIERE, C. et SAPIN, C.: Etude biomécanique de la liaison tête-thorax. Report of the
    Biomechanics Laboratory, Renault-Peugeot, France (1968).
10 TERZUOLO, C.A. and POPPELE, R.E.: Myotatic reflex. Its input-output relation.
    Science *159:* 734–745 (1968).
11 TERZUOLO, C.A.; MCKEAN, T.; ROBERTS, W.J., and ROSENTHAL, P.L.: Gain of the
    stretch reflex in extensor muscles of decerebrate cats; in ANDERSEN and JANSEN
    Excitatory synaptic mechanisms, pp. 327–332 (Blindern, Norway 1970).
12 TISSERAND, M.: Comportement du rachis cervical lors de chocs dorsaux. Report No.73
    INRS, Nancy-Vandoeuvre 1966).
13 WISNER, A.; DONNADIEU, A. et BERTHOZ, A.: Etude biomécanique de l'homme soumis
    à des vibrations de basse fréquence. Trav. hum. *1* (2): 18–56 (1965).

Author's address: Dr. A. BERTHOZ, Laboratoire de Physiologie du Travail, 41 rue
Gay Lussac, *75 Paris 5ème* (France)

Medicine and Sport, vol. 8: Biomechanics III, pp. 261–267 (Karger, Basel 1973)

Equivalence between
Positive and Negative Muscular Work

H. MONOD and J. SCHERRER

Laboratoire de Physiologie du Travail du CNRS
Faculté de Médecine Pitié-Salpétrière, Paris

It is not possible to compare positive and negative muscular work using mechanical criteria. For the physicist 'negative work' has no sense, as he considers that the muscle does not perform any mechanical work when, after it has been raised, a weight is put down in its initial place; in this case the weight, not the muscle, performs the work.

The physiologist cannot resolve the question so easily, following EPIC-TETE's recommendations. Concentric, as well as eccentric contractions, involve physiological changes of the same nature, i.e. pulmonary gas exchanges, minute volume, heart rate, EMG, ionic transfers between muscle fibers and blood, which differ only in magnitude.

Since the first experiments by CHAUVEAU at the end of the last century, many studies have been devoted to the measurement of the energy cost of negative work for general exercise. For instance, ASMUSSEN [1952], ABBOTT *et al.* [1952] have shown that for bicycling negative work is not as expensive as positive work (from 3 to 125 times less, depending on the speed of contraction).

The aim of this study has been to consider the same problem for local muscle work. By local work, we mean work performed by only one muscle or one synergistic muscle group [SCHERRER and MONOD, 1960]. Positive and negative work is compared in 2 different ways, by muscle work capacity and by energetic cost of work for different kinds of *biceps brachii* contractions.

I. Method

The experiments were carried out on 3 male subjects, 25–38 years old, examined during a 2-year period; 2 successive series of measurements were carried out.

---

1   With the technical assistance of Mrs. HUART.

Series A (195 measurements). Oxygen consumption, heart rate and EMG activity were studied during dynamic work of low or moderate intensity performed at a constant power without exhaustion. The muscle contractions were of 3 types: either alternating motor and resistant (mixed work), as is generally used in ergometric testing, or strictly motor (positive work) or strictly resistant (negative work).

Series B (72 measurements). Local dynamic work capacity, heart rate, and EMG activity were studied during dynamic work, either mixed or positive, of sufficiently high intensity to entail local muscular exhaustion relatively soon.

1. Dynamic work on the ergometer. In the 2 series, the work was performed by the biceps brachii and synergistic muscles. The contractions were given on a specially designed ergometer, including pulleys and cables, a strain gauge, an electromagnetic clutch and an electric motor [Du PASQUIER and MONOD, 1967]. This ergometer allows the subject to perform 3 different kinds of work: (1) mixed work: the apparatus is used as an ordinary ergometer; (2) positive work: the subject, his forearm extended, contracts his biceps and lifts the weight; at the end of the movement, a contact orders the excitation of the clutch; the motor resists and slows down the fall of the weight, and (3) negative work: the subject, his forearm extended, presses a contact, the clutch is excited and the motor raises the weight slowly at the same time that the subject flexes his forearm; at the end of the course, the subject releases the contact and, cutting the excitation of the clutch, contracts his biceps to slow the fall of the weight.

2. Physiological data. Oxygen consumption was measured by an open circuit method from the minute volume and an analysis of the expired gases. The oxygen consumption was calculated using the Margaria's nomogram [MONOD *et al.*, 1963]. Heart rate was obtained from electrodes placed at the lower limbs or on the precordial area and connected either to a polygraph (Alvar) for series A, or to a cardiotachometer (Rood) for series B. An EMG was obtained through skin electrodes on the biceps brachii. The integrated EMG was measured by an integrator functioning on the principle of condenser charges and discharges.

3. Procedure. The dynamic work was carried out in a supine position by the flexor muscles of the forearm in order to obtain a maximal relaxation of the posture muscles.

In series A, the work was bilateral. The external force opposite the right and left biceps contraction was of 5.7–17 kg. The contractions were always carried out at a spontaneous speed, at a frequency of 10/min. The power, which was kept constant during the test, varied from 14 to 60 kgm/min. These values are below the critical power of the muscles concerned, as pointed out by the series B measurements. During each session the oxygen consumption was measured in periods of 4 min. In most cases, 2 tests of different types were carried out during the same session.

In series B, the work was unilateral. The weights mobilized at the rate of 12/min varied from 5.7 to 17 kg as in series A, but applied only to one side. The work, either mixed or purely positive, was carried out at a constant power until local exhaustion. Following SCHERRER *et al.* [1954], this was achieved when the amplitude or the initial frequency of the movements could no longer be continued by the subject. Local exhaustion is defined, for each power, by the length of the test (time-limit, t) and the amount of work carried out (work-limit, W). The values of t and W are used to obtain the critical power of the muscle, i.e. the maximal power for nonexhaustive work [MONOD and SCHERRER, 1965; MONOD, 1972].

During each session, the subjects generally carried out 2 successive tests of different types with one arm, and then with the other. In 5 cases the tests were voluntarily stopped after about 45 min, before exhaustion.

II. Results

Oxygen intake, EMG activity and maximal time for the tests until exhaustion were compared with the amount of work done by the biceps brachii. But it should be recalled that the same number of kilogrammeters relates to either the positive mechanical work of lifting the weight carried out by the muscle (W_p), the negative work done by this muscle (W_n) when the weight is lifted by the ergometer, or the mixed work ($W_m = W_p + W_n$) when the muscle performs positive and negative work.

A. Energetic Cost of Work

Net oxygen intake ΔV_{O_2} was calculated for the local exercises performed in steady-state conditions (series A). Each of the 3 subjects carried out the 3 types of dynamic work, 5 times with 4 different weights. A linear relationship was established between the 60 individual values of ΔV_{O_2} and the power of the test for each of the 3 types of activity (positive, negative and mixed work). In addition, the oxygen cost for unloaded movements carried out 5 times by each subject was calculated (fig. 1).

B. Critical Power of Active Muscles

The relationship $W = a + bt$ was determined for each of the 3 subjects, for both the right and left arm, for mixed and for positive work. In this relationship, W designates the work-limit and t the time-limit. The W for a given t is always greater for positive work than for mixed work. In order to eliminate individual differences in local work capacity, the W has been expressed as the percentage of the maximal work done in 25 min (fig. 2).

In the above relationship, b represents the critical power corresponding to the maximum rate of energy supply. A regular difference exists between the 2 types of activity. For positive work the mean critical power was 31.3 kgm/min for the right arm and 27.3 kgm/min for the left arm. It appears, from the comparison, that the critical power for positive work is greater than

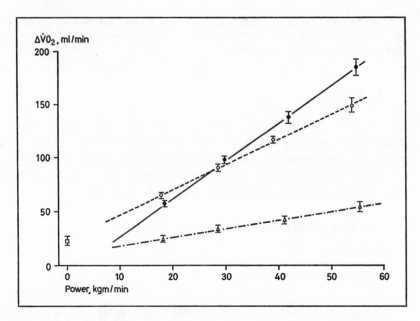

Fig. 1. Energy expenditure for positive, negative and mixed work. Net oxygen intake (in ml/min, STPD) is plotted against the power for 3 types of local exercise. In addition, the energetic cost of unloaded work is represented. Each point is the mean (with the SD) of 15 experiments (5 tests on 3 subjects). ●—● = mixed work, $y = 3.6 x - 8.6$; ○---○ = positive work, $y = 2.4 x + 23.7$; △–·–△ = negative work, $y = 0.80 x 10.4$; □ = unloaded work.

the critical power for mixed work. The supplementary positive work that can be accomplished in a given time, when the negative work normally included in the mixed work has been excluded, is + 30.7% for the left side and + 33.2% for the right side with a mean of + 32%.

C. Electromyogram

A corrected integrated EMG was obtained for 54 tests of series A and 19 tests of series B.

The integrated EMG is still greater for positive work than for negative work for an equal mobilized weight. The comparison can be made by calculating the ratio of the EMG for positive work with the EMG for negative work. This presents a considerable variability; for the unloaded movements

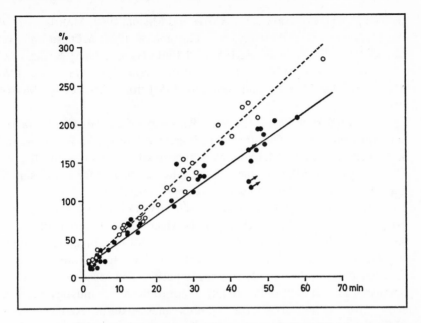

Fig. 2. Critical power for positive and mixed work. The work-limit as the percentage of the amount of work done in 25 min is plotted against the time-limit in minutes, for positive work (O, ----) and for mixed work (●, ——). The slope of the curves represents the critical power of the muscle. Three subjects were tested. Points with a small arrow denote tests voluntarily stopped before exhaustion.

it is 3.6 and for mixed work with weight it varies from 1.3 to 2.7 with a mean of 2.1 for 21 tests. This ratio decreases when the mobilized weight increases, from 2.3 at 9.5 kg to 1.6 at 17 kg.

An increase of the integrated EMG occurs, as is usually seen with muscle fatigue [SCHERRER and MONOD, 1960]. This increase of the EMG activity is proportionally the same in the 2 parts of the movement in mixed dynamic work with exhaustion. But the ratio between the EMG activity of positive work and that of negative work varies irregularly between 1.5 and 5.0 with a mean of 2.8 for 19 tests.

III. Discussion

The energy expenditure of positive and negative work has been studied for various muscular activities for a large amount of muscles: *walking* on an inclined plane [MARGARIA, 1938; COTES and MEADE, 1960; MARGARIA et al., 1963]; *climbing* stairs [BENEDICT and PARMENTER, 1928; MÜLLER and HETTINGER, 1952; BANERJEE et al., 1959 and 1961; HESSER, 1965; NAGEL et al., 1965; RICHARDSON, 1966; KAMON, 1970); *bicycling* [ABBOTT et al., 1952; ASMUSSEN, 1952; ABBOTT and BIGLAND, 1953] (for references see MONOD, 1973).

In all these experiments the ratio (R) *energy expenditure of positive work to energy expenditure of negative work* is greater than 1, with very large variations between 1.3 and 8.4. In one case (high-speed contractions) this R equals 125 [ASMUSSEN, 1952], but generally the values calculated for R by using the data of the authors refered to above are approximately 3.

The same figure has been obtained for local work in the present experiment using 2 different experimental methods. Heart rate measurement, minute volume and EMG activity did not quite so certainly confirm an R equal to 3.

Many factors could explain an R which differs from 3: an asymmetric activity pattern of the muscles engaged in positive and negative exercise, percentage of the net oxygen intake linked not to the active muscles but to the extra activity of the heart and the respiratory muscles during exercise or inertia of the body segment which must be added to or deducted from the charge moved, depending on the type of muscular activity.

Whatever the value of R in investigations dealing with comparisons of negative and positive work, it seems that Chauveau's hypothesis (supported later by numerous other authors including HILL and FENN) of a mechanical energy absorption by the muscle during eccentric contraction, cannot be considered as verified. If the idea of an R equal to 3 is retained, it remains to explain the meaning of this figure from a mechanical point of view.

Summary

A comparison between positive and negative work, performed on a special ergometer, was made using physiological criteria. Oxygen consumption, pulmonary ventilation and heart rate were measured on 3 subjects during 195 various ergometric tests. The 'critical power' (maximal work capacity without fatigue) was determined considering the results of 72 tests with exhaustion. The EMG activity was recorded in both series of tests.

The results show that, at a spontaneous speed of movement, the net energy expenditure of negative work is equal to one–third of that of positive work. In the same way, suppressing the negative work allows a 33-percent increase of the critical power of positive work. EMG variations, minute volume and heart rate changes confirm, to a lesser extent, the lower energy expenditure in local negative work.

References

Abbott, B. C.; Bigland, B., and Ritchie, J. M.: The physiological cost of negative work. J. Physiol., Lond. *117:* 380–390 (1952).

Asmussen, E.: Positive and negative muscular work. Acta physiol. scand. *28:* 364–382 (1952).

Kamon, E.: Negative and positive work in climbing a laddermill. J. appl. Physiol. *29:* 1–5 (1970).

Margaria, R.: Positive and negative work performances and their efficiencies in human locomotion. Int. Z. angew. Physiol. *25:* 339–351 (1968).

Monod, H.: How muscles are used in the body; in 'The structure and function of muscle', ed. by Bourne, Acad. Press, 3 vol. (1972, in press).

Monod, H.: Physiological equivalent of local negative work (to be published, 1973).

Monod, H.; Bouisset, S. et Laville, A.: Etude d'un travail musculaire léger. III. Influence de la charge. Arch. int. Physiol. *71:* 441–461 (1963).

Pasquier, P. E. Du et Monod, H.: Un dispositif pour l'étude du travail moteur ou du travail résistant. Trav. hum. *29:* 323–328 (1967).

Scherrer, J. et Monod, H.: Le travail musculaire local et la fatigue chez l'homme. J. Physiol., Paris *52:* 419–501 (1960).

Author's address: Prof. H. Monod, Laboratoire de Physiologie du Travail, CNRS, Faculté de Médecine Pitié-Salpétrière, *Paris* (France)

III. EMG

Medicine and Sport, vol. 8: Biomechanics III, pp. 270–274 (Karger, Basel 1973)

EMG—Force Relationship during Voluntary Static Contractions (M. Biceps)

G. Rau and J. Vredenbregt

Institute for Perception Research, Eindhoven

In biomechanical studies, the EMG activity picked up by surface electrodes can be used as a tool in two essentially different ways: (1) as an indication of time periods during which a muscle is activated or is in an inactive state; these periods can be measured with high accuracy and are very important, especially for investigation of muscle coordination, and (2) as an indication of the force exerted by the muscle momentarily; a greater EMG activity (characterising amplitude and frequency of discharges of the activated muscle cells) can be expected to cause a higher value of force, and *vice versa*.

The aim of our experiments on the human biceps muscle was to find a quantitative and reproducible relationship between the EMG activity and the value of force during static contractions. If such a relationship exists we could use it, at least under static conditions, to estimate the exerted force by only recording the EMG signal. The relationship should also give insight into the peripheral functional mechanisms of the muscle as, for example, fatigue; in addition it could be used as a diagnostic tool for disorders in the muscular function.

Methods

In our experiments the subject was seated on a chair, his right upper arm resting upon a horizontal platform. The angle between the upper arm and the forearm was 90°; the force in the direction of the longitudinal axis of the biceps muscle was measured at the wrist. The EMG signal was picked up by a bipolar electrode system placed transversely above the belly of the biceps. The subject sustained a force of constant value controlled by an indicating meter. As shown by figure 1, the force, the EMG signal and the rectified and integrated signal, i.e. the area between the signal and the zero line, were recorded. As a measure of the EMG activity we used the slope of the rectified and integrated signal; thus, a steeper slope means a greater EMG activity, and *vice versa*.

At the beginning of each experiment a reference measurement was carried out. The subject exerted a force of 18 kg for about 2 sec and if the EMG value had changed in respect to previous ones, the EMG values of this experiment were corrected by a factor. In this way, differences which might have been caused by changes in the position of the electrodes or of the subject himself were diminished [for details see Rau and Vredenbregt, 1970].

Results

As an example, the results of one subject are shown in figure 2. There is the EMG activity as a function of time for different constant forces as parameter.

We would refer to 3 characteristics: (1) the relationship can be approximated by straight lines; (2) the slopes of the lines are much steeper for high forces than for low forces, and (3) the lines show a well-defined EMG activity at the beginning.

The next question is: Is it really possible to estimate the momentarily sustained force by the slopes of the curves even if the force is not kept constant during the whole experiment? In other words: Is the relation invariant in time? To answer this question we asked the subject to change the initial force during an experimental run suddenly to another definite one. The results are shown in figure 3. Here again the EMG activity is a function of time. There are various examples available: in one series of measurements the initial force of high value was sustained, and changed after 10 sec to a lower one. In another case, a small force was sustained at the beginning and changed after 10 or 20 sec to the previous high level. The slope appeared to be only dependent on the force sustained momentarily. We repeated this experiment also at lower levels of force (mean values 3–5 measurements). In the range of forces between 20 and 100% F_{max}, we are therefore able to estimate the sustained force by measuring the EMG activity only.

We recently began to investigate the range below 20% F_{max} and observed also an increase in EMG activity dependent on the force at very low values, for example even at 5% of the maximum force. This seems to be an unexpected result, because it is commonly assumed that forces lower than 15% F_{max} can be sustained without a time limit and without fatigue. This assumption was derived from measurements of the maximum endurance time as reported by Rohmert [1960] and Monod and Scherrer [1965], for example.

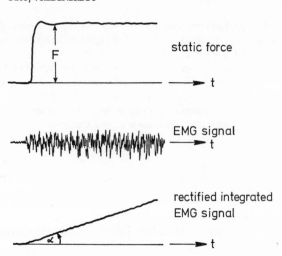

Fig. 1. Registration of the force, the EMG signal and the rectified and integrated EMG signal (schematically). The angle was evaluated as a measure of the total EMG activity.

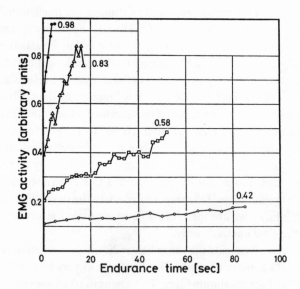

Fig. 2. Variations of EMG activity during continuous static contractions of constant forces exerted by the arm flexor (m. biceps); the angle of the joint was 90°. Example of one subject (V. R.). The curves are mean values of 3–5 measurements; the relative deviations were within 10% at the beginning of the curves and about 20% at the end. Parameter: static force F/F_{max}.

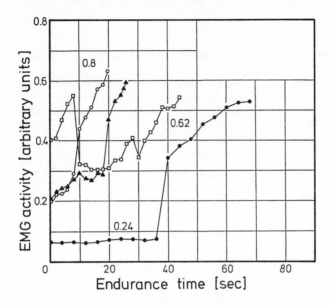

Fig. 3. EMG activity as a function of endurance time. The static forces (80, 62 and 24% F_{max}) were changed suddenly to another value after certain time periods. Curves are mean values of 3–4 measurements. Parameter: static force F/F_{max} (changed). Subject V.R.

We think there are changes in the muscle tissues also at very low forces, indicating a sort of peripheral fatigue. This is supported by the subject's observations, who reported a local feeling of fatigue in a small portion of the muscle; perhaps the muscle is used in a different way at low forces than at high forces.

Considering the results of figures 2 and 3 and especially the relatively small deviations of the data, we assume that the methods of quantifying surface EMG signals (as shown in fig. 1, together with the reference measurement) are appropriate for studying properties of the muscle during contractions. The EMG activity as a function of time can be considered as a representation of peripheral fatigue, which depends on the exerted value of force. Taking into account the changes of the muscular function due to fatigue as mentioned, a quantitative estimation of the force is possible by considering the EMG only, at least under the conditions we defined. The question whether certain disturbances of the normal muscular function could be described by these procedures demands further investigations.

References

MONOD, H. and SCHERRER, J.: The work capacity of a synergic muscular group. Ergonomics *3:* 329–338 (1965).

RAU, G. and VREDENBREGT, J.: Electromyographic activity during voluntary static muscle contractions. Report 192 (Institute for Perception Research, Eindhoven 1970).

ROHMERT, W.: Ermittlung von Erholungspausen für statische Arbeit des Menschen. Int. Z. angew. Physiol. *18:* 123–164 (1960).

Authors' addresses: Dr. G. RAU, Forschungsinstitut für Anthropotechnik, *D-5309 Meckenheim* (FRG); J. VREDENBREGT, Institute for Perception Research, Insulindelaan 2, *Eindhoven* (The Netherlands)

Medicine and Sport, vol. 8: Biomechanics III, pp. 275–277 (Karger, Basel 1973)

Some Problems of the Electrical Activity of Elementary Human Movements

L. Lukács, Clara Lukács and M. Nemessúri

Pediatric Hospital of Buda, and Hungarian Institute of Physical Education, Budapest

In the special literature, the majority of EMG examinations refer to pathological cases; however, EMG examinations of movement processes of healthy people are becoming more frequent. More and more proceedings are published about the realization of EMG examinations of simple and more complicated movement patterns. They deal with EMG examinations of walking swimming, gymnastic exercises, moreover, with movements of work and calisthenics.

In general, contact electrodes were used for the examination of healthy muscles, but sometimes graphs were also taken by means of needle electrodes, furthermore telemetric EMG examinations have appeared.

Examining the various human movement processes such as grasping, throwing, walking, weight-lifting, etc., we have observed that EMG graphs taken under the same conditions from different persons have shown, more or less, sometimes significant deviations. It has occurred that when taking repeated EMG graphs from the same person, the bio-electric activity showed a significant change even in the case of identical movement patterns.

We decided to examine simple movements of healthy human muscles with 1° of joint freedom. The range of variety was analysed on a series of photographs and we tried to find out the cause of it.

We examined the biopotential of serial isometric functions of several muscles. In addition to this, the extension, flexion and pronation of the wrist has been examined during function of the m. extensor digitorum communis, m. flexor digitorum sublimis and m. pronator teres. On the lower limb the flexion and extension of the joints of the knee and hip, as well as the abduction, adduction, rotation outward and inward were analysed while the m. biceps, m. quadriceps femoris and m. adductor longus were in operation.

Table I. Ratio between the biological movement analysis and the electric activity

	Number of experiments	Number of experiments, %
In accordance	13 ⎞ 28	35 ⎞ 76
Partly in accordance	15 ⎠	41 ⎠
Discordant	9	24
Total	37 =	100

Apart from simple movements, we have taken in every case the biopotentials of movements done against resistance. When applying DISA equipment for EMG graphs, needle electrodes were used.

The results of our experiments have shown that the bioelectric activity was, in only one-third (35%) of the cases, in good agreement with the results that were expected from the biological point of view of movement; in 41% of the cases smaller deviations occurred and, as was anticipated, based upon our former examinations, the EMG graphs were different in 24% of the cases (table I).

In the following, we describe the development of the ratio between the biological movement analysis and the bio-electric activity based on some functional forms.

During the isometrical functioning we expected that biopotentials could be measured at every muscle. This presented itself in the case of the examined person, No. 1, where in the quadriceps as well as in the biceps the peak potential activity was 900 μV (with an interference curve), and in some potentials of action a peak potential of 300 μV appeared. However, examining subject No. 7 (during all 6 actions) the peak potential of the biceps appeared to be much higher (4,000 μV) than that of the quadriceps (2,000–1,600 μV), although we expected a higher peak potential at the quadriceps because of the isometric muscle work done in a horizontal position with stretched knees. Relatively high discharges (1,400–800 μV) were measured in the adductors. The examined subject, No. 2, showed a very different picture from that expected; here, only in the adductor longus were single action potentials measured of 300, 300 and 800 μV, respectively, whilst the m. quadriceps and biceps femoris did not show any reaction.

In 2 cases it was found that from the muscles of the forearm no potentials of action could be conducted from the m. pronator teres during isometrical action.

Examining the extensions in every subject, the highest peak potentials were derived from the extensors, moreover in the other 2 muscles bioelectric activity could also be measured except in 2 cases, in which the biceps femoris (opposing activity), adductor longus and pronator teres remained silent.

The EMG graph of wrist flexion was formed, as we anticipated, at the flexion of the knee. The picture was different from what we had expected. This manifested itself in higher peak potentials measured on the m. quadriceps and adductor longus, as there were generating lines of force on the biceps in the direction of displacement (kinetor activity). On the examined subject No. 7, for instance, we measured during a simple flexion some action potentials on the biceps and quadriceps with a peak potential of 800 μV, whilst on the adductor an interference curve with a peak potential of 4,200 μV took shape. Loading the biceps and the quadriceps the peak potentials increased to 1,800 μV each, whilst the adductor unexpectedly decreased to 3,600 μV.

In examining the other displacements, we met with similar symptoms. For instance, at the adduction of one of our subjects the adductor showed a peak potential of only 500–600 μV, but in the biceps just as much, however, as in the quadriceps we measured a peak potential of 1,400 μV.

Looking for the cause of the described symptoms based upon our observations, we think that the inexactitude of the movement direction of the central nervous system manifests itself in a bio-electrical activity and, according to this, the tension distribution diameter is unsuitable and differs from the program in its realisation.

This inexactitude is rather significant, as it concerns nearly one-quarter of the examined cases. Its appearance is individual, but on the occasion of serial examinations of the same person it may change with time.

We continued our examinations in this direction, but on the basis of our work to date, it seems that before carrying out complicated EMG examinations the bio-electric activity of simple movement processes should be analysed on every person in order to state the exactness of the individual time and the motoric reaction.

Author's address: Dr. L. Lukács, Pediatric Institut of Buda, Budai Területi Gyermekkórház, Bólyai-u. 9, *Budapest II* (Hungary)

Medicine and Sport, vol. 8: Biomechanics III, pp. 278–287 (Karger, Basel 1973)

An Analysis of Muscle Coordination in Walking and Running Gaits[1]

B. R. BRANDELL

Department of Anatomy, University of Saskatchewan, Saskatoon, Saskatchewan

I. Introduction

The objectives of the present study have been to study details of the coordination between lower limb movements and muscle contractions during natural walking and running gaits, and to critically compare the functional phases of these two basic forms of locomotion. In spite of the extensive research on locomotion by earlier workers, including photographic studies by MAREY [9], MURRAY et al. [10], BRAUNE and FISCHER [3], the combined force plate and muscle tension studies of SCHWARTZ et al. [11], combined cinematographic and EMG studies by EBERHARDT et al. [4], SUTHERLAND [12] and GRAY and BASMAJIAN [5], and the use of radio-telemetry with electromyography by JOSEPH [8], no earlier EMG study of running appears to have been reported, and it was felt by the author that a combined EMG-cinematographic study of free locomotion performed for great enough distances to ensure a natural gait would define aspects of motion and muscle coordination overlooked in previous studies.

II. Materials and Methods

EMG potentials were recorded from 12 lower limb muscles by means of fine indwelling wire electrodes and a 7-channel tape-recorder, an earlier model of which has been described by BRANDELL et al. [2]. Of the 7 channels, 6 were electromyographic and one was a synchronizing marker channel. The recorder was carried on the subject's back and allowed him complete freedom of movement during walking and running.

Movements of the subject were tracked by means of an electrically driven Mitchel motion picture camera at speeds of 64 frames/sec during walking and 80 frames/sec during running. Alternate flashings of a small blinking light taped to the subject were synchronized with simultaneous upward deviations of the tape-recorder marking channel, so that the first motion picture frame in which a light blink occurred corresponded (within 3 msec) to

1   This study was supported by the Fitness and Amateur Sport Directorate, Department of National Health and Welfare, Canada.

the moment when the marker channel began to deviate. Therefore, all the frames of the motion pictures can be very exactly correlated with the corresponding points on the EMG tape-recordings. The tapes were played back onto an oscilloscope screen, which was photographed by a continuously moving 35-mm linegraph camera film. Each frame of the motion picture was represented by 1 horizontal unit of a one-tenth of an inch unit graph paper, and the 35-mm films were enlarged to bring all points on the EMG and the motion graphs into register with each other.

The vertical coordinate of the graphs was determined by the angles formed between: (1) the thigh and the vertical; (2) the leg (crus) and the vertical; (3) the thigh and leg (knee joint); (4) the foot and the vertical; (5) the foot and leg (ankle joint), and (6) the trunk and the vertical; and by linear measurements of: (1) the vertical oscillations of the trunk, and (2) the horizontal progression of the trunk. All measurements were made by means of a Vanguard Motion Analyzer – angles to the nearest degree and linear measurements to the nearest Analyzer unit of 0.001 in.

Each of 7 male and 3 female subjects walked, and after a rest, ran at an easy pace for 30 laps of a 22-yd course. Subjects were asked to walk or run at what seemed to them a natural speed and cadence. They were photographed during laps 2, 4, 15, 17, 28 and 30 as they travelled in a straight line for about 45 ft in front of a gridded background on one side of the course. The subject pressed a small hand-carried switch at the start of each photographed lap to simultaneously illuminate a second light and depress the seventh channel of the recorder to identify the photographed part of the otherwise uninterrupted EMG record.

Of the 12 muscles recorded, the 3 2-joint hamstring muscles, vastus medialis, vastus lateralis and the triceps surae of the calf muscles are considered in this paper.

III. Phasic Occurrence of Muscle Activity (figs. 1–3)

A. Two Joint Hamstring Muscles

1. Walking
Last half of swing-phase knee extension, after the thigh has completed its forward swing, and is being held almost stationary.

2. Running
Last three-quarters of swing-phase knee extension after the thigh has completed its forward swing, and started backwards.

B. Vastus Muscles

In both walking and running gaits, during knee flexion after the heel (or toe) strikes the ground, often reaching maximum activity at the transition

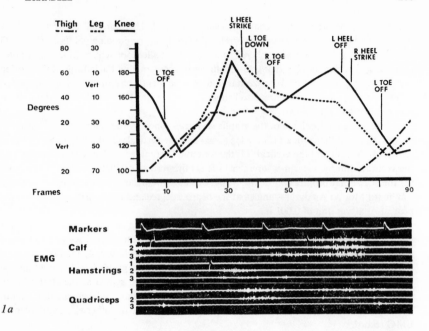

Fig. 1. Muscle coordination during limb motions of walking and running gaits. Calf muscles: (1) medial head of gastrocnemius; (2) lateral head of gastrocnemius, and (3) soleus. Hamstring muscles: (1) semitendinosus; (2) semimembranosus, and (3) long head of biceps femoris. Quadriceps muscles: (1) vastus medialis; (2) vastus lateralis, and (3) rectus femoris.

between knee flexion and extension, but never continuing into the extension of a constant speed gait. These muscles often begin their activity, when the knee starts to flex prior to heel strike, and occasionally are active near the end of the preceding knee extension.

C. Calf Muscle Group (Triceps Surae)

1. Walking

Strongest after the heel lifts from the ground, as the trunk is falling and gaining speed and as the *knee is flexing*. Activity often begins before the heel lifts, while the forward motion of the leg (crus) is decreasing and the knee is extending.

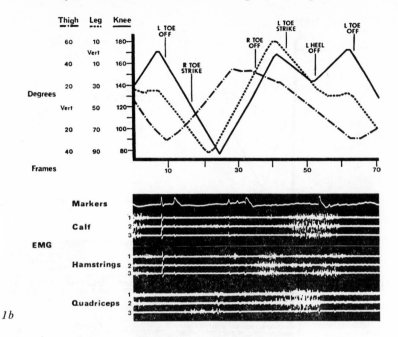

1b

2. Running

Strongest after the heel lifts, as the trunk is rising and gaining speed and the *knee is extending*. At the same time, the forward motion of the leg (crus) is abruptly slowed and briefly reversed. Calf muscle activity usually *begins* shortly after the heel strikes, while the knee is flexing and the forward motion of the leg is gradually reduced.

IV. Discussion (figs. 1–3)

A. Forward Propulsion

1. Walking

A gain of forward speed seems to depend upon the force of gravity lowering the trunk from the ball of one foot onto the heel of the opposite foot *while the knee is flexing*. The calf muscles are active at this time, but the heel is simply rising from the ground, while the foot maintains a constant

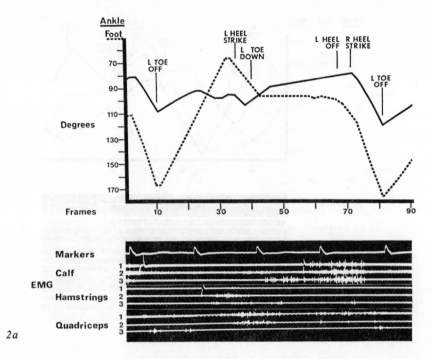

Fig. 2. Muscle coordination during limb motions of walking and running gaits. Calf muscles: (1) medial head of gastrocnemius; (2) lateral head of gastrocnemius, and (3) soleus. Hamstring muscles: (1) semitendinosus; (2) semimembranosus, and (3) long head of biceps femoris. Quadriceps muscles: (1) vastus medialis; (2) vastus lateralis, and (3) rectus femoris.

angular relationship to the leg and, as a result, *the ankle joint is not plantar flexing*. The moment that the opposite heel contacts the ground and begins to assume the weight of the body, the calf muscle activity ceases, and the ankle joint begins to plantar flex.

2. Running

Increasing speed to produce forward propulsion seems to result from the strong contraction of the calf muscles, which plantar flexes the ankle joint *at the same time that the knee is extending* and the trunk is rising. Surprisingly, active propulsion by the calf muscles is limited to about the first third of the plantar flexion of the ankle joint and extension of the knee joint.

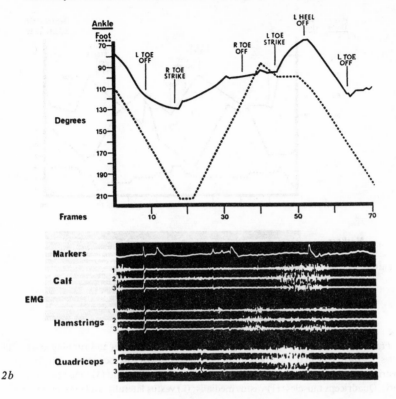

2b

B. Knee Extension of the Stance Phase

This movement of the knee joint always corresponds to: (1) calf muscle activity; (2) a rapid forward motion of the thigh, and (3) *a marked slowing or reversal of the forward motion of the leg*. The sudden slowing of forward leg motion might be compared to the snapping of a whip. The knee rapidly extends because the thigh is carried forward by momentum after the leg is brought up short by the restraining action of the calf muscles.

1. Walking

Knee extension of the stance phase serves to bring the femur about 20° forward of a vertical position and places the trunk in front of its base of support in the foot. It is *after this* that the trunk undergoes its most rapid gain in speed, as the knee flexes and the heel rises from the ground.

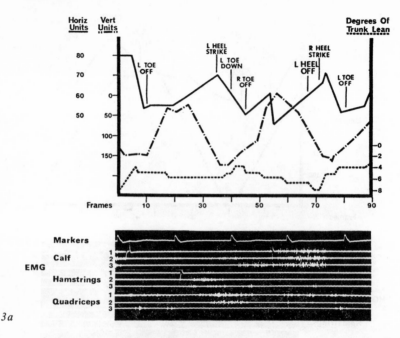

Fig. 3. Muscle coordination during trunk motions of walking and running gaits. Calf muscles: (1) medial head of gastrocnemius; (2) lateral head of gastrocnemius, and (3) soleus. Hamstring muscles: (1) semitendinosus; (2) semimembranosus, and (3) long head of biceps femoris. Quadriceps muscles: (1) vastus medialis; (2) vastus lateralis, and (3) rectus femoris.

2. Running

Knee extension during the stance phase of running *is part of the propulsion mechanism.* At the start of knee extension the femur is almost vertical and, since the leg is leaning forward at an angle of 30–40°, the trunk is already well ahead of its base of support. Knee extension serves to propel the proximinal end of the femur and trunk forward, while the trunk is elevated by the rising heel and plantar flexion of the ankle joint, both movements being produced by the calf muscles. It is, perhaps, significant that calf muscle activity ceases before the ankle has plantar flexed to 90°. At acute angles, the leg (crus) is propelled forward more than upward by the rising heel and, therefore, at these angles calf muscle action on the heel would be the most effective in producing forward propulsion.

The role, which the vastus muscles may play in knee extension of the stance phase is problematical. Although their activity ceases before the start of knee extension, these muscles often reach their most intense activity at the transition from knee flexion to knee

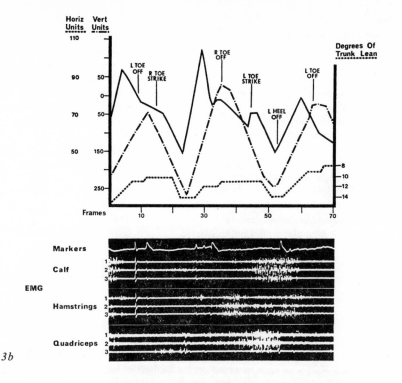

3b

extension, and it is tempting to suggest that the force from their contractions is prolonged sufficiently to initiate knee extension. This idea seems to be supported by HUBBARD's description [6] of ballistic motions, and by the findings of INMAN *et al.* [7] and AHLGREN and ÖWALL [1] that peak electrical energy precedes maximum mechanical force of muscle contractions by an interval of between 40 and 80 msec.

C. Knee Extension of the Swing Phase

Knee joint extension during the swing phase is the reverse analogue of knee joint extension during the stance phase. In the swing phase, knee extension is caused by slowing, or reversing, the forward motion of the thigh, while momentum continues to carry the leg rapidly forward. *In both walking and running*, this part of the stride is *always* correlated with strong activity of the 2-joint hamstring muscles, especially semimembranosus. These muscles seem to exert a differential control on the thigh and leg, producing a detectable restraint on the forward movement of the thigh at a distinctly earlier time than when the forward movement of the leg is slowed and reversed.

Table I

	Type of race	
	1,500 metre	100 metre
Total duration of stride, sec	0.72	0.59
Stance phase duration, sec	0.20	0.11
Speed of body, m/sec	6.2	8.9
Length of stride, m	4.44	5.25
Amplitude of knee extension, degrees	42	26
Knee joint angle at toe-off, degrees	175	171
Rate of knee joint extension, sec/degree	0.0025	0.0023
Amplitude of ankle joint plantar flexion, degrees	72	64
Ankle joint angle at toe-off, degrees	135	131
Rate of ankle joint plantar flexion, sec/degree	0.0015	0.0010
Amplitude of thigh movement, degrees	58	50
Rate of thigh movement, sec/degree	0.0022	0.0018

D. A Comparison of Slow and Sprint Running Strides in Competition

Table I, which compares a relatively slow stride from a 1,500-metre race with a very fast stride from a 100-metre dash, is based on the analysis of motion pictures taken at 64 frames/sec. Rate of forward motion has been calculated by reference to stripes on the track, and fence posts at a known distance in the background. The races were run at the same track meet by two of the top athletes in the province of Saskatchewan, Canada.

The most striking difference to be noted between the two strides is the extremely short duration of the stance phase in the 100-metre dash, which was 0.18 of the total stride duration as compared to 0.28 in the 1,500-metre race, and the noticeably greater amplitude but slower velocity of knee joint, ankle joint and thigh motions in the longer duration stride. The greater speed and distance achieved in the sprint stride is clearly due to the relatively shorter but quicker angular motions of the limb which provide greater thrust, reduce the relative duration of the support phase and increase the relative duration of the unsupported 'float' phase. A very important difference between the two strides was that in the slower stride initial ground contact was made with the heel, which caused a momentary reversal in the forward direction of the thigh and a knee flexion of about 26° at a rate of 0.078 sec/degree. In the sprint stride initial contact was made with the ball of the foot so that the forward motion of the thigh was only slowed but not reversed, and the knee joint only flexed 10° but at the very fast rate of 0.016 sec/degree. This very brief knee flexion corresponds exactly to the slight hesitation of forward thigh motion. Apparently, in the sprint stride a more continuous forward thigh motion was achieved by keeping the body weight on the ball of the foot, and absorbing contact shock at the ankle joint by the calf muscles, rather than at the knee joint by the quadriceps femoris muscles. The interruption of forward leg motion and correlated knee joint extension of the stance phase, which was the period of most intense calf muscle activity in the experimental tests of running, lasted 4 times as long in the 1,500-metre race as in the 100-metre dash. It is suggested that the more gradual interruption of forward leg motion in the slower stride, as compared to the sprinting stride, corresponded to less intense calf muscle activity, and greater dependence upon the vastus muscles to initiate the knee extension of the stance phase.

References

1 AHLGREN, J. and ÖWALL, B.: Muscular activity and chewing force. A polygraphic study of human mandibular movements. Arch. oral Biol. *15:* 271–280 (1970).
2 BRANDELL, B.R.; HUFF, G.J., and SPARK, G.J.: An electromyographic-cinematographic study of the thigh muscles using M.E.R.D. (Muscle Electronic Recording Device). I. Proceedings of the First International Congress of Electromyographic Kinesiology. Electromyography *8:* suppl. 1, pp. 67–75 (1968).
3 BRAUNE, C.W. und FISCHER, O.: Der Gang des Menschen. I. Versuche unbelasten und belasten Menschen. Abh. math.-phys. Kl. Sächs. ges. Wiss. *21:* 153–322 (1895).
4 EBERHART, H.D.; INMAN, V.T., and BRESLER, B.: The principal elements in human locomotion; in KLOPSTEG Human limbs and their substitutes, pp. 437–471 (Hafner, New York 1968).
5 GRAY, E.G. and BASMAJIAN, J.V.: Electromyography and cinematography of leg and foot (normal and flat) during walking. Anat. Rec. *161:* 1–16 (1968).
6 HUBBARD, A.W.: An experimental analysis of running and of certain fundamental differences between trained and untrained runners. Res. Quart. amer. Ass. Hlth phys. Educ. *10* (3): 28–38 (1939).
7 INMAN, V.T.; RALSTON, H.J.; SAUNDERS, J.B. DE C.M.; FEINSTEIN, B., and WRIGHT, E.W.: Relation of human electromyogram to muscular tension. Electroenceph. clin. Neurophysiol. *4:* 187–194 (1952).
8 JOSEPH, J.: The pattern of activity of some muscles in women walking on high heels. Ann. phys. Med. *9:* 295–299 (1968).
9 MAREY, E.J.: Movement (translated by E. PRITCHARD) (Appleton, New York 1895).
10 MURRAY, P.M.; DROUGHT, A.B., and KORY, R.C.: Walking patterns of normal men. J. Bone Jt Surg. *46:* 335–360 (1964).
11 SCHWARTZ, R.P.; TRAUTMAN, O., and HEATH, A.L.: Gait and muscle function recorded by electrobasograph. J. Bone Jt Surg. *18:* 445–454 (1936).
12 SUTHERLAND, D.H.: An electromyographic study of the plantar flexors of the ankle in normal walking on the level. J. Bone Jt Surg. *48:* 66–71 (1966).

Author's address: Dr. BRUCE R. BRANDELL, Department of Anatomy, University of Saskatchewan, *Saskatoon, Saskatchewan* (Canada)

Medicine and Sport, vol. 8: Biomechanics III, pp. 288–293 (Karger, Basel 1973)

Discussion of the Paper of Dr. Brandell

W. T. LIBERSON

Veterans Administration Hospital and University of Miami, Miami, Fla.

It is a privilege to open the discussion of the inspiring paper of Dr. BRANDELL on 'Comparison between walking and running'. I have not studied the latter but have spent a great deal of time in investigating the former. Our results are quite comparable; however, since some discrepancies remain I would like to go over my own data in order to better restate my position.

In my technique I remained faithful to the use of accelerometers which I introduced to this field in the early thirties. However, as was stated during this conference, the use of accelerometers leaves undetermined the position of the segments of the body in motion. I have, therefore, been using simultaneously electrogoniometers and moving pictures. The latter shows a split image: on the left the movement itself and on the right the corresponding tracings of different parameters of the gait.

Lately, I have been using at times video-tape recorders and on-line computers. The latter permit one to display stick figures of the walking subject, computed from the electrogoniograms and the vectors of linear (the trunk in the frontal and horizontal planes) accelerations.

I have been interested in both the general organization of the gait and the action of some specific muscles, which appeared in my study to be somewhat different from the generally accepted views. A typical tracing (fig. 1) represents:

1. A 'vertical' accelerogram showing the double hump of Berenstein at the time of the double stance.

2. A 'horizontal' accelerogram showing a 'dip' of backward acceleration at about the same time.

3. A hip electrogoniogram, showing its flexion (upward deviation) with adiscontinuity (sudden slowing indicated by arrow) at the time of the resto-

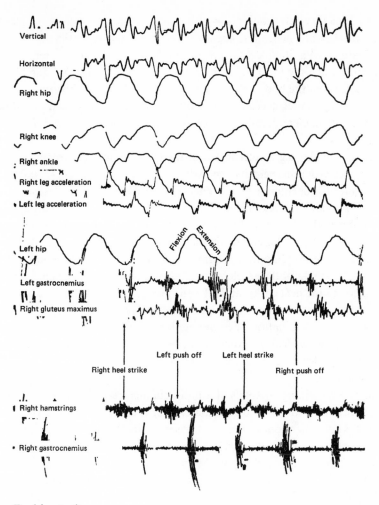

Fig. 1 (see text).

ration of the horizontal forward acceleration of the trunk; this second ascending phase is followed by a plateau at the time of the heel strike, succeeded in turn by a descending phase corresponding to hip extension.

4. A knee electrogoniogram, more complex, showing at first its flexion followed by its extension at the time of the above-mentioned slowing of the hip flexion. The extension of the knee is marked by the classical 'double lock' pattern.

5. An ankle electrogoniogram, still more complex, showing (a) a plantar flexion (downward deflexion) beginning at about the same time as hip and knee flexion; (b) a dorsiflexion (upward deflexion of the curve), terminating at the time when the knee begins its extension, and (c) a second dorsiflexion at the time when the body rolls over the ankle in the sagittal plane.

6. An angular leg acceleration of the ipsilateral leg, recorded by a differential accelerometer, permitting one to eliminate the acceleration due to gravity.

7. An angular leg accelerogram of the opposite lower extremities. Each of these accelerograms permits one to determine several important landmarks of the gait: (a) beginning of the knee flexion (downward deflexion of the curve; (b) maximal acceleration of the leg (the top of the curve) always corresponding within a few milliseconds only, to the beginning of the ankle dorsiflexion; (c) the crossing of the base line by the downward deflexion of the curve, corresponding to the vertical position of the opposite leg; (d) the lowermost position of the accelerogram corresponding to the extension of the knee, prior to the heel strike, and (e) return of the curve to the base line (heel strike) where it remains until the time of the 'heel off' just before the 'toe off'.

Thus, the combination of both angular accelerograms provides the observer with a great deal of information. Both may be recorded on the same channel of an electrocardiograph (which permits one to have quite inexpensive equipment).

8. The remaining channels of a multi-channel recorder may be used for the electrogoniograms of the contralateral leg and a number of EMG.

The above-mentioned precise correlation in time (within a few milliseconds) of such seemingly unrelated events as maximal acceleration of the entire leg (depending mostly upon the contraction of the iliopsoas) and of the beginning of the dorsiflexion of the ankle (depending upon the contraction of the tibialis anterior) shows that even locomotion is probably carried out, to a certain degree, according to an 'open loop' mechanism.

During our studies we could find that, contrary to the usually held view, the gastrocnemius is silent during the plantar flexion proceeding the 'toe off'. Exceptionally, its activity was present at the plantar flexion as in the cases of Dr. BRANDELL. One can see in his illustrations that even during the running this muscle remains silent during the second half of the plantar flexion. We believe that in ordinary walking this muscle prevents the subject from falling on his face until the other leg is ready to support the body weight. Far from being a 'push off' muscle it deserves the name of a 'let go' muscle at that particular moment (fig. 2a).

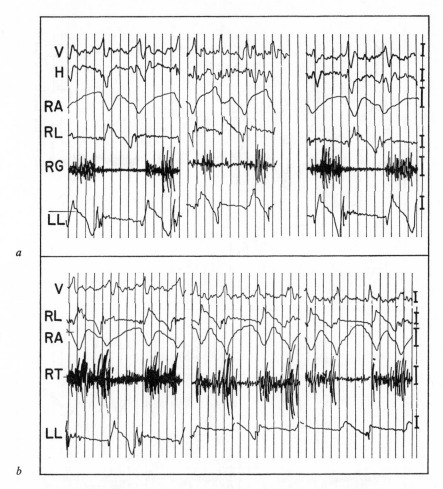

Fig. 2. a From top to bottom: V = vertical accelerogram; H = horizontal accelerogram; RA = right ankle angular electrogoniogram; RL = angular accelerogram of the right
lower leg; RG = EMG, right gastrocnemius; LL = angular accelerogram of the left lower
leg. *b* Same abbreviations as above; RT = right tibialis anticus. Each tracing (a and b)
represents samples from 3 different walking subjects. The thin vertical lines indicate intervals of one-fifth of a second; the short heavy vertical lines indicate 1 *g* acceleration or 5 mV.

Again contrary to the general view, the second burst of the contraction
of the tibialis anterior occurs at the very end of the second plantar flexion of
the ankle and, therefore, this muscle does not seem to present an eccentric
contraction at that time. It may pull the body slightly forward, thus con-

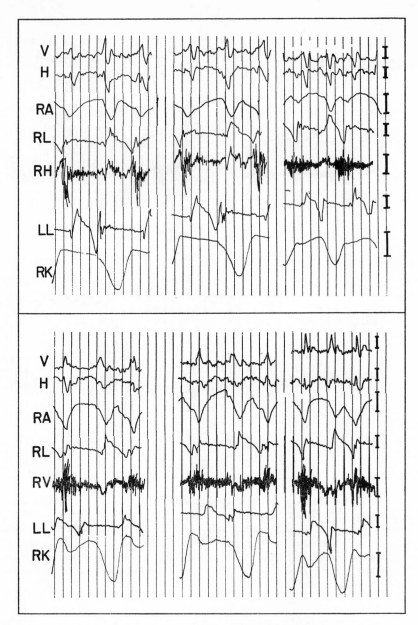

Fig. 3. Same abbreviations as in figure 1: RK = right knee electrogoniogram. RV = EMG of the right vastus lateralis. RH = EMG of the right hamstrings.

tributing to a momentary loss of balance. Gluteus maximus, hamstrings and quadriceps all contract almost simultaneously at the time of the heel strike. Thus, they stabilize the leg and also the pelvis permitting the contralateral iliopsoas to perform the major function of the gait, namely to move the opposite leg forward, thus placing the center of gravity ahead of its normal position. In addition, the hip extensors perform at that very time the 'oar' action, pulling the body forward in the same time as the pendulum represented by the contralateral leg contributes to the same effect (fig. 1 and 2b).

It is obvious from our data that hamstrings *do not* contract during the knee flexion and the quadriceps *does not* contract during the knee extension until the phase of double lock. Thus, in addition to the tibialis anticus, the following sequence of muscle action could be established from our data (fig. 3).

1. Gluteus maximus (together with ipsilateral hamstrings, quadriceps and contralateral iliopsoas).
2. Ipsilateral gastrocnemius.
3. Contralateral gluteus maximus and the above-mentioned 'synergists'.
4. Contralateral gastrocnemius.

References

LIBERSON, W. T.: Une nouvelle application du quartz piezoélectrique. Piezoélectrographie de la marche et des mouvements volontaires. Trav. hum. *4:* 1–7 (1936).
LIBERSON, W. T.: Biomechanics of gait. A method of study. Arch. phys. Med. *46:* 37–48 (1965).
LIBERSON, W. T.: Démarches normales et pathologiques. Une méthode d'études, pp. 428–435 (Excerpta Medica, Amsterdam 1966).
LIBERSON, W. T.; HOLMQUEST, H., and HALLS, A.: Accelerographic study of gait. Arch. phys. Med. *43:* 547–551 (1962).

Author's address: Dr. W. T. LIBERSON, M.D., Ph.D., University of Miami, Veterans Administration Hospital, 1201 Northwest 16th Street, *Miami, FLA 33125* (USA)

Medicine and Sport, vol. 8: Biomechanics III, pp. 294–300 (Karger, Basel 1973)

Electromyography of the Erector Spinae Muscle[1]

B. Jonsson

Department of Anatomy, University of Göteborg

Introduction

The function of the erector spinae muscle in various movements and postures of the body has been the subject of several investigations. In most of these studies the erector spinae muscle has been regarded as one single muscle, without taking into consideration the fact that it is a large group of different muscles, most of which have different functions.

The most common method in the study of the function of the erector spinae muscle is to place surface electrodes above the erector spinae [among others, ALLEN, 1948; FLOYD and SILVER, 1951; ASMUSSEN, 1960; CHAPMAN and TROUP, 1969]. In such studies the whole muscle mass is regarded as one single unit. Surface electrodes, however, mainly record the EMG activity from the most superficial part of the muscle group. This means that the electrodes will mainly record activity from the multifidi muscles when placed over the lower lumbar spine, while they will mainly record activity from the longissimus muscle when placed over the lower thoracic spine. Differences in EMG activity recorded by surface electrodes from the lower lumbar part of the spine and from the lower thoracic part of the spine may, therefore, be due either to a difference in function between the two levels or to a difference in function between the two muscles or both.

As a matter of fact, functional differences exist between the individual muscles of the erector spinae as well as between different levels [MORRIS et al., 1962; PAULY, 1966; JONSSON, 1970 b, c; WALTERS and MORRIS, 1970].

It is, thus, important in future studies of the erector spinae muscle either that the different muscles be studied separately, or that such a technique be used that the investigator can control which muscle he is studying. The aim of this paper is to describe and discuss how examinations of the erector spinae muscle should be performed in order to obtain a reliable and standardized technique.

1   This work was supported by the Swedish Medical Research Council, Project No. 12 X–2711.

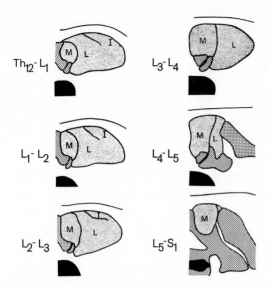

Fig. 1. Topography of the multifidi (M), longissimus (L) and iliocostalis (I) muscles at different lumbar levels as seen on cross-sections of a lumbar spine.

Topography of the Erector Spinae Muscle

The erector spinae muscle can be subdivided into a large number of small muscles. However, the only muscles in the thoracic and lumbar spine which can be assumed to be available to electromyographic studies are the multifidi, the longissimus, the iliocostalis and the spinalis muscles [JONSSON, 1970a]. The muscles of the neck are associated with certain special aspects, anatomically as well as functionally, and will not be discussed here.

The topography of the multifidi, the longissimus and the iliocostalis muscles in the lumbar spine is indicated in figure 1. The multifidi muscles constitute the main part of the erector spinae in the lower lumbar spine, usually reaching the surface of the erector spinae laterally to the midline of the body. In the upper lumbar spine and in the thoracic spine the multifidi constitute a small muscle group. Above the level of the iliac crest the longissimus muscle is the largest muscle of the erector spinae. It is located laterally to the multifidi muscles. The iliocostalis muscle is mainly located in the thoracic spine, where it forms the most lateral part of the erector spinae muscle. In the upper lumbar spine the longissimus is laterally covered with

the iliocostalis muscle. The spinalis muscle is a narrow muscle located super-
ficially in the thoracic spine close to the midline of the body.

Insertion of EMG Electrodes

The type of electrode which seems most useful for electromyographic
studies of the function of the erector spinae muscle is wire electrodes. They
can be used in dynamic as well as in static conditions.

The most reliable technique for insertion of EMG electrodes into individual muscles of the erector spinae muscle appears to be one in which the
insertion is guided by fluoroscopy, preferably TV-fluoroscopy [JONSSON and
REICHMANN, 1970]. It is then possible to locate the electrode exactly in relation
to the skeleton.

Fig. 2. The technique for insertion of EMG electrodes into the multifidi (M), longissimus (L) and iliocostalis (I) muscles in the upper lumbar spine.

The multifidi muscles of the thoracic and upper lumbar spine are best
reached if the electrodes are inserted approximately 10 mm laterally to the
midline and with an antero-medial inclination of the electrode (fig. 2). In the
lower lumbar spine the multifidi will be reached if the electrode is inserted
15–20 mm laterally to the midline and perpendicularly to the skin. The depth
of insertion should be 20–30 mm below the skin surface.

The longissimus muscle is best reached if the electrode is inserted perpendicularly to the skin surface at a distance of 35–40 mm laterally to the
midline. The depth of insertion should be 20–30 mm below the skin surface.

The iliocostalis muscle is best reached if the electrode is inserted 1–3 cm
medially to the palpable border of the erector spinae muscle. The depth of
insertion should be 5–10 mm below the muscle fascia. The iliocostalis muscle
will not be reached below the level of the L_3 spinous process.

The spinalis muscle will be reached if the electrode is inserted super-
ficially a few millimetres laterally to the midline.

In those cases, too, where the insertion of the electrodes is guided by fluoroscopy, it happens now and then that the electrodes are placed in the wrong muscles. It is, therefore, advisable to verify the location of the electrodes after insertion. This can be done radiologically after injection of carbon dioxide through the electrode [REICHMANN and JONSSON, 1967; JONSSON and REICHMANN, 1970]. The carbon dioxide remains inside the muscle fascia and can be visualized radiologically. The location of the gas and the direction of those gas streaks which are located along the bundles of muscle fibres indicate in which muscle the gas has been injected.

Function of Individual Muscles of the Erector Spinae Muscle

The functions of individual muscles of the erector spinae muscle in different lumbar levels were studied electromyographically in 2 series of investigations, one being performed on 13 [JONSSON, 1970b] and the other on 10 healthy young subjects. The results from the two investigations are mainly the same. Both investigations clearly indicated that there exist differences in function between the multifidi, longissimus and iliocostalis muscles. The results also showed that functional differences of one and the same muscle between different lumbar levels may exist. The results presented here in figures 3–5 were obtained from the last of the two investigations.

The experiments were conducted both with the subject in the erect sitting position and in the standing position. In the first position only attempted movements against a controlled resistance were studied. In the latter position

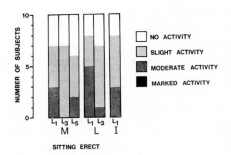

Fig. 3. Number of subjects showing no, slight, moderate or marked activity, respectively, in the multifidi (M), longissimus (L) and iliocostalis (I) muscles at different lumbar levels in erect sitting position.

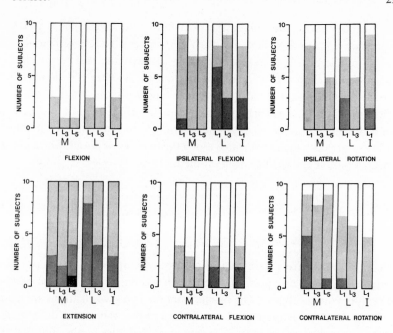

Fig. 4. Number of subjects showing no, slight, moderate or marked activity, respectively (cf. fig. 3), in the multifidi (M), longissimus (L) and iliocostalis (I) muscles at different lumbar levels in attempted movements of the trunk against a resistance.

only symmetric standing postures with and without weights held in the hands were studied.

In the unloaded erect sitting posture the majority (60–80%) of the subjects showed weak or moderate activity in all muscles studied and on all lumbar levels (fig. 3). Attempted flexion of the trunk resulted in a marked decrease of the activity in all muscles tested (fig. 4). In attempted extension, on the other hand, all muscles in all subjects were active. Ipsilateral flexion of the trunk resulted in no obvious change in the activity as compared to that in sitting erect. Contralateral flexion, on the other hand, resulted in a marked decrease in the activity of all muscles. Ipsilateral attempted rotation of the trunk gave a tendency to a decrease in the degree of activity in the multifidi and the longissimus. Contralateral rotation gave rise to a tendency towards an increase in activity in the multifidi muscles and a decrease in the iliocostalis muscle.

In standing at ease, most subjects showed slight or moderate activity in all muscles studied (fig. 5). A very marked decrease in activity was observed

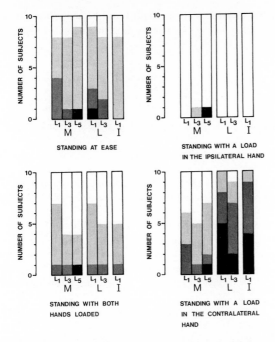

STANDING AT EASE

STANDING WITH A LOAD
IN THE IPSILATERAL HAND

STANDING WITH BOTH
HANDS LOADED

STANDING WITH A LOAD
IN THE CONTRALATERAL
HAND

Fig. 5. Number of subjects showing no, slight, moderate or marked activity, respectively (cf. fig. 3), in the multifidi (M), longissimus (L) and iliocostalis (I) muscles at different lumbar levels in standing with or without loads in one or both hands.

in standing with a load in the ipsilateral hand. Only in one subject was some activity recorded from the multifidi muscles. When the subject stood with the contralateral hand loaded, the multifidi showed less activity and the iliocostalis and longissimus showed more activity than in standing at ease. In standing with both hands loaded, less activity than in standing at ease was recorded in all muscles.

Concluding Remarks

The erector spinae muscle is an important group of muscles, the function of which has often been the subject of electromyographic studies. It is obvious that the different muscles of which it is built up have different functions. It must, therefore, be emphasized that it is important to standardize the technique for locating the electrodes and to take into consideration which muscle of the erector spinae is actually being studied. It is not sufficient to study the erector spinae as one single unit. In those cases, too, where the investigator is not interested in the differences in function between the individual muscles, he should use a standardized technique so as to make sure that he is examining the same muscle in all subjects.

References

ALLEN, C.E.L.: Muscle action potentials used in the study of dynamic anatomy. Brit.J phys. Med. *11:* 66–73 (1948).

ASMUSSEN, E.: The weight-carrying function of the human spine. Acta orthop.scand. *29:* 276–290 (1960).

CHAPMAN, A.E. and TROUP, J.D.G.: The effect of increased maximal strength on the integrated electrical activity of lumbar erectores spinae. Electromyography *9:* 263–280 (1969).

FLOYD, W.F. and SILVER, P.H.S.: Function of erectores spinae in flexion of the trunk. Lancet *260:* 133–134 (1951).

JONSSON, B.: Topography of the lumbar part of the erector spinae muscle. An analysis of the morphologic conditions precedent for insertion of EMG electrodes into individual muscles of the lumbar part of the erector spinae muscle. Z.Anat.EntwGesch. *130:* 177–191 (1970a).

JONSSON, B.: The functions of individual muscles in the lumbar part of the erector spinae muscle. Electromyography *10:* 3–19 (1970b).

JONSSON, B.: The lumbar part of the erector spinae muscle. A technique for electromyographic studies of the function of its individual muscles. Thesis, Göteborg (1970c).

JONSSON, B. and REICHMANN, S.: Radiographic control in the insertion of EMG electrodes in the lumbar part of the erector spinae muscle. Z.Anat.EntwGesch. *130:* 192–206 (1970).

MORRIS, J.M.; BENNER, G., and LUKAS, D.B.: An electromyographic study of the intrinsic muscles of the back in man. J.Anat., Lond. *96:* 509–520 (1962).

PAULY, J.E.: An electromyographic analysis of certain movements and exercises. I. Some deep muscles of the back. Anat. Rec. *155:* 223–234 (1966).

REICHMANN, S. and JONSSON, B.: Contrast radiography with carbon dioxide for the location of intramuscular EMG electrodes. Electromyography *7:* 103–124 (1967).

WALTERS, R.L. and MORRIS, J.M.: Effect of spinal supports on the electrical activity of muscles of the trunk. J. Bone Jt Surg. *52:* 51–60 (1970).

Author's address: Dr. BENGT JONSSON, Department of Anatomy, University of Umeå, *S-901 87 Umeå* (Sweden)

Medicine and Sport, vol. 8: Biomechanics III, pp. 301–308 (Karger, Basel 1973)

EMG Action Potentials of Rectus Abdominis Muscle during Two Types of Abdominal Exercises

Y. GIRARDIN[1]

Département d'Education Physique, Université de Montréal, Montréal

Contraction of the psoas muscle produces thigh flexion and/or flexion of the lumbar column on the thigh. Under special circumstances, such as when an individual with weak abdominal muscles is in a supine position and attempts to sit up, psoas contraction can provoke hyperextension of the lumbar spine. This reversal of function has been referred to as the 'psoas paradox' by RASCH and BURKE [10].

In our hypokinetic era, it has been established that lower back pain is one of the most frequent medical complaints [2]. Muscle training of the abdominal is recommended to counteract this affliction. Abdominal exercises initiated from the supine position are commonly employed to fulfil this purpose. However, since there is a tendency for abdominal strength to be functionally inferior to psoas strength [10], strengthening of the latter through such exercises may accentuate lumbar hyperextension which has frequently been associated with lower back pain. Therefore, it seems advisable to emphasize abdominal strengthening while minimizing the role of the psoas.

EMG studies reported during the last 2 decades indicate that the sit-up and curl-up exercises, performed according to various patterns, elicit a high intensity of muscle action potentials (MAP), particularly from the upper and lower segments of the rectus abdominis [3–5, 7–9, 11, 12].

The rectus abdominis involvement during a modified hook sit-up, along with other forms of abdominal exercises, was studied by LIPETZ and GUTIN [7]. The authors have described the exercise as follows: 'The lower leg rests across the top of a bench 14 inches high. The back of the knee remains in contact

1   The author is indebted to Mr. PAUL MARTIN and Mr. PIERRE SIMARD for their technical assistance in the pursuit of this work.

with the corner of the bench. The hands are clasped behind the neck and the elbows are held forward throughout. The subject sits up until the chest touches the quadriceps' [7]. This modified hook sit-up elicited greater MAP from the upper and lower rectus abdominis than most of the abdominal exercises studied.

It is a well established fact that, at the time of contraction, the relative length of the muscle fiber affects the maximal tension that the muscle fiber can develop. According to ÅSTRAND and RODAHL [1] the maximal tension that the fiber can exert occurs at a relative length of about 1.2:1 and decreases at lower and higher length. In the performance of an exercise similar to the modified hook sit-up previously referred to, it would appear that, due to the position of the subject, the origin and insertion of the psoas muscle would be closer to each other than during the execution of any other abdominal exercise, thereby minimizing its action on the lumbar spine.

Therefore, the modified hook sit-up of LIPETZ and GUTIN [7] might be improved if the bench height could be adjusted and the lower body segments immobilized in order to obtain an exercise which would prevent hyperextension of the lumbar spine while exercising the abdominal muscles.

The purpose of this study was to compare the intensity of the EMG action potentials elicited from the upper and lower segments of the left rectus abdominis during the execution of 2 types of abdominal exercises. The first exercise, referred to as the conventional sit-up (CSU), was used as a yardstick against which the second exercise, the bench sit-up (BSU), was compared.

Methodology

16 male volunteers participated in this study. Their biometric characteristics are presented in table I. The subjects were randomly divided into 2 equal groups. Members of group I performed the CSU followed by the BSU. The reverse order was employed for members of group II.

Table I. Biometric characteristics of subjects

	Age, months	Height, cm	Weight, kg	Surface area, m²
Mean	258.4	177.0	72.0	1.85
SD	18.6	6.8	6.4	0.11

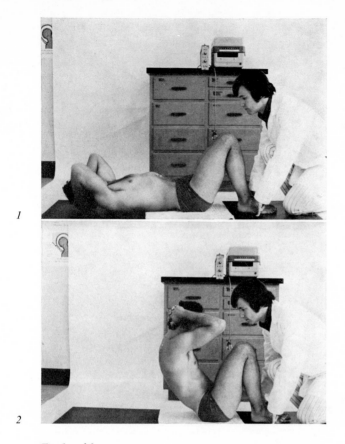

Fig. 1 and 2

The two abdominal exercises were performed as follows.

1. For the CSU, the subject assumed a supine position. His lower legs were flexed so that a 45-degree angle was obtained at the knee joint. His feet were flat on the floor and his ankles were held down by an assistant (fig. 1). The movement started with a flexion of the head on the trunk and was followed by a curling of the trunk until the head was close to the knees (fig. 2). The downward phase of the movement was performed in the same manner.

2. For the BSU, the subject assumed a supine position. The posterior surface of his lower legs rested on an adjustable bench so that a 90-degree angle was obtained at the knee joint. The subject's shanks and thighs were

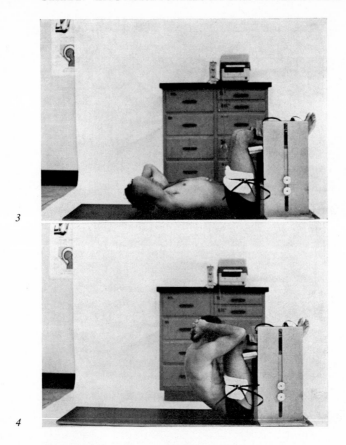

Fig. 3 and 4

immobilized by means of adjustable belts (fig. 3). The movement was executed as for the CSU except that the range of upward motion was restricted due to the position of the lower segments (fig. 4).

In both exercises, the hands were clasped behind the neck and the elbows were kept forward throughout. Both sit-ups were performed according to constant time intervals as given by a metronome. The upward phase was completed in 3 sec and, following a 2-sec static hold, the subject returned to the initial position in 3 sec. Prior to the experimental sessions all subjects were trained in the proper execution of both exercises.

Paired 16-mm biopotential skin electrodes were secured to the skin covering the muscles studied. The center of the umbilicus served as a point of reference for electrode positioning [6, 11]. Two electrodes, 2.5 cm apart, were placed over the upper segment of the left rectus abdominis following an imaginary line 3 cm lateral to the umbilicus. Another set of electrodes was secured above the lower segment of the same muscle according to the same specifications. These sites for electrode placement have been used by other investigators [5, 6, 9, 11]. A fifth electrode was placed over the supero-lateral surface of the ilium and served as a ground. In order to more accurately compare the CSU and BSU action potentials the electrodes remained on the subjects from one exercise to the other.

The EMG was recorded by means of a Honeywell 906 C Visicorder. The paper speed was set at 25 mm/sec and prior and after the execution of each movement a calibration signal was recorded such that a pen deflection of 20 mm corresponded to 1 mV input.

The intensity readings of the MAP were computed for each phase of each exercise: the upward phase, the static hold and the downward phase. The readings were derived by averaging the amplitudes of the pen displacements at 0.5 sec intervals during the recording. The calculated values were expressed in millimeters and were transformed into millivolts before being submitted to statistical treatment.

Results

For each phase of the movement, as well as for the 3 phases combined, the BSU elicited a significantly greater intensity of contraction from the left upper rectus abdominis than the CSU (table II).

Similar results were found for the left lower rectus abdominis. The BSU consistently provoked a significantly greater intensity of muscle contraction than did the CSU (table III).

Discussion

From other investigations it was demonstrated that the CSU, used as a yardstick in this study, compared favorably with other forms of abdominal exercises in terms of MAP elicited from the rectus abdominis [3, 8, 12]. This particular sit-up could, therefore, be regarded as a typical representative of the most commonly used abdominal exercises.

In most investigations of various forms of sit-up exercises, the authors

Table II. Intensity (mV) of map in left upper rectus abdominis during the two types of sit-up (N = 16)

	Mean ± SD	Δ − mean ± SD	Paired t-ratios
1. Upward phase			
BSU	1.68 ± 0.89	0.34 ± 0.27	5.04[1]
CSU	1.34 ± 0.75		
2. Static hold			
BSU	1.38 ± 0.83	0.89 ± 0.57	6.29[1]
CSU	0.49 ± 0.33		
3. Downward phase			
BSU	1.17 ± 0.62	0.16 ± 0.25	2.47[3]
CSU	1.01 ± 0.51		
Mean (1, 2, 3)			
BSU	1.41 ± 0.77	0.46 ± 0.32	5.75[1]
CSU	0.95 ± 0.51		

1 $p < 0.001$.
2 $p < 0.01$.
3 $p < 0.05$.

Table III. Intensity (mV) of map in left lower rectus abdominis during the two types of sit-up (N = 16)

	Mean ± SD	Δ − mean ± SD	Paired t-ratios
1. Upward phase			
BSU	1.48 ± 0.65	0.56 ± 0.43	5.26[3]
CSU	0.92 ± 0.32		
2. Static hold			
BSU	1.39 ± 0.83	1.05 ± 0.80	5.24[3]
CSU	0.34 ± 0.16		
3. Downward phase			
BSU	0.94 ± 0.36	0.26 ± 0.24	4.39[3]
CSU	0.68 ± 0.29		
Mean (1, 2, 3)			
BSU	1.27 ± 0.56	0.63 ± 0.42	5.90[3]
CSU	0.64 ± 0.23		

1 $p < 0.05$.
2 $p < 0.01$.
3 $p < 0.001$.

do not report a static hold period between the upward and downward phases. It is quite understandable, since minimal contraction is needed to maintain the body in this position during most sit-up exercises. However, for the BSU, at the point where the subject reaches the end of the upward phase, the gravitational line of his upper body falls outside his base of support. This occurrence is quite interesting since intense contraction is required in order to resist the gravitational force. Moreover, as pointed out by FLINT [3], during sit-ups, the first 45° of trunk flexion as well as the last 45° of trunk extension are the responsibility of the abdominal muscles, while the middle phase of the exercise is primarily the responsibility of the hip flexors [3]. In the BSU, as can be observed from the results in tables II and III, the rectus abdominis is solicited throughout the complete range of the movement.

For both the BSU and the CSU, consistently higher MAP readings were obtained during concentric contraction (upward phase) as opposed to eccentric contraction (downward phase). These findings concur with those of previous studies [3, 11].

Higher electrical activity was observed in the upper rectus abdominis than in the lower rectus abdominis; this was also reported by LIPETZ and GUTIN [8].

The reduction in the distance between the origin and insertion of the hip flexors and the psoas has most probably decreased their participation and therefore enhanced the rectus abdominis involvement. However, an electromyographic study of these muscles during the BSU should be conducted to verify this hypothesis.

During the experimental sessions, it was observed visually that hyperextension of the lumbar spine during the BSU was virtually impossible. Since the CSU was performed like a trunk curl, the subjects did not hyperextend the lumbar spine during the execution of this exercise either.

References

1 ÅSTRAND, P. O. and RODAHL, K.: Work physiology (McGraw-Hill, New York 1970).

2 COBEY, M. C.: Postural back pain (Thomas, Springfield 1956).

3 FLINT, M. M.: Abdominal muscle involvement during the performance of various forms of sit-up exercise. Amer. J. phys. Med. 44: 224–234 (1965).

4 FLINT, M. M.: An electromyography comparison of the function of the iliacus and the rectus abdominis muscles. Amer. J. phys. Ther. Ass. 45: 248–253 (1965).

5 FLINT, M. M. and GUDGELL, J.: Electromyographic study of abdominal muscular activity during exercise. Res. Quart. amer. Ass. Hlth phys. Educ. 36: 29–37 (1965).

6 FLOYD, W. F. and SILVER, P. H. S.: Electromyographic study of patterns of activity of the anterior abdominal wall muscles in man. J. Anat., Lond. 84: 132–145 (1950).

7 LIPETZ, S. and GUTIN, B.: An electromyographic investigation of the rectus abdominis in various abdominal exercises. AAHPER Convention, Boston 1969.

8 LIPETZ, S. and GUTIN, B.: An electromyographic study of four abdominal exercices. Med. Sci. Sports 2: 35–38 (1970).

9 PARTRIDGE, B.S. and WALTERS, C.E.: Participation of the abdominal muscles in various movements of the trunk in man. An electromyographic study. Phys. Ther, Rev. 39: 791 (1959).

10 RASH, P.J. and BURKE, R.K.: Kinesiology and applied anatomy, 3rd ed. (Lea & Febiger, Philadelphia 1967).

11 SHEFFIELD, F.J.: Electromyographic study of the abdominal muscles in walking and other movements. Amer. J. phys. Med. 41: 142–147 (1962).

12 WALTERS, C.E. and PARTRIDGE, M.J.: Electromyographic study of the differential action of the abdominal muscles during exercise. Amer. J. phys. Med. 36: 259–268 (1957).

Author's address: Dr. YVAN GIRARDIN, 5440 Decelles 303, Montreal, Que. (Canada)

Medicine and Sport, vol. 8: Biomechanics III, pp. 309–314 (Karger, Basel 1973)

Electromyographic Study of Quadriceps and Hamstring Involvement in Knee Stability[1]

H. H. MERRIFIELD and C. G. KUKULKA

Ithaca College, Ithaca, New York, N.Y.

Introduction

Stability of the knee joint depends almost entirely on the integrity of its muscles and ligaments. Following trauma or surgery for knee instability, muscle strengthening extension and flexion exercises have been empirically recommended. Various opinions exist regarding the effectiveness of the quadriceps and hamstring muscle groups and the particular role of each muscle within the group on maintaining joint stability.

The purpose of this study was to investigate the function of the quadriceps and hamstring muscles in producing leg realignment from valgus and varus stresses.

Methodology

Subjects

Six male Ithaca College students, between the ages of 20 and 27, with no previous injuries to the right knee, volunteered to serve as subjects. Prior to the test sessions, each subject was asked to practice isolated contractions of the quadriceps and hamstring muscle groups.

Instrumentation

A Teca N8 and SDC Electrophysiological Recording System was utilized to record the action potentials. For intersubject standardization, pairs of surface electrodes were placed parallel with the long axis and straddled the motor points of the rectus femoris, vastus lateralis, vastus medialis, biceps femoris,

1   A paper presented at the 3rd International Seminar on Biomechanics in Rome, Italy, on September 28, 1971.

semitendinosus and semimembranosus. Permanent records of the EMG were obtained on light sensitive paper by use of an F03 Recorder.

The Sprague-Walters table was employed for the testing of knee abduction (valgus) and adduction (varus). With the knee joint at approximately 15° of flexion, the joint line was aligned with the front surface of a vice in order to prevent femoral rotation. The distal end of the leg was placed in a metal cuff that could travel across the table. To produce valgus and varus stress, a one-eighth inch steel cable with a weight pan at its end was attached to each side of the cuff.

In order to project a 35-mm slide composed of an angular grid vertically downward on the knee and leg, a Viewlex slide projector, Model V-500-P, was mounted on a platform 5.5 ft above the knee joint. The apex of this projected grid was placed over the proximal reference point between the femoral condyles, and a line representing 0° of the grid was aligned with the middle of the leg.

Procedure

Values for muscular contractions and valgus and varus stresses were obtained from testing of the subject's right leg. With six sets of EMG electrodes attached, the thigh was carefully placed in the vice.

The knee stability testing and recording of action potentials was conducted under 2 conditions: the thigh musculature (1) relaxed, and (2) contracted. Before testing of the contracted state, the subject was checked by EMG monitoring for isolation of the individual muscle group contractions. When no apparent action potentials were observed from the antagonistic muscle group, the subject was considered ready for testing. This procedure was repeated at the conclusion of testing.

In testing during the relaxed state, the deviation value was obtained 3 sec after the weight was placed on the weight pan; and for the following 5 sec, the action potentials, if any, were recorded. In testing during muscular contraction, the subject relaxed while the weight was added, and after 3 sec, the deviation value was observed. The subject was then instructed to try to bring his leg back into alignment with the 0-degree line of the projected grid by contracting the particular muscle group being monitored.

Each contraction lasts for 5 sec with continuous EMG recordings. After 4 sec of the contraction, another recording of the deviation was obtained to observe how near to the 0-degree line the subject had aligned his leg. The testing sequence was with 4.5, 6.8 and 9.1 kg of stress, first for valgus and then the varus.

During the testing in both conditions, caution was taken to keep the knee joint at approximately 165° and to see that leg rotation did not occur.

Treatment of Data

Amplitude and frequency of the action potential responses were the parameters utilized for determining muscle activity. The mean electrical activity based upon the 3rd and 4th sec of the EMG monitoring was expressed in microvoltage. From subjects' EMG, means were determined for each valgus and varus stress. Direct readings expressed in degrees of the angular displacements of the leg for the stress conditions were recorded. Means and standard deviations were calculated.

Results

The angular displacement measurements on the 6 subjects for both valgus and varus stresses increased with additional weight increments as shown in figure 1. It may be noted with muscle contraction each displacement value decreased. For example, with 9.1 kg of valgus stress, the mean resting value was 6.9°, and with muscle contractions 2.58 and 3.75° were registered for the quadriceps and hamstring groups, respectively. It appeared that the quadriceps group produced more leg realignment with valgus stress, whereas the hamstring group indicated only a slight advantage in varus stress.

During the relaxed phase for valgus and varus stresses with the 3 weight increments, no action potential recordings were produced for the 6 muscles monitored. With volitional contraction the microvoltage for the 3 quadricep muscles followed a similar pattern for varus stress and demonstrated different patterns with valgus stress (fig. 2). The 3 muscles generally increased in microvoltage with larger loads. The vastus lateralis displayed the largest microvoltage output for both stresses. However, as the valgus stress was increased, the vastus lateralis output decreased indicating a reversal pattern with the other muscles. The rectus femoris demonstrated only a slight increase with additional valgus stress whereas a large increase in microvoltage output for the vastus medialis occurred.

In Figure 3, the semitendinosus is shown to elicite the most activity under

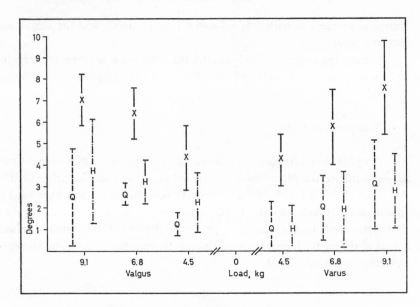

Fig. 1. Mean and standard deviation with valgus and varus stress: muscles relaxed (―――); quadriceps contracted (– – – –); hamstrings contracted (―·―·).

all stress conditions, with the semimembranosus second and biceps femoris the least. Increased valgus stress appeared to have little effect on hamstring activity. With increased varus stress, the semitendinosus output increased and the biceps femoris displayed a slight opposite trend.

Discussion

The maximum valgus and varus angular displacement measurements in this study were similar to the findings reported by others using a similar technique [5, 9].

It appears that no reflexes were produced during valgus and varus stresses with external loads up to 9.1 kg. If motor neurons were excited, the muscular responses must have been minimum. Thus, our findings agreed with PETERSEN and STENER [7], who reported that no reflex effects were detected during tension of the medial collateral ligament. In fact, reflex inhibition of at least the quadriceps muscle group has been found in response to various stimuli including capsular compression [2] and joint distension [1]. Perhaps the technique of gradual loading, type of EMG electrodes employed and/or

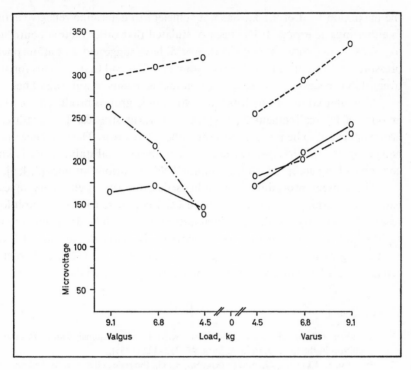

Fig. 2. Mean readings with valgus and varus stress for rectus femoris (———); vastus medialis (—·—·); vastus lateralis (– – – –).

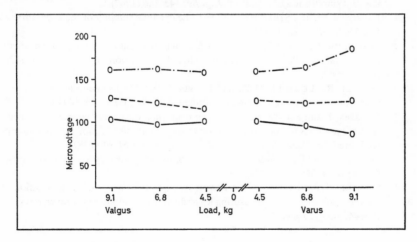

Fig. 3. Mean readings in valgus and varus stress for biceps femoris (———); semi-tendinosus (—·—·); semimembranosus (– – – –).

the maximum load of 9.1 kg was not sufficient to determine reflexive action in either muscle group. It has been postulated that other factors contribute to joint stability. SEMLAK and FERGUSON [8] have suggested that atmospheric pressure is an important force aiding joint stability and that a valgus force of over 10 lb is necessary to cause force on the ligaments in a normal knee.

Muscular contractions from the two muscle groups produced a certain amount of leg realignment from valgus and varus stresses. The quadriceps group appeared to be more important with valgus stress. The influence of the hamstring and quadriceps muscle groups on the coronal stability of the knee confirms a long accepted clinical opinion held by various authors [3, 4, 6].

The present procedure and findings describe our preliminary investigations concerning the knee that have been conducted in our kinesiology research laboratory. Studies in which knee stability will be determined in the sagittal and transverse planes have been already planned. Additional EMG monitoring of muscles comprising the pes anserinus and iliotibial tract in reference to all 3 planes will be included in these future studies.

References

1   ANDRADE, J. R. DE; GRANT, C., and DIXON, A. ST. J.: Joint distension and reflex muscle inhibition in the knee. J. Bone Jt Surg. *47:* 313–322 (1965).
2   EKHOLM, J.; EKLUND, G., and SKOGLUND, S.: On the reflex effects from the knee joint of the cat. Acta physiol. scand. *50:* 167–174 (1960).
3   KITTLESON, A. C.; O'CONNOR, G. A., and RODRIGUEZ, H. P.: Evaluation of knee stability and stabilizing systems. Radiology *89:* 145–146 (1967).
4   LAURENCE, M. and STRACHAN, J. C. H.: The dynamic stability of the knee. Proc. roy. Soc. Med. *63:* 34, 759 (1970).
5   MOREHOUSE, C. A. and SODERBERG, G. L.: An instrument for the objective evaluation of knee stability. Paper presented at National Convention of Amer. Ass. of HPER, Boston 1969.
6   OUELLET, R.; LEVESQUE, H. P., and LAURIN, C. A.: The ligamentous stability of the knee. An experimental investigation. Canad. med. Ass. J. *100:* 45–50 (1969).
7   PETERSEN, I. and STENER, B.: A study in man using the medial collateral ligament of the knee joint. Experimental evaluation of the hypothesis of ligamento-muscular protective reflexes. Acta physiol. scand. *48:* suppl. 166, pp. 51–61 (1959).
8   SEMLAK, K. and FERGUSON, A. B.: Joint stability maintained by atmospheric pressure. Clin. Orthop. *68:* 294–300 (1970).
9   SPRAGUE, R. B.; TIPTON, C. M.; FLATT, A. E., and ASPREY, G. M.: Evaluation of a photographic method for measuring leg abduction and adduction. J. amer. phys. Ther. Ass. *46:* 1068–1078 (1966).

Author's address: Dr. H. H. MERRIFIELD, Ph.D., Associate Professor, Division of Physical Therapy, Ithaca College, *Ithaca, NY 14850* (USA)

Medicine and Sport, vol. 8: Biomechanics III, pp. 315–321 (Karger, Basel 1973)

An Electromyographic Study of Selected Shoulder Muscles during Arm Support Activity

DOROTHY A. LEGGETT and JOAN C. WATERLAND

University of Wisconsin, Madison, Wisc.

I. Introduction

Much of what is known about the action of muscle has been determined in the past from anatomical knowledge of muscle attachments and the line of pull with relation to the joints. As electromyography became more commonly used, muscle actions were tested through movements of flexion, extension, abduction, adduction and rotation while body parts were weighted, with the assumption that resisted movements would facilitate the observations.

The present study was concerned with describing muscle action during skilled movement, performed while supported by the hands. Since much of the movement was performed against gravity, the resistance was generally the weight of the body and, thus, was greater than that in most other studies of muscle function. Because the subjects were skilled, one would expect to find only those muscles active which were necessary to the performance of each movement [8, 9, 11, 13, 14]. In order to avoid confusion in terminology, the movements will be described as normally performed; that is, flexion of the trunk on the arms will be described as arm extension.

II. Experimental Procedures

Muscle action potentials were recorded during selected gymnastic movements from the sternal portion of the pectoralis major, the anterior and posterior sections of the deltoid, the upper and lower fibers of the trapezius and the latissimus dorsi muscles. The recordings were made with surface electrodes and a visicorder optical polygraph. The subjects were 5 male members of the University of Wisconsin intercollegiate gymnastics team, chosen for their skill in performing the movements studied. These movements included the hip pullover (fig. 1a) on the single parallel bar, swing to handstand (fig. 1b) on the parallel bars and the back up-rise (fig. 1c) on the stationary rings. The EMG were recorded continuously during each test movement and the performance was simultaneously filmed by a 16-mm camera. By means of a timing device, the EMG subsequently could be synchronized with the film, thus permitting comparison of the overt movement with the covert muscle patterns (fig. 2).

III. Results and Discussion

A. Pectoralis Major Muscle (Sternal Portion)

Activity in the sternal portion of the pectoralis major muscle was noted during movements in which arm flexion occurred from a position of hyperextension to one in line with the trunk. Action potentials were also noted during arm extension from a position of high forward flexion. At the point when pectoralis major ceased to function during arm extension, a slight increase in activity was noted in posterior deltoid. This interaction was pointed out by WRIGHT [17] in 1928, and also noted by BRUNNSTROM [4]. Pectoralis major along with anterior deltoid was noted also to be active in bracing the body when it assumed a prone position as the arms moved into adduction and extension from high forward flexion.

B. Anterior and Posterior Portions of the Deltoid Muscle

The standard textbook description of deltoid action is that the muscle as a whole abducts the arm, the anterior portion flexes and medially rotates and the posterior portion extends and laterally rotates. The above description is a simplified version of a muscle over which much controversy has taken place. DUCHENNE [6] stated that the posterior fibers of the deltoid act both during abduction and adduction, the transition taking place at 45° [6, p. 47]. BEEVOR [3] claimed that the posterior deltoid acts in adduction but not during abduction. WRIGHT [17] agreed with BEEVOR that the posterior deltoid does not take part in abduction, but is active during adduction, although with less force than during hyperextension.

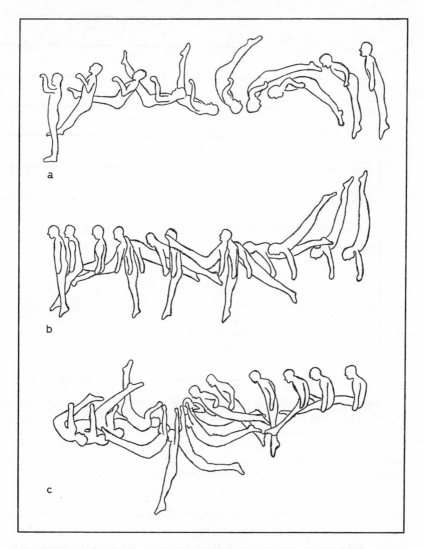

Fig. 1. Film tracings of 3 gymnastic movements performed while supported on the hands: *a* the hip pullover, *b* swing to handstand, *c* the back up-rise.

In the present study, the posterior deltoid showed some form of activity at all times. This was not true of the anterior fibers, although YAMSHON and BIERMAN [19], SCHEVING and PAULY [15] and MORTENSEN [10] found all parts of the deltoid active during abduction, flexion and extension. In the inverted

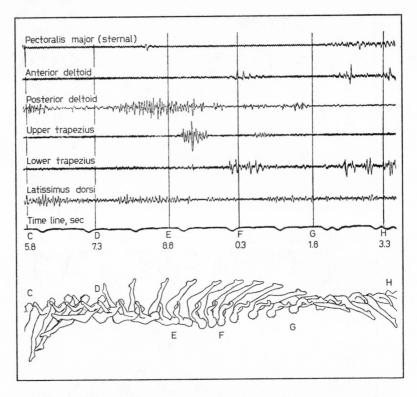

Pectoralis major (sternal)

Anterior deltoid

Posterior deltoid

Upper trapezius

Lower trapezius

Latissimus dorsi

Time line, sec

| C | D | E | F | G | H |
| 5.8 | 7.3 | 8.8 | 0.3 | 1.8 | 3.3 |

Fig. 2. Sample recording of the muscle action potentials, and tracings taken from a film of the same test performance. The figures lettered C, D, E, F, G and H were traced from film frames at 1.5-sec intervals apart. The corresponding positions on the muscle activity recordings are indicated by the vertical lines lettered identically and showing the time read from the time line. The unlettered tracings were taken from film frames, equally spaced between those lettered, in order to provide additional information.

hang in pike position at the beginning of the back up-rise, the posterior deltoid of all 5 subjects was active. In one subject, it appeared to be the only muscle active of the muscle group tested. BASMAJIAN and BAZANT [1], BEARN [2] and SCHEVING and PAULY [15] found that the posterior deltoid functioned in an antigravity capacity when its movement was resisted. Probably, in the present situation, where the resistance was the weight of the body (approximately 150 lb), the posterior deltoid was acting to prevent subluxation.

During two movements in which arm abduction occurred with medial rotation, the posterior rather than the anterior fibers recorded action potentials in coaction with the upper fibers of trapezius. WRIGHT [17] reported that the lateral portion of the posterior deltoid, and not the anterior fibers took part in the movement of abduction of the humerus while in a position of medial rotation. The recordings made by SCHEVING and PAULY [15] show marked contraction of posterior deltoid and noticeable recession of the recordings made by the anterior fibers during resisted medial rotation of the humerus. They do not comment on this behavior other than to note that all parts of the deltoid, because they are active in all arm movements, are possibly another 'articular muscle of Winslow'. It is suggested here, instead, that the resistance to medial rotation provoked abduction, for a noted increase in activity occurred in the middle fibers of the deltoid. If abduction did occur at the time that medial rotation was tested, this might very well have been the reason for posterior deltoid activity.

The anterior portion of the deltoid generally appeared quiescent except when called upon to perform the function of flexion of the humerus. There was only slight indication that anterior deltoid was called on to perform either abduction or medial rotation. WRIGHT [17] observed that this portion of the deltoid, although it undoubtedly had inward rotator action, did not function in this capacity unless in conjunction with another movement. Action potentials in anterior deltoid were always accompanied by lower trapezius activity, the timing and pattern shape of which were surprisingly similar to that from the deltoid.

C. Upper and Lower Fibres of the Trapezius Muscle

Like the posterior deltoid, the lower fibers of trapezius were active in some degree throughout all test movements. Muscle action potentials were more noticeably consistent during arm flexion than abduction and appeared in advance of those from the arm flexors. WIEDENBAUER and MORTENSEN [16,] on the contrary, reported more activity during the early phase of abduction than of flexion.

The present study concurs with the WIEDENBAUER and MORTENSEN [16] findings that lower trapezius is not important in scapular depression. When approximately 150 lb were supported on the hands during stationary upright position on the parallel bars, only minimal activity was noticeable. This is not in accord with the opinions of other writers [4, 6, 7, 17, 18].

Both the upper and lower portions of trapezius were active during scapular retraction, which concurs with other EMG studies [16, 18]. When the subject was in an inverted hang position supported by the hands, there was noticeable activity in upper trapezius. This observation coincides with the experimental findings of BEARN [2] and NAJENSON and SOLZI [12] that upper trapezius, when resisted, takes part in antigravity activity. YAMSHON and BIERMAN [18] judged upper trapezius to be an accessory muscle in head extension. In the present study it was noticeably active during the handstand on the parallel bars when there was strong hyperextension of the neck and head. Activity was much less noticeable when the body was in a prone position, and the head was not extended beyond the line of the trunk.

D. The Latissimus Dorsi Muscle

The latissimus dorsi muscle, in addition to arm extension, possible medial rotation and adduction, was active consistently toward the lowest point of all trunk and leg swings which, due to centrifugal force, caused it to function in an antigravity capacity on the hips.

References

1 BASMAJIAN, J.V. and BAZANT, F.J.: Factors preventing downward dislocation of the adducted shoulder joint. An electromyographic and morphological study. J.Bone Jt Surg. 41: 1182–1186 (1959).

2 BEARN, J.G.: An electromyographical study of the trapezius, deltoid, pectoralis major, biceps and triceps muscles during static loading of the upper limb. Anat. Rec. 140: 103–107 (1961).

3 BEEVOR, C.E.: The Croonian lectures on muscular movements and their representation in the central nervous system. Lancet i: 1715–1724 (1961).

4 BRUNNSTROM, S.: Clinical kinesiology (Davis, Philadelphia 1966).

5 DANIELS, L.; WILLIAMS, M., and WORTHINGHAM, C.: Muscle testing (Saunders, Philadelphia 1956).

6 DUCHENNE, G.B.: Physiology of motion (translated by EMANUEL B. KAPLAN) (Saunders, Philadelphia 1959).

7 GRAY, H.: In Goss Anatomy of the human body, 28th ed. (Lea & Febiger, Philadelphia 1966).

8 HINSON, M.M.: An electromyographic study of the push-up for women. Res. Quart. amer. Ass. Hlth phys. Educ. 40 (2): 305–311 (1969).

9 KAMON, E. and GORMLEY, J.: Muscular activity pattern for skilled performance and during learning of a horizontal bar exercise. Ergonomics 2 (4): 345–357 (1968) (Taylor & Francis, London 1968).

10 MORTENSEN, O.A.: Personal communication (1970).

11 LUNDERVOLD, A.J.S.: Electromyographic investigations of position and manner of working in typewriting. Acta physiol. scand. 24: suppl. 84, pp. 1–171 (1951).

12 NAJENSON, T. and SOLZI, P.: Antigravitatory function of the trapezius muscle. Electromyography 9 (2): 215–218 (1969).

13 OKAMOTO, T.; TAKAGI, K., and KUMAMOTO, M.: An electromyographic study on the process of the acquisition of proficiency in gymnastic kip; in KATO Proc. Int. Congr. of Sport Sciences, 1964, pp. 422–424 (University of Tokyo Press, Tokyo 1966).

14 ONO, M.: Study on effect of muscle training; in KATO Proc. Int. Congr. of Sport Sciences, 1964, pp. 428–429 (University of Tokyo Press, Tokyo 1966).

15 SCHEVING, L. E. and PAULY, J. E.: An electromyographic study of some muscles acting on the upper extremity of man. Anat. Rec. 135 (3): 239–245 (1959).

16 WIEDENBAUER, M. M. and MORTENSEN, O. A.: An electromyographic study of the trapezius muscle. Amer. J. phys. Med. 31: 363–372 (1952).

17 WRIGHT, W. G.: Muscle function (Hoeber, New York 1928).

18 YAMSHON, L. J. and BIERMAN, W.: Kinesiologic electromyography. II. The trapezius. Arch. phys. Med. 29: 647–651 (1948).

19 YAMSHON, L. J. and BIERMAN, W.: Kinesiologic electromyography. III. The deltoid. Arch. phys. Med. 30: 286–289 (1949).

Authors' addresses: Dr. DOROTHY A. LEGGETT, School of Physical and Health Education, Queen's University, Kingston, Ont, (Canada); Dr. JOAN C. WATERLAND, Department of Physical Education for Women, Lathrop Hall, University of Wisconsin, Madison, WI 53706 (USA)

Medicine and Sport, vol. 8: Biomechanics III, pp. 322–327 (Karger, Basel 1973)

Potentials Recorded from the Semiorbicularis Oris as a Result of Contralateral Facial Nerve Stimulation

P. Tonali and D. Gambi

Clinic of Nervous and Mental Diseases, Catholic University, Rome

From the anatomical point of view, the existence of transversal intrinsic fascicles extending from joint to joint in the orbicularis oris has been indicated by some authors. However, this opinion is in opposition to the view expressed by some EMG-graphists [Passerini et al., 1968a] who state that no action potential can be recorded from a portion of the orbicularis oris as a result of contralateral facial nerve stimulation. Our aim is to clarify if this anatomical and physiological condition could be verified by EMG investigation techniques.

Methods and Techniques

Four healthy subjects ranging from 30 to 65 years of age were examined. The facial nerve was stimulated at the exit from the stylomastoid foramen by surface electrodes delivering 0.1–0.2 msec impulses at maximal intensity. In each subject the right and left nerves were stimulated one after the other. Muscle action potentials were recorded from the upper semiorbicularis and, subsequently, from the lower semiorbicularis by means of concentric needle electrodes (with a surface area of 0.00193 mm^2 and an outer diameter of 0.305 mm) placed on both sides of the median line in double simultaneous recordings.

At first, the two electrodes and the median line were 3 cm apart; afterwards the electrode placed contralaterally to the stimulated nerve was set at 2, 1.5 and 1 cm distance, respectively, from the median line. The same method was applied for nerve stimulation on either side. On the whole, 8 recordings for each semiorbicularis were accomplished. Care was taken to place needle electrodes always at the same depth (2–3 mm) along parallel lines on the cutaneous-mucous edge of the upper and lower lip, 5 mm apart. The distance from the stimulation site to the outer recording points was in the range of 9.5–10.5 cm. In one subject action potentials of the homolateral semiorbicularis to the nerve were also studied and evoked from points 2.5, 2 and 1 cm distant from the median line.

Muscle action potentials were recorded by means of a rack-mounted 4-channel 'SDC 3R' model Medelec EMG. As far as physical diffusion of evoked potentials and selectivity

Table I. M. semiorbicularis oris superior

	Right				Left			
	Distance of median line, cm	Evoked muscle potentials			Distance of median line, cm	Evoked muscle potentials		
		latency, msec	ampli-tude, mV	dur-ation, msec		latency, msec	ampli-tude, mV	dur-ation, msec
Right facial stimulus	3	3.2	4.5	10	3	–	–	–
					2	7.5	0.7	6
					1.5	5.5	1	6.5
					1	3.7	1.2	7
Left facial stimulus	3	–	–	–	3	4	6	9
	2	8	0.03	4				
	1.5	6	2	6				
	1	4.7	2.5	7.3				

of recording are concerned, we refer to PINELLI's papers [1949, 1951]. According to these results, controls of our investigations were made with bipolar concentric needle electrodes.

The potentials in simultaneous recordings were (1) recorded by searching the optimal point of needle insertion, and (2) were not synchronous and not equal in shape to contralaterally evoked potentials, were considered as potentials originating in the site of derivation [PINELLI and SALA, 1969].

Results

In all subjects, facial nerve stimulation evokes muscle action potentials in the upper or lower homolateral semiorbicularis, as well as in the contralateral semiorbicularis. This happens when the sites of recordings and the median line are not more than 2 cm apart (with the exception of a few cases in which a maximum distance of 2.5 cm was reached).

Contralaterally evoked potentials increase in amplitude as the recording is placed more proximally to the site of stimulation and closer to the median line; also latency time decreases under such conditions. A simple (di- or triphasic) shape usually prevails.

Table I shows various parameters of evoked potentials of the right and left upper semiorbicularis following homo- and contralateral stimulation of

Table II. M. semiorbicularis oris inferior

	Right				Left			
	Distance of median line, cm	Evoked muscle potentials			Distance of median line, cm	Evoked muscle potentials		
		latency, msec	amplitude, mV	duration, msec		latency, msec	amplitude, mV	duration, msec
Right facial stimulus	3	3.5	7	9.5	3	–	–	–
					2	–	–	–
					1.5	4.5	1.2	8.5
					1	4	1.5	13
Left facial stimulus	3	–	–	–	3	3	7	6.7
	2	6.5	0.1	4				
	1.5	5	2	7.5				
	1	4	2.2	8				

the facial nerve in one of the cases under investigation. Differences in latency times of contralaterally evoked potentials from points placed 1 cm apart are in the range of 3.3–3.8 msec. Greater differences have never been recorded for equal distances (see table II, concerning the lower semiorbicularis in a different case), whereas a value of 1 msec was recorded for only one subject (table III). This leads to the conclusion that in the contralateral muscle excitation travels at velocities ranging between 2.6 and 10 m/sec. How can the difference in latencies be interpreted between the beginning of the action potentials recorded at different points from the orbicularis oris?

A low amplitude of phases reducing differences in latency can be expected, this is due to physical diffusion of the action potentials arising from the end-plate zone. However, if this zone is very circumscribed and few muscle fibres are placed extending at a relatively long distance, an interval can occur owing to the conduction of muscular action potentials from the end-plate zone to the recording point when this is located very near to the active muscle fibres [BUCHTHAL *et al.*, 1955]. On the basis of table III, we may point out that a conduction velocity of 15 m/sec (over 1.5 cm distance) has been attained in the subject in whom action potentials, evoked from several points of the homolateral semiorbicularis, have been investigated.

Table III. M. semiorbicularis oris superior

Right					Left			
Right facial stimulus	Distance of median line, cm	Evoked muscle potentials			Distance of median line, cm	Evoked muscle potentials		
		latency, msec	ampli-tude, mV	dur-ation, msec		latency, msec	ampli-tude, mV	dur-ation, msec
	2.5	3	2	7.7	2.5	–	–	–
	2	3.2	2.5	8.8	2	6.3	0.15	10.2
	1	4	1.5	11	1	5.3	0.4	8.3

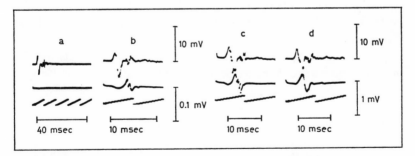

Fig. 1. Stimulation of right facial nerve: simultaneous recordings from upper semi-orbicularis oris (above, left side; below, right side). The right electrode is placed 3 cm from the median line; the left electrode is placed in *a* 3 cm, in *b* 2 cm, in *c* 1.5 cm, and in *d* 1 cm from the median line.

Fig. 2. Stimulation of left facial nerve: simultaneous recordings from upper semi-orbicularis oris (above, left side; below, right side). The left electrode is placed 3 cm from the median line; the right electrode is placed in *a* 3 cm, in *b* 2 cm, in *c* 1.5 cm, and in *d* 1 cm from the median line.

Discussion

In our study we can refer to the term 'conduction velocity', which is meant very generally, since it is not measured between 2 points of stimulation, but between 2 recording points. In such cases the procedure established for the investigation of conduction velocity of motor fibre cannot be applied; as a result, our measurements may include nerve components (having a conduction velocity of 30–50 m/sec) as well as muscle components (conduction velocity of 3–5 m/sec). Special mention should also be made of the two electrodes, which are sometimes placed in connection with groups of different muscle fibres. This results in a broader scattering of values. Hence, the data gathered in our cases may only be considered by way of orientation.

Our results lead to the conclusion that most muscle fibres of the semiorbicularis oris stop at the median line, whereas a small number of them cross it. Some of the latter fibres stop a few millimetres after the median line, others reach a maximal distance of 2 cm. This would result in an anatomical structure in which muscle fibres are closely intertwined.

Our results concur rather well with the opinion that the motor end-plate zone hardly extends beyond the median line. On the other hand, the considerable scattering and polyphasic character of homolaterally evoked muscle action potentials induce the supposition that in the semiorbicularis oris the motor end-plates cover quite an extensive area on the homolateral side. These polyphasic characters may also be due to other factors, such as the different arrangement in the pattern of muscle fibres. The reduced scattering and the lower number of phases of contralaterally evoked potentials may be accounted for on the basis of selective recording (which excludes electric potentials being transmitted to the opposite side) and of the probably poor number of plates, pertaining to the contralateral nerve, which are located beyond the median line.

Further investigation can be performed, including stimulation of facial nerve at 2 points according to the method by PASSERINI et al. [1968 b].

Summary

The authors, by EMG investigation techniques, have studied whether in normal subjects the facial nerve participates in the innervation of the contralateral section of the upper and lower semiorbicularis oris muscles.

In all subjects facial nerve stimulation evokes a muscle action potential in the upper or lower homolateral semiorbicularis, as well as in the contralateral semiorbicularis. This happens when the sites of recordings and the median line are not more than 2 cm apart. The collected values show that conduction in the contralateral section of muscle takes place partly along nervous fibres and partly along muscle fibres. This means to say that the zone of the motor end-plates extends beyond the median line.

References

BUCHTHAL, F.; GULD, C., and ROSENFALCK, P.: Propagation velocity in electrically activated muscle fibres in man. Acta physiol. scand. *34:* 75 (1955).

GAMBI, D. et TONALI, P.: Sur la réinnervation faciale par le nerf controlatéral. Paper of the 4th Int. Congr. of Electromyography, Brussels 1971.

PINELLI, P.: Premesse all'analisi clinica dei valori del singolo potenziale d'azione muscolare. Riv. Neurol. *21:* 3 (1949).

PINELLI, P.: La forme des potentiels d'action étudiée chez l'homme par différentes électrodes. Congr. Neurol., Paris *3:* 98 (1951).

PINELLI, P. e SALA, E.: Ricerche elettromiografiche nei processi distrofici muscolari. Sulla modalità topografica di atrofia e sulle caratteristiche funzionali delle fibre muscolari. Atti 13th Congr. Naz. Soc. Ital. di Neurol., Messina 1969.

PASSERINI, D.; SALA, E., and VALLI, G.: Controlateral reinnervation in facial palsy. Electromyography *8*(2): 115 (1968a).

PASSERINI, D.; SALA, E. e VALLI, G.: Possibilità e limiti di una nuova metodica per la determinazione della velocità di conduzione del nervo facciale. Riv. Neurol. *38:* 320 (1968b).

Author's address: Prof. PIETRO TONALI, Clinica delle Malatte Nervose e Mentali, Università Cattolica, Via della Pineta Sacchetti, 526, *Rome* (Italy)

Medicine and Sport, vol. 8: Biomechanics III, pp. 328–333 (Karger, Basel 1973)

Electromyographic Study of the Learning Process of Walking in 1- and 2-Year-Old Infants

T. OKAMOTO

Kansai Medical School, Department of Physical Education, Hirakata-shi, Osaka

The learning process of walking in 1- and 2-year-old infants was studied for more than 1 year in terms of the functional mechanism of the muscles. The EMGs were recorded from the time when the 1-year-old infant was unable to walk without support and acquired the ability to walk for the first time until the time when it arrived at mastery to some extent. In case of the 2-year-old infant, who was already able to walk, until the time when she acquired even more skillful walking ability. In addition, myograms were recorded of 30 subjects from infants to adults in order to examine the process of walking as it develops with age. These EMG recordings were made with an 8-channel ink-writing oscillograph using surface electrodes of 10 mm in diameter.

Figure 1 shows the typical myograms in the initial period of the learning process of the 1-year-old infant. From the basogram recorded, a walking cycle is divided into a swing and a stance phase.

Figure 1a is the myogram recorded at the time when the infant walked 2 or 3 steps for the first time on the 380th day after birth. It shows strong discharges throughout the stance phase in many muscles, when compared with the adult pattern. From the forms, the strong discharge pattern may be considered to come from the squatting posture which needs a strained effort with hip and knee flexed. That is, the biceps femoris, the gluteus maximus and the sacrospinalis would act as the antagonist of the iliopsoas, known as deep hip flexor, and would be concerned in fixing the hip joint and the pelvis. The co-contraction of the tibialis anterior and the gastrocnemius shows the stabilization of the ankle joint. The marked discharge of the rectus femoris during the swing phase would seem to show that this muscle acts as hip flexion to lift up the leg.

On the 1st day of walking, the rolling motion of the sole of the foot and

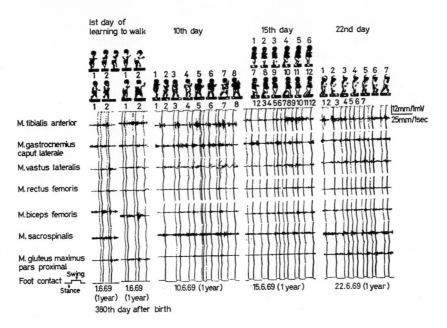

Fig. 1. Learning process of a 1-year-old infant.

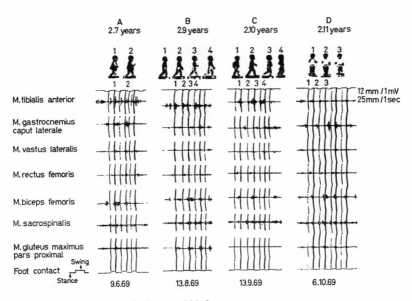

Fig. 2. Learning process of a 2-year-old infant.

the pushing-off motion of the foot in the stance phase, and the arm swing as a reflex movement, all of which are shown in an adult, are not perceived. That is, the heel and the toe touched the ground almost simultaneously, and the hip and the knee extension after the shock absorption of the leg in the stance phase were not well performed as compared with that of adults. The width between both feet at the stage of double support was very wide and the form of the spread arms served to maintain the balance. It seems certain that in order to propel its body, the infant obtains its strength by the forward swing of the leg from the squatting position in leaning forward. These seem to be the characteristics of the walking form on the 1st day of learning to walk.

On the 10th day after learning to walk, the infant walked more than 10 steps. The discharge pattern of the tibialis anterior tended to decrease during the stance phase, and on the 15th day a part of the discharges was becoming similar to the adult pattern. That is, it shows that the unnecessary stabilization of the ankle joint has begun to disappear. The discharge of the gastrocnemius was recorded not only throughout the stance phase but also at the latter part of the swing phase. In this case the co-contraction of the tibialis anterior and the gastrocnemius indicating unnecessary stabilization of the ankle joint is also perceived in the swing phase. The discharge pattern of the biceps femoris and of the vastus lateralis appeared not only in the stance phase, but also in the swing phase. That is, the discharge of the biceps femoris in the early part of the swing phase seems to mean that this muscle acts for knee flexion, while the discharge of the vastus lateralis in the latter half of the swing phase indicates knee extension. In this period, the infant is unable to stand on one leg only. The swing phase should, therefore, be as short as possible.

From the 50th to the 85th day after learning to walk, the discharge patterns of these muscles have begun to show a reciprocal relationship. From the 85th day, the discharge of the biceps femoris and the vastus lateralis in the swing phase which appeared in the initial period of learning to walk has decreased extremely. As to the change in the discharge patterns of these muscles, it would seem to show that the infant has become fairly accustomed to walking. In this period, the width between both feet at the stage of double support has been narrowed, and the position of the hip joint is higher than in the initial period of learning to walk. Even when the infant was still unable to walk alone from the 11th to the 12th month after birth, the discharges of both the tibialis anterior and the gastrocnemius already showed a nearly reciprocal relationship when it walked using a supporting frame. From this finding, the appearance of the reciprocal patterns of both these muscles from the 50th to 85th day seems to show that the infant has acquired a comparatively stable walk during this period.

From the 174th to 365th day after learning to walk, though there is some variation in the discharge patterns of the muscles, there is no obvious change in the activity of the muscles. Only the discharge pattern of the tibialis anterior was comparatively similar to the adult one. At about 1 year after learning to walk, the pushing-off motion of the foot is not yet seen, and the characteristics of infant walking still partially remain. Figure 2 is the myogram of the learning process of the other 2 year-old infant.

During the latter half of the second year (fig. 2a, b), the discharge of the vastus lateralis showed a marked reduction during the stance phase, and this pattern was frequently similar to the adult one. It shows that the unnecessary stabilization of the knee joint has begun to disappear. At the end of the second year (fig. 2c, d), the discharge of the gastrocnemius in the first half of the stance phase has disappeared and the discharge showed a burst which is similar to the adult one. In this period, the pushing-off motion of the foot was also recognizable from its forms, but the knee extension after the shock absorption of the leg in the stance phase was not well performed, as compared with that of adults. Therefore, it seems that the strong kicking motion of the leg has not yet been developed. The unnecessary contractions of the biceps femoris, the gluteus maximus and the sacrospinalis during the stance phase have tended to decrease or disappear, and these patterns have generally come to be similar to the adult pattern. From the forms, this seems to come from the decrease of the forward body sway.

From the beginning of 3 years of age, the discharge pattern of the rectus femoris and of the deltoideus pars spinata of the shoulder girdle was frequently similar to the adult pattern. In this period, it seems that hip flexion during the swing phase is not performed actively and the strength to propel the body has to be obtained by the pushing-off motion. The arm swing as a reflex movement has already been found. From about the end of the second year, the walking form has begun to lose the characteristics of infant walking and to acquire an adult walking form by use of the kicking motion.

When we observe the process of walking as it develops with the age of the subjects, many muscles still kept unnecessary partial contractions until the 6th year. But the pattern after the 7th year has become almost similar to the adult one, and the adult walking form is nearly accomplished.

Briefly speaking, at the period of transition to the adult pattern and the period of accomplishment, individual muscles indicated some degree of difference. It would seem that the period of transition to the adult pattern takes place from about the 3rd to the 6th years and the period of accomplishment at about the 7th year. Especially at about the 3rd year, the discharge pattern of many muscles participating in the movement of the leg, the spinal column and the shoulder girdle suddenly begin to be similar to the adult pattern. Therefore, we can conclude that the period around the 3rd year is the most important for the transition to the adult pattern.

Thus, from these data on the learning process of walking in 1- and 2-year-old infants and the process of walking developing with age, the following conclusions can be summarized.

1. As to the periods of transition to the adult pattern, that of the tibialis anterior was at about the 1st month after learning to walk, that of the vastus lateralis was at about the middle of the 2nd year, those of the gastrocnemius, the biceps femoris, the gluteus maximus and the sacrospinalis were at about the end of the 2nd year, those of the rectus femoris and the deltoideus pars spinata were at the beginning of the 3rd year. The period of accomplishment of the deltoideus pars spinata was at about the 4th year and those of the other muscles were at about the 7th year.

2. The stance percentage of a single walking cycle showed $61.9 \pm 10.5\%$ in the 1st month after learning to walk, and after that it decreased in accordance with the mastery of walking and the standard deviation became smaller than that of the initial period of learning to walk. At the 3rd and the 4th month after learning to walk, it became about 50% and after that it did not show any remarkable change. From the end of the 2nd year, the stance percentage increased with age. Over the 7th year of age, it was more than 50% and became almost 60% as in an adult.

3. Even when the infant was still unable to walk alone, from the 11th to the 12th month after birth the discharge patterns of both the tibialis anterior and the gastrocnemius already showed a reciprocal relationship when it walked using a supporting frame.

4. When, on the 380th day after birth, the infant succeeded in walking 2 or 3 steps for the first time, the discharge pattern which indicated the stabilization of the ankle joint and the maintenance of the squatting posture leaning forward with the knee and hip joint flexed, was observed in the stance phase.

The discharge pattern recorded in the swing phase indicated that the strength to propel the body was produced by the forward swing of the leg from this squatting position.

5. On the 10th day after learning to walk, the infant became able to walk more than 10 steps. In this period, the continuous discharge of the tibialis anterior during the stance phase began to decrease or disappear. On the 15th

day after learning to walk, a part of the discharge patterns began to be similar to the adult pattern and the unnecessary stabilization of the ankle joint began to disappear.

6. From the 50th to the 85th day after learning to walk, the infant became relatively accustomed to walking and the discharge patterns of both the tibialis anterior and the gastrocnemius began to show a reciprocal relationship.

7. At about the middle of the 2nd year, the unnecessary stabilization of the knee joint began to disappear.

8. At about the end of the 2nd year of age, the arm swing, the pushing-off motion of the foot and the decrease of the forward body sway, which are found in an adult, could be recognized. From this period, the walking form began to lose the characteristics of infant walking and to acquire those of the adult one by use of the kicking motion.

9. The adult walking form was nearly accomplished by the 7th year.

Author's address: Dr. Tsutomu Okamoto, Department of Physical Education, Kansai Medical School, *Hirakata-shi, Osaka* (Japan)

IV. Applied Biomechanics in Sport

Medicine and Sport, vol. 8: Biomechanics III, pp. 336–341 (Karger, Basel 1973)

Biomechanics of Sport

Emerging Discipline[1]

R. C. NELSON

Biomechanics Laboratory, Pennsylvania State University, University Park, Pa.

Biomechanics has become a well developed area of research within a number of disciplines concerned with human motion. Pioneering work in physical rehabilitation, anatomy, aerospace science, automobile safety, industrial engineering (human factors) and biomedicine has greatly expanded our knowledge of human biomechanics. Within the field of sport and physical education, however, biomechanics has progressed slowly and today remains relatively underdeveloped. This lack of development is characterized by a limited number of qualified researchers, a shortage of well equipped, productive laboratories, a small number of graduate programs training doctoral students and a general lack of identity in the scientific community. In spite of its current status, the writer believes that recent progress indicates that biomechanics of sport will develop rapidly in the 1970s and reach full maturity during this period.

I. Recent Advances

The 1st International Seminar on Biomechanics held in Zurich, Switzerland, in 1967 was a milestone for the emerging discipline of sport biomechanics. This seminar, which focused international attention on biomechanics of sport, provided the opportunity for many interpersonal contacts leading to increased communication among researchers throughout the world. As an outgrowth of these contacts, the 2nd Seminar was conducted 2 years later in Eindhoven, Holland, resulting in further interchange of information and ideas

1   Introductory paper presented at the 3rd International Seminar on Biomechanics in Rome, Italy, on September 30, 1971.

on an international basis. It became evident that other seminars dealing with specific areas of sport were needed. The Symposium on Biomechanics of Swimming in Brussels followed a year later. In October 1970, a biomechanics symposium was conducted at Indiana University, Bloomington, Indiana, USA: the first such meeting on biomechanics of sport in North America. This 3rd International Seminar in Rome, which represents a continuation of the tradition begun in Zurich is noteworthy for the number and quality of papers concerned with sports. Undoubtedly, continued expansion of research in sport will be evident on the program of the 4th Seminar to be held at the Pennsylvania State University, University Park, Pennsylvania, USA, in August, 1973.

In addition to increased communication, outstanding progress has been made during the past 4 years in the quality and number of research institutes and graduate programs which have been established. In the USA alone, doctoral programs supported by research facilities have been implemented at 8 universities while several other institutions are in the process of beginning such programs. In Canada, where no programs existed 4 years ago, there are a number of excellent programs being currently established. Active research programs are now in operation in Japan and research institutes are expanding rapidly in England, Belgium, East and West Germany, Switzerland, Austria, Finland, the Netherlands, Poland, Hungary, Russia and Czechoslovakia.

Although instrumentation systems and research methodology are not fully developed, recent progress has simplified many measurement problems in the study of sport performances. New intermittent, 16-mm cameras such as the LoCam, manufactured by Redlake Laboratories, Inc., Santa Clara, California, USA, provide adequate frame rates (up to 500/sec) for filming most sports movements. Interchangeable shutters, internal timing lights, portability and compactness make such a camera ideal for filming human movements. Modern film analysis systems such as the Vanguard Film Analyzer, made by the Vanguard Corporation, Melville, Long Island, USA, make possible accurate measurement of X, Y coordinate data, angle, and frame count from film. The more advanced systems now provide for automatic recording of these data on paper tape or punch cards in preparation for computer processing.

Digital computer technology has greatly changed many facets of biomechanics research. Calculations formerly done on rotary calculators are routinely performed by computer in much less time. Through skilled programming, it is possible to increase the number of experimental variables under investigation. These time-saving aspects also permit the inclusion of more subjects and conditions in experiments. Computer simulation of sports movements, currently in the early stages of development, offers a new dimension to research in biomechanics. The paper presented at this seminar on computer simulation of diving exemplifies the potential of this new technique. The application of 3-dimensional computer graphics display media makes possible even more effective use of simulation techniques.

The expansion of seminars, conferences and symposia on biomechanics, development of research laboratories and graduate programs, improvement in instrumentation and utilization of computer technology are representative of the recent developments in biomechanics of sport. It is the author's opinion that these factors have been of primary importance in stimulating interest in and expansion of the study of the biomechanical aspects of sports and will form the basis for continued development during the next decade. What direction and form the maturation of this young discipline will take is uncertain. The following ideas are a few possibilities based on the author's experience and his discussions with colleagues.

II. Future Development

It is essential that the number of research laboratories and graduate programs be expanded over the next few years. Of special importance is the development of a large number of highly qualified researchers who will form the basis for continued development. Modern, completely equipped laboratories will be needed to stimulate research and provide the necessary facilities for training of researchers.

As the quantity of research in biomechanics of sport increases it will be essential that results and findings become more easily available. Unfortunately, research in biomechanics is widely dispersed throughout many scientific journals, making it difficult and time-consuming to locate the information. This problem could be somewhat alleviated if an international journal of sport biomechanics were to be established. Publication in many languages of general textbooks on biomechanics of sport and specialized books of biomechanical aspects of selected sports are needed to improve communication and greatly assist teaching in this field.

Scientific meetings which have been so essential to past growth must be expanded. International meetings should continue on a biannual basis while regional and national meetings should be scheduled during alternate years. Some of these smaller meetings should focus on specific sports with inter-

action among practitioners and researchers. Workshop and conference programs should also be devoted to recent advances in instrumentation, provide
instruction and practical experience in utilization of research equipment and
stimulate interest in teaching biomechanics of sport. Certainly this discipline
should become an integral part of the scientific program of the Olympic
Games. Tentative plans have already been initiated to include biomechanics
in the program of the scientific congress at the 1976 Games to be held in
Montreal, Canada.

As the biomechanist becomes accepted among human performance researchers, he must be prepared and willing to participate in team research.
This team approach will lead to a better understanding of the 'whole man'
engaging in sport activity since the physiological, psychological, biomechanical, sociological and even anthropological aspects can be simultaneously
investigated. The contribution to be made by the biomechanist will be of
primary importance to the success of such cooperative research.

As research in biomechanics of sport becomes more theoretical and sophisticated it is imperative that close interaction be maintained with the applied
field of sport and physical education. It is through dissemination of research
findings and application of results to improve human performance that biomechanics can make a significant contribution to sport. If, in the future,
researchers work more closely with trainers and coaches it is the author's
opinion that the next major improvements in sports performance will occur
as a consequence of applying the results obtained from biomechanics research.

III. Olympic Archives

As the research activity in biomechanics of sport increases, it will be
possible to conduct team research involving scientists from a number of countries. This approach will be of special value when international sports competitions are the sites for the collection of data. It is the writer's hope that
international cooperation will ultimately lead to a research program conducted in conjunction with the Olympic Games. Such a project would make
a significant contribution to the scientific documentation of Olympic performances and focus attention on biomechanics of sport.

This concept of studying Olympic athletes is not new, as they have served
as subjects for extensive research by sports medicine specialists, anthropol-
ogists, psychologists, sports sociologists, physiologists and sport historians.
This work has led to a variety of scientific publications and to a better under-

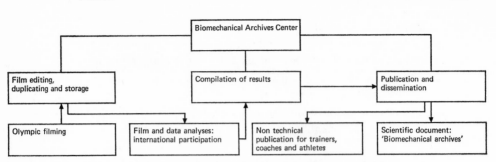

Fig. 1. Proposed organizational plan for a biomechanical archives of Olympic performance.

standing of these outstanding athletes. Although this research has been of obvious scientific merit, it is often of questionable value to the athletes or their coaches since they seldom read research publications and the results are frequently of little or no practical value to performance.

Since biomechanics research deals specifically with technique and the quantitative aspects of performance, a collection of films and results derived from analyzing the films should be of primary interest to coaches and trainers as well as to the sportsmen. A general plan for a biomechanical archives, shown schematically in figure 1, would include high-speed 16-mm films taken during the actual competition, if possible, or in some instances during practice sessions. The events would necessarily be limited to those which can be analyzed with cinematographic procedures. The most obvious would be gymnastic and diving events, field events in track and field and some running events.

The original films could be housed at a designated location and duplicates produced for distribution. Ideally, a number of research laboratories throughout the world would cooperate and participate in the film analysis and data reduction aspects of the program. Perhaps each research group could specialize in the study of certain events. The results of these analyses would be collected and stored at the archives center. A scientific document might be completed which, along with the films, would constitute the official *archives*. In addition, practical findings would be made available to coaches, trainers and athletes of all countries for use in improving performances. As these biomechanical records are accumulated over the years, it will be possible to assess the changes in techniques and style which have occurred. By the time such an archives is well established, biomechanics of sports will be fully recognized as a discipline which makes significant contributions to certain aspects of society.

Summary

In summary, the young discipline of biomechanics of sport has developed very rapidly in the past few years and it is suggested that accelerated growth will occur in the 1970s. The first steps in this process will be the expansion and development of research facilities and graduate training programs. This will produce the scholars needed to increase the research activities and implement results of this research. Further stimulation will result from an increased number of seminars, conferences and symposia which will lead to the complete acceptance of biomechanics of sport as an area of scientific study. Symbolic of this acceptance might be the establishment of an official biomechanical archives of Olympic performance.

Author's address: Dr. RICHARD C. NELSON, Biomechanics Laboratory, The Pennsylvania State University, *University Park, PA 16802* (USA)

Medicine and Sport, vol. 8: Biomechanics III, pp. 342–348 (Karger, Basel 1973)

Analysis of Running Pattern in Relation to Speed

T. Hoshikawa, H. Matsui and M. Miyashita

Department of Physical Education, University of Nagoya, Nagoya

The principal aim in running is to move one's body as fast as possible. One of the limiting factors in running speed is the energy output mechanism in man; therefore, many investigators have studied oxygen consumption during running. On the other hand, Fenn [1930] pointed out that the greater part of energy expended in running went into accelerating and decelerating the limbs, particularly the lower limbs. Taking this conclusion into consideration, it is important to know how to move one's limbs whilst running.

The purpose of this study is to investigate the running patterns adapted to the changes of progressive speed from the biomechanical point of view. The subjects in this experiment were 8 male adults, 1 excellent runner who took part in a 400-meter race of the Olympic Games in 1964, 4 average runners who had participated in interschool games and 3 poor runners who were not sportsmen. The experiments were carried out on a treadmill. The subjects were tested at 7 different speeds ranging from 200 m/min to 500 m/min with speed intervals of 50 m/min.

I. The Analysis of Temporal Pattern

A special transducer was developed for recording a pressure exerted by the foot. From the recordings a swing phase time, a support phase time and a stride phase time were measured. As the speed was increased, those phase times were reduced, with the support phase time showing the greatest reduction. Ratio of the swing to the support phase time was approximately 0.2 at 500 m/min. The time for one stride of the excellent runner was longer than the time of the poor runners at every speed.

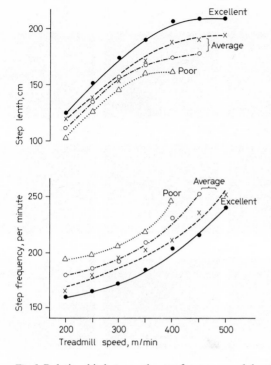

Fig. 1. Relationship between the step frequency and the step length and running speed.

A step frequency and step length in relation to running speed were calculated (fig. 1). A progressive speed (V m/min) in running is the product of step frequency (F steps/min) and step length (Lm) as is shown below:

$$V \text{ (m/min)} = (F \text{ steps/min}) \cdot (Lm) \tag{1}$$

Furthermore, in order to consider the case of increasing speed, if one expresses an increment of progressive speed by Δv, that of step frequency by Δf and that of step length by Δl, respectively, equation 1 may be expressed as follows:

$$(V + \Delta v) = (F + \Delta f) \cdot (L + \Delta l) \tag{2}$$

$$\Delta v = F \cdot \Delta l + L \cdot \Delta f + \Delta f \cdot \Delta l \tag{3}$$

Equation 3 means that the increment of progressive speed in running is supplied by $F \cdot \Delta l$, $L \cdot \Delta f$ and $\Delta f \cdot \Delta l$. The responsibility of each factor for Δv was calculated. At relatively lower speed to the subject, Δl was dominant; at higher speed, however, a greater part of Δv was compensated with Δf.

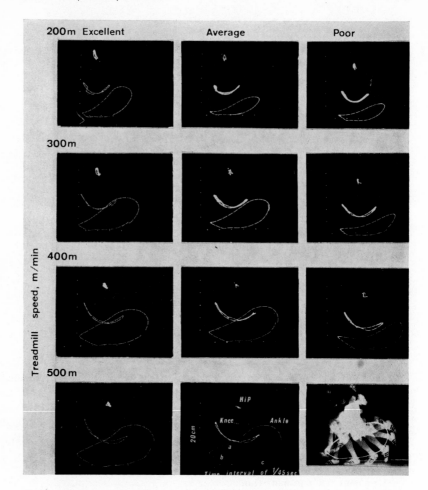

Fig. 2. Effect of speed on the pattern of leg movement.

II. The Analysis of Leg Movement

Running is a series of rotatory movements of one segment upon another. We tried to analyze leg movement during running. Firstly, the displacement of hip, knee and ankle were measured by a photographic method aided by the interrupted and stroboscopic procedures (fig. 2).

As the speed was increased, the displacements of knee and ankle became larger, but that of the trunk decreased. Comparing the displacement between

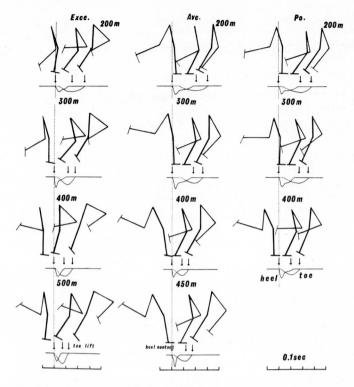

Fig.3. The leg movement during support phase.

the subjects the excellent runner's displacement was larger than the poor runner's at a given speed.

Furthermore, the velocities of the foot at the reference points a, b and c during one cycle were measured. As the speed increased, the velocity of the foot at a, b and c increased considerably with speed. Making a comparison between subjects, there was a clear difference in the speed of the foot. The excellent runner had a higher speed than the poor runner, especially at the moment when the heel makes contact with the ground.

Secondly, the range of movement and the angular velocity of the knee joint were measured by electrogoniometers. The angle range became considerably wider with speed, with the exception of the ankle joint. In the case of the excellent runner the angle movement was larger than in the case of the poor runner. The angular velocities of the leg joint were calculated from the goniograms. The angular velocities increased remarkably with speed, especially

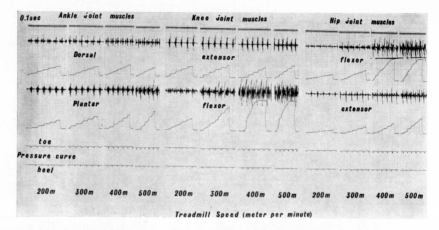

Fig. 4. The EMG and its integration of the leg muscles during running at various speeds for the excellent runner.

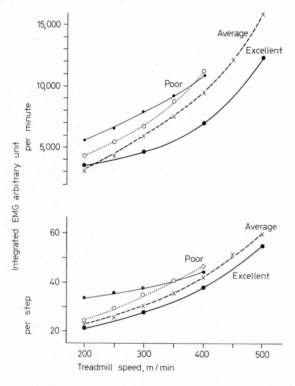

Fig. 5. Integrated EMG showing arbitrary units in relation to speed.

the ankle extension at the toe lift, the knee flexion immediately after the toe lift, the hip flexion at the opposite foot in the contact and the knee extension at the toe lift of the opposite foot. The excellent runner had a higher angular velocity than the poor runner at a given speed.

Based on data from the goniogram, the leg movements at the support phase were illustrated (fig. 3). This drawing shows visually the surprising fact that there is a difference in the leg movement of the excellent runner and of the poor runner even at a lower speed which could be run easily by every subject. Namely, the knee of the driving leg is pushed up high in front of the body and the recovery foot makes rapid progress. These changes in the angle action were in accordance with the observations of the photographic investigation.

III. The Analysis of Muscular Activity

A leg movement depends upon the muscular activities of leg muscles. The direct relation of muscular activity to running speed was revealed quantitatively by means of EMG integrators. Figure 4 is a typical recording of the excellent runner. From the recording it can be seen that the integrated EMG increased remarkably with speed. Figure 5 shows that there is a nonlinear correlation between the muscular activity and speed. Furthermore, this graph represents the difference of muscular activity among individuals at a given speed. Namely, the muscular activity for the excellent runner is lower than that of the poor runner at a given speed.

References

1   BIGLAND, B. and LIPPOLD, O.C.J.: The relation between force, velocity and integrated electrical activity in human muscles. J. Physiol., Lond. *123:* 214–224 (1954).

2   CAVAGNA, G.A.; SAIBENE, F.P., and MARGARIA, R.: Mechanical work in running. J. appl. Physiol. *17:* 249–256 (1964).

3   FENN, W.O.: Frictional and kinetic factors in the work of sprint running. Amer.J. Physiol. *92:* 583–611 (1930).

4   FENN, W.O.: Work against gravity and work due to velocity change in running. Amer. J. Physiol. *93:* 433–462 (1930).

5   HUBBARD, A.: An experimental analysis of certain fundamental differences between trained and untrained runners. Res. Quart. amer. Ass. Hlth phys. Educ. *10:* 28–38 (1937).

6   GOLLNICK, P.D. and KARPOVICH, P.V.: Electrogoniometric study of locomotion and

of some athletic movements. Res. Quart. amer. Ass. Hlth phys. Educ. *35:* 357–369 (1964).

7   FINLEY, F. R. and KARPOVICH, P. V.: Electrogoniometric analysis of normal and pathological gaits. Res. Quart. amer. Ass. Hlth phys. Educ. *35:* 379–384 (1964).

8   LIPPOLD, O. C. J.: The relation between integrated action potentials in a human muscle and its isometric tension. J. Physiol., Lond. *117:* 492–499 (1952).

9   SARGENT, R. M.: The relation between oxygen requirement and speed in running. Proc. roy. Soc. B *100:* 10–22 (1926).

Author's address: TAMOTSU HOSHIKAWA, Department of Physical Education, Aichi Prefectural University, *Nagoya City* (Japan)

Medicine and Sport, vol. 8: Biomechanics III, pp. 349–353 (Karger, Basel 1973)

A Study on Relations between Physical Performance and Physical Resources

M. Miyashita, M. Miura, K. Kobayashi and T. Hoshikawa

Department of Physical Education, University of Nagoya, Nagoya

The present study was intended to examine experimentally the relations between physical performance and physical resources in running 5,000 m. The speed during the 5,000-meter run is one of the indices of physical performance in endurance events. Since the total energy required to run 5,000 m depends to almost 80% upon the oxygen intake during running, the maximum oxygen intake is a direct index of the physical resources. Therefore, the values of the maximum oxygen intake of the long-distance runners were analyzed in relation to the speed in their best 5,000-meter run.

Procedure

The subjects in this experiment were 10 untrained males between 22 and 35 years of age, and 45 trained males between 19 and 23 years of age. The trained subjects had undertaken running training for several years (the total running distance is more than 10,000 m every day).

The maximum oxygen intake was measured during the maximum run on the treadmill. Expired air during running was collected in Dauglas bags each minute of the run and analyzed by a Beckman oxygen analyzer. The heart rate was calculated from an ECG during running, and the respiratory rate was determined with the aid of a thermister attached to the inside of the mask. All subjects ran at least 4 min before they reached exhaustion.

Results

Since heart rate calculated from the ECG was over 180/min and the respiratory rate was 50 to 70 during the minute preceding exhaustion, the

Table I. The maximum oxygen intake in order of the best record of a 5,000-meter run

Record	N	Speed, m/sec	Height, cm	Weight, kg	Max. oxygen intake l/min	ml/kg·min
14:00 ~ 14:59	10	5.65 (0.08)	168.0 (5.2)	55.9 (3.4)	3.89 (0.23)	70.1 (4.4)
15:00 ~ 15:59	13	5.36 (0.07)	169.1 (5.1)	56.8 (5.0)	3.71 (0.41)	65.2 (5.1)
16:00 ~ 16:59	14	5.04 (0.09)	166.4 (5.7)	56.4 (4.8)	3.48 (0.32)	62.1 (5.2)
17:00 ~ 17:59	8	4.78 (0.08)	168.8 (6.4)	56.2 (6.6)	3.28 (0.36)	58.6 (3.4)
19:56 ~ 28:41	10	3.63 (0.48)	165.4 (5.4)	62.6 (6.3)	3.15 (0.28)	50.6 (4.8)

Standard deviations in parentheses.

maximum oxygen intake achieved in this experiment can be considered to be the true maximal value.

There was not much difference in age, body height and body weight among the subjects engaged in this experiment. The mean values of maximum oxygen intake were presented in the order of the best record of the 5,000-meter races (table I). There is a clear trend that the greater the maximum oxygen intake, the better the best record of a 5,000-meter run.

Going into details, however, there is a certain amount of individual difference in the mean speed of 5,000 m runs among subjects who showed the same ability of oxygen intake.

The values obtained for trained subjects in this experiment are shown together with data reported by the other investigators [SALTIN and ÅSTRAND, 1967; KAGAYA, 1970; AOKI, 1970] in figure 1. The ordinate (Y) is the speed of the best 5,000-meter run, and the abscissa (X) is the maximum oxygen intake per body weight. The regression equation and its standard deviation were calculated between the mean speed and the maximum oxygen intake as is shown below.

$$Y = 0.0431X + 2.50 \pm 0.232.$$

These results suggest that the individual difference in performance among those runners who have the same physical resources is within approximately

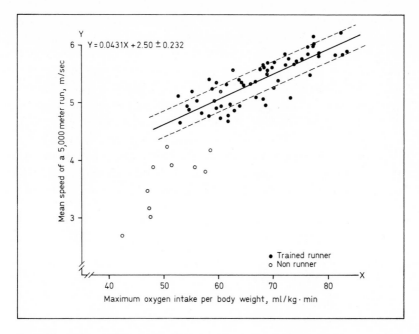

Fig.1. The relationship between the mean speed of a 5000-meter best run and the maximum oxygen intake per body weight.

0.46 m/sec. The values obtained for the untrained subjects are illustrated in the same figure (fig. 1). Their running performance is lower than that of the trained subjects.

Discussion

KARPOVICH [1959] pointed out the possibility of predicting how long strenuous work can be continued; namely, it depends on the man's ability to take up oxygen during work and his ability to accumulate an oxygen debt. On the other hand, he indicated that the availability of oxygen was the decisive factor in the winning of a race. The availability of oxygen is the important factor in running performance. In this study, the authors tried to show the effect of the techniques quantitatively in long-distance running under the condition that the subjects were well trained.

In principle, man converts chemical energy into kinetic energy through movement. Therefore, the amount of chemical energy must be taken into consideration in order to know the effect of the energy convertion through movement. The physiological (chemical) energy is divided into two items; oxygen intake (Oi) and oxygen debt (Od). CAVAGNA *et al.* [1964] reported that the work done against gravity in running seemed to be constant and the work due to speed changes in forward direction increased with speed. Those works were expressed in kcal/kg/min. Therefore, it was assumed in this study that the mechanical work done in running at a certain speed increases in proportion to the body weight. If the man (weight, W) runs a certain distance (L) within a given time (T), the mechanical work is given as $k \times W \times L$ (k is constant) and physiological energy is given as $Oi \times T + Od$. Since the physiological energy used in running is equal to the external work, the following formula is obtained.

$$k \times W \times L = Oi \times T + Od.$$

Dividing by T·W gives,

$$k \times L/T = Oi/W + Od/W/T,$$

where L/T is speed (V). Then, in the other words,

$$V = 1/k \times (Oi/W + Od/W/T).$$

Therefore, the maximum oxygen intake per body weight is an appropriate index of the physical resources for a 5,000-meter run.

In this experiment, the linear relation was shown between maximum oxygen intake and mean speed at 5,000 m best run. There were, however, large individual differences among runners on the same level of physical resources. For instance, there were two runners, A and B, of the same ability to take in oxygen (72 ml/kg/min). 'A' had a best time of 14 min 42 for the 5,000-meter run, but 'B' had 16 min 23. This difference in performance is influenced by the so-called 'technique', 'skill' or 'art' of running. One coach said that the art of distance running lies in distributing one's energies as efficiently, as completely and as rapidly as possible [DOHERTY, 1960]. Logically there must be an ideal form which produces a perfect balance between maximum speed and maximum economy of energy. But each runner has his own style with his natural and acquired habits. The style of running consists of 6 elements: the overall action, body angle, arm swing, foot placement, rear-leg lift and length of stride. Therefore, when little difference in each element is accumulated, it will remarkably affect the performance, as has already been mentioned in this experiment. Performance can be improved by a certain amount of training or practice, but there still exist some differences in performance. The results obtained in this experiment revealed the amount of differences in running performance under the condition that the subjects were well trained runners. The maximum difference in this study is approximately 0.46 m/sec.

The motion of running is composed of various kinds of factors. Therefore, a theoretical investigation of a means to systematize such complicated phenomena is important. The authors believe that the results of this study will be useful for further investigations on running.

References

AOKI, J.: Personal communication (1970).

CAVAGNA, G. A.; SAIBENE, F. P., and MARGARIA, R.: Mechanical work in running. J. appl. Physiol. *19* (2): 249–256 (1964).

DOHERTY, J. K.: Modern track and field (Prentice Hall, Englewood Cliffs 1960).

KAGAYA, H.: Personal communication (1970).

KARPOVICH, P. V.: Physiology of muscular activity (Saunders, London 1959).

SALTIN, B. and ASTRAND, P. O.: Maximal oxygen uptake in athletes. J. appl. Physiol. *23* (3): 353–358 (1967).

Author's address: Dr. MITSUMASA MIYASHITA, Department of Physical Education, University of Nagoya, *Nagoya-City* (Japan)

Medicine and Sport, vol. 8: Biomechanics III, pp. 354–358 (Karger, Basel 1973)

Mechanics of Distance-Running during Competition

MARLENE ADRIAN and ELLEN KREIGHBAUM

Biomechanics Laboratory, Washington State University, Pullman, Wash.

Since research on the mechanics of running has been conducted primarily under non-competitive situations and the published data are descriptive in nature or quantified from films of the side view of runners, the following investigation was conducted to determine differences in foot placement, changing angles of the leg and asymmetrical movement of body parts. Front views of first and last phases of competitive running were used.

13 members of the United States 1970 Olympic Long Distance Training Center were selected for subjects. Runners were competing in a 24-hour relay in which each member ran a total of 27–31 mi in 1-mi intervals. Average times for these runners were 4:42.14 to 5:10.7 min.

Anthropometric data were collected as shown in table I. Comparisons with research of TANNER [3] and COSTILL et al. [1] showed that these runners were taller and heavier but otherwise typical of distance-runners. Somatotypes were obtained using the Carter-Heath anthropometric scale. The endomorphic component of these runners was less than that reported by Tanner using another technique. Observation of the running pattern provided no evidence that style of running was related to the anthropometric data. No predictive relationship was found between height and length of stride in this study.

All subjects were filmed with a 16-mm camera having a telephoto lens and operating at a speed of 64 frames/sec. Front views were taken for condition I (miles 10–12) and condition II (miles 28–30). Since the runners were performing in competition, no body landmarks were artificially used; however, the shapes of muscles, bones and other body landmarks were readily identifiable.

Differences were recorded by subjective categorization of deviations using a stop-action projector and a film reader, and a computerized graphic tablet system consisting of a film projected onto a tablet connected to a miniature computer. Points touched with the tablet pen were stored in the computer as coordinate points. All analysis was done by the computer. Reliability for

Table I. Anthropometric measures

Measures	Mean (N = 13)	Standard deviation
Body and skeletal		
height, dm	17.8605	0.4876
weight, kg	68.2938	3.8362
Ponderal Index	13.2292	0.3234
Body density	1.0848	0.00906
Body fat, %	6.9868	1.4765
Lean body weight, kg	61.45	4.594
Biacromial, cm	41.1962	1.3080
Bitrochanteric, cm	33.2000	1.4448
Bi-iliac, cm	29.8769	2.9529
Elbow, cm	7.1615	0.2507
Knee, cm	10.8385	0.4775
Thigh length, cm	41.3727	1.9221
Leg length, cm	45.2969	2.5509
Calf girth, cm	35.7985	3.4072
Biceps girth, cm	30.1392	3.5889
Skinfolds, mm		
Subscapular	6.7615	0.9305
Abdomen	6.1154	1.0442
Triceps	5.0385	1.0055
Calf	4.0308	0.6694
Suprailiac	4.1615	0.5919
Somatotype	1.5, 5, 3.5	

both forms of analysis was found to be satisfactory by a second analysis 2 months later.

Head rotation, head lateral flexion, shoulder rotation and trunk lateral flexion were classified as slight, moderate or heavy. Generally, results showed that the head rotation did not change from conditions I to II. Right head flexion increased in half of the subjects, but only two showed an increase in left head lateral flexion. Shoulder rotation showed increases, decreases and no change, with more subjects in the last category. The same was found for trunk lateral flexion. Foot and arm placements were measured according to whether or not each was short of, at or beyond the midline of the body. A majority of runners did not change their arm swing, and observed changes did not follow a pattern. The runners all placed their feet on or beyond the line of progression in condition I. This agrees with the findings of SLOCUM and JAMES [2]. Four runners who were at the line of progression overstepped

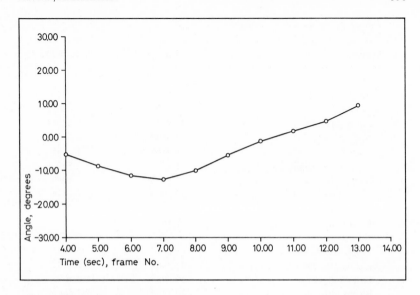

Fig. 1. Typical plot of computer print-out for foot angle during the support phase of running. Frame 4: landing; frame 13: take-off.

the midline with their right foot during condition II. One increased the lateral distance between his feet.

Computer analysis scores were plotted to show angular changes of front view right and left head, leg, thigh and foot angles with the perpendicular during the support phase (fig. 1). In general, the foot was perpendicular or everted at the time of take-off. The thigh and lower leg were perpendicular at take-off but showed an angle with the vertical at landing. Leg-thigh relationships at landing showed no typical pattern. In some cases the leg was closer to the vertical than the thigh; in some cases the reverse was true, and in other cases the angle was the same. Differences between right and left angles were noted in the majority of runners.

Results from the plots of 8 runners showed that during condition II, 4 had greater head deviation from the vertical than during condition I. The thigh angle increased in 3 runners, decreased in 2 and remained the same in 3 runners. The leg angle increased in 2, remained the same in 4. The foot position at take-off showed greater eversion in 4 runners during condition II. Contours of 1 runner with angular measurements and force lines through the hip are shown in figure 2. The greater deviation from the vertical of the right thigh can be noted for this runner.

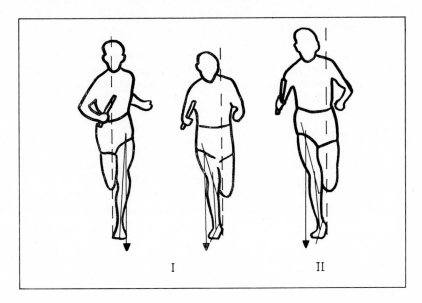

Fig. 2. Angles at initial weight bearing for conditions I and II. Leg, thigh, head, foot-line ———; vertical - - - - - - -; line of force through hip ——▶ .

And during condition II, greater shoulder rotation and lesser angle of lower leg and thigh with the vertical were measured. The change in lower leg-thigh angle decreased the torque acting on the knee, lower leg and ankle and corresponded with a decrease in trunk lean. In decreasing the torque, less stress was placed upon the bones and joints of the lower limb. A shoulder rotation may have facilitated this decrease in torque as well as the shortening of the stride. With a reduction in speed of running there is less effort required to receive the impact of foot contact and, therefore, a more vertical leg is advantageous.

One runner displayed an increase in bowing of the lower legs during condition II which placed a greater moment on the bones of the lower leg. SLOCUM and JAMES [2] indicated that minor bowing is consistent with good running. The left foot of 2 runners and both feet of 1 runner showed flat-footedness.

The causes of individual adjustment variations as well as individual asymmetry in running patterns noted from conditions I and II need to be investigated. The relationship of leg dominance, postural abnormalities and the condition of the running clockwise on a circular track should not be excluded from further study.

Acknowledgements

Acknowledgements are given to DENNIS HUSTON and ROBIN FRY for their assistance with the collection of data. The computerized graphic tablet system was developed from initial funding by the Washington Association for Retarded Children Trust Fund.

References

1   COSTILL, D.L.; BOWERS, R., and KRAMER, W.F.: Skinfold estimates of body fat among marathon runners. Med. Sci. Sports 2 (2): 93–95 (1970).
2   SLOCUM, D. and JAMES, S.: Biomechanics of running. J. amer. med. Ass. 205 (11) (1968).
3   TANNER, J. M.: The physique of the Olympic athlete (Allen & Uniwin, London 1964).

Author's address: Dr. MARLENE ADRIAN, Biomechanics Laboratory, Washington State University, *Pullman, WA 99163* (USA)

Medicine and Sport, vol. 8: Biomechanics III, pp. 359–363 (Karger, Basel 1973)

The Treadmill used as a Training and a Simulator Instrument in Middle- and Long-Distance Running

A. Dal Monte, S. Fucci and A. Manoni

Institute of Sports Medicine, Italian Olympic Committee, Rome

Introduction and Purpose of the Work

The athlete's functional evaluation seems more useful when carried out during performances which are as similar as possible to those of specific training or of competition. The purpose of this experiment is to ascertain whether the motor-driven treadmill may simulate middle- and long-distance running in the laboratory under the double aspect of energy cost and form of movement.

For the kinematic study of movement the following instruments were used: (1) high-speed cine-camera: 100 photograms/sec; (2) mobile mirror, size 2 × 1.50 m²; (3) motor-driven treadmill with variable inclination and speed, and (4) chronometer.

For the study of energy cost the following instruments were used: (1) cardio-frequencymeter with biopotential electrodes fixed to the individual; (2) portable gas meter (Max-Planck type), with bags in which to collect samples of expelled breath at a percentage of 0.6; (3) equipment for the quantitative analysis of breathing gas; (4) chronometer; (5) mask with reduced dead space volume, equipped with inhalation and exhalation valve of very low resistence to the air flux, and (6) flexible tube of constant diameter for the canalisation of the exhaled air going into the gas meter.

The method used in the operation was the same both for the treadmill and track tests. The test on track were effected with a van to which the necessary modifications for the use of the scientific equipment were carried out.

The experiment was performed with 3 middle-distance athletes, of average athletic performance, who in addition to the specific training had 24 weeks of treadmill running for a total of 30 h of training each. Tracing spots were marked on the 3 athletes in correspondence with the articulation links of the

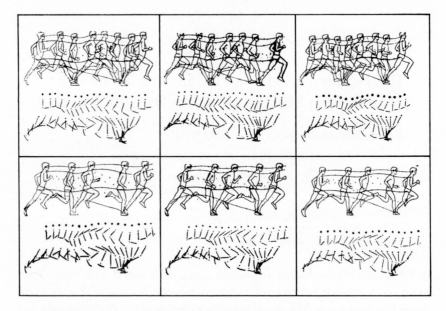

Fig. 1. Curves of movement analysis obtained at 20 km/h. In the upper 3 positions, subjects *a*, *b* and *c* are running on a track. In the lower 3 positions, the same subjects are running on tread mills. Speed in all cases, 20 km/h.

shoulder, elbow, hand, hip, knee and ear lobe. To get a correct evaluation of the angular movement of the limbs, adhesive tapes were applied along the bone segments.

The evaluation of data on track was made by using a cine-camera, carried on a car which was driven beside the athlete at the same speed, to avoid parallax errors. Timing control was used to ensure that the speed was as previously established (athletes had been trained beforehand to maintain the required rhythm). The speeds adopted were: 15, 18 and 20 km/h which correspond to training and competition speeds for athletes whose performance is 4 min on the 1,500 m and 15 min on the 5,000 m. The evaluation of the data was carried out in the laboratory using the reflection of the image on a mirror placed at an angle of 45° to the running direction. This artificial method was used to have the same shooting distance and to be able to film the athlete both side- and frontwards simultaneously.

For the evaluation of the energy cost, cardiocirculation, breathing and metabolic tests were carried out both in the laboratory and on the track at the 3 adopted speeds: 5, 18 and 20 km/h.

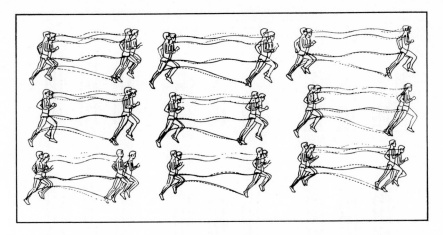

Fig. 2. Comparison between the curves obtained from subjects *a*, *b* and *c* at 15, 18 and 20 km/h.

The parameters measured were as follows: (1) cardiac frequency (f_h); (2) ventilation per minute ($\dot{V}$); (3) oxygen uptake ($\dot{V}O_2$); (4) quantity of carbon dioxide exhaled ($\dot{V}CO_2$); (5) respiratory quotient (R); (6) pulse of oxygen ($\dot{V}O_2/f_h$), and (7) ventilation equivalent (REO_2).

Analysis of the Data

The two running types were compared by using a graphical method. The trajectories of two chosen points during the stride were compared and the difference was recorded. To do this, a moviola with reflector was used with which the necessary photograms were obtained to gain an exact idea of the movement. In order to compare the two curves, the drawings of these two trajectories were laid the one upon the other and the difference observed. The duration was measured in hundredths of a second using a camera which took 100 photograms/sec.

The running phases were conventionally defined as follows:

1. *Aerial phase.* The period during which the body is not in contact with the ground.

2. *Dumping phase.* The period from the moment the foot touches the ground to the moment when the hip articulation passes the external malleolus perpendicular.

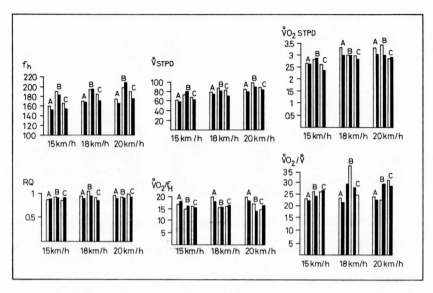

Fig. 3. The histograms show differences between the parameters explored on track (□) and treadmill (■), respectively. Upper left, the heart frequency; upper centre the ventilation; upper right, the oxygen consumption. Lower left shows respiration quotient (RQ). Lower middle shows oxygen pulse. Lower right shows oxygen, $\dot{V}O_2/\dot{V}$.

3. *Extension phase.* The period from the moment when the hip articulation passes the malleolus perpendicular, to the moment in which the foot touches the ground.

The data thus obtained were represented in table form in order to make the results easier to compare. As regards the analysis of the cardio-circulatory, respiratory and metabolic values, these parameters were then put into graphic form for comparison.

The 3 tests on the track and in the laboratory were carried out with the same instruments and on the same day in order to avoid variations in the athletes' physical condition.

Comment and Conclusion

In both cases, differences were noticed in the form of movement at the considered speed: 15, 18 and 20 km/h; however, the running technique on the treadmill is not worse than that on the track because on the treadmill the movement in the vertical plane is reduced.

The differences between the two types of running tends to diminish with the increase of speed during the stride (probably a prolonged training on the treadmill could make the trajectory similar during the stride).

During the various running phases, it was observed that the extension phase and the stride were shorter on the treadmill than on the track. The energy used and the progress of the various cardiocirculatory, respiratory and metabolic parameters studied do not present relevant differences in the two cases. This is acheived through maintaining the tape at an inclination of 5% at the lowest speeds and at an inclination of zero at higher speeds. The similarity between track and treadmill running, both in energy cost and in movement, increases as the speed increases. Naturally, this is acheived after sufficient training on the treadmill.

It can be concluded that the treadmill may be considered, although within certain limits, a specific simulator of middle-distance running, as there is no outstanding difference, kinematic or energy cost, from the track running as long as it is adopted at a speed close to that of competitions.

The instrument can be used for training, for it allows control of proper breathing mechanics through the recording of lung ventilation and of pneumotachograms, heart frequency, ECG and oxygen consumption, which can only be measured in a laboratory and not on the track.

Author's address: Prof. ANTONIO DAL MONTE, Istituto di Medicina dello Sport, Via dei Campi Sportivi, 46, *I-00197 Rome* (Italy)

Medicine and Sport, vol. 8: Biomechanics III, pp. 364–369 (Karger, Basel 1973)

Comparison of the Kneeling and Standing Sprint Starts

A Kinematographical Analysis Incorporating Electromyography

M. Desiprés

Research Laboratory, Department of Physical Education,
University of the Orange Free State, Bloemfontein

The kneeling start is basically very stable because of the following factors: (1) the support base is large (2 hands and 2 feet well spread), and (2) the center of gravity is low and well within the base of support.

During excessive body lean with the center of gravity closer to the hands, rapid movement of the feet is necessary to catch up the falling bodyweight. Short strides bring about optimal mechanical advantage of the leg joints through a small range of motion. This experiment is limited to acceleration during the first 1.0 sec only. Although the kneeling start replaced the standing start in 1887, it has been varied. A modernized standing start technique has been used in the Republic of South Africa during the past few years.

The aim of this experiment was to determine the merits and shortcomings of the kneeling start *versus* the standing start. 17 athletes (10 male and 7 female) were used in the experiment. All were sprinters. Two male athletes had been credited with best times of 10.4 sec (kneeling start) and one female with 11.4 sec (also kneeling start). The others were sprinters of average performance having participated in sprints for a number of years.

Kinematography and Reaction Time

Reaction time of foot release was determined with the use of electronic timing equipment consisting of starting blocks, electric stop-watches and a starter's pistol which activated the watches which were tripped after release of pressure on block. Kinegrams were drawn of the movement of the center of gravity which was determined using Dähne's method. Body links were measured anthropometrically prior to the tests. Link centers of gravity were determined with stencils using Dempster's determinations.

Electromyography

Surface electrodes were attached to the musculi quadriceps femoris, transversely (vastus medialis and vastus lateralis, 15 cm above the patella); hamstrings on the biceps femoris and semi membranosis above the tendons, and gastrocnemius, transversely (medial and lateral bulges, 5 cm above the start of tendo Achilles).

Findings

In the following table we see a summary of the findings:

Table I

Parameter:	Kneeling	Significance
Reaction time rear foot	faster	$p = 0.05$
Reaction time front foot	faster	$p = 0.40$
Length first stride	shorter	$p = 0.02$
Time for first stride	shorter	$p = 0.01$
Length second stride	shorter	$p = 0.01$
Time second stride	shorter	$p = 0.05$
Length third stride	shorter	$p = 0.01$
Time third stride	longer	$p = 0.80$
Time to cover fixed distance	less	$p = 0.01$
Average velocity of C.G. at 1.0 sec	greater	$p = 0.01$

It appears that the reaction time of the rear foot is shorter in the kneeling start. This can be attributed to the center of gravity being closer to the hands which is a support not to be found in the standing start. The same holds good for the front foot. For the same reason the time for the first stride is shorter kneeling so as to catch up with the body mass.

Rectilinear horizontal average velocity of the center of gravity is greater. Time and distance as well as velocity of the second stride has the same pattern, in favor of the kneeling start. In the kinegram depicting rotary motion of the ball of the foot in relation ot the hip joint center during the kneeling start, the rear foot is in contact with the block for a significantly shorter period than in the standing start.

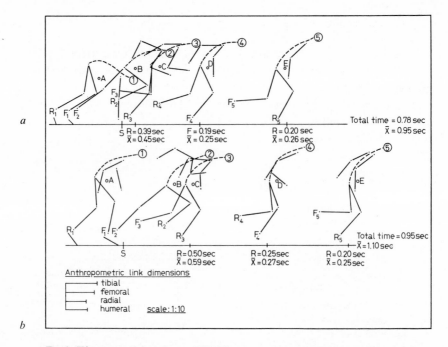

Fig. 1. Kinegram sprint start. *a* Kneeling: reaction time 0.13 sec; R = rear foot, 0.187 sec, x̄ = 0.221 sec; F = front foot, 0.390 sec x̄ = 0.407 sec. *b* Standing: reaction time 0.16 sec; R = rear foot, 0.193 sec, x̄ = 0.265 sec; F = front foot, 0.485 sec, x̄ = 0.427 sec. S = start.

The front foot remains in contact with the block significantly longer in the kneeling start and, therefore, force can be applied longer to facilitate more forceful reaction. During the first stride the rear foot is in contact with the ground for 0.13 and 0.11 sec, respectively.

Quantitative electromyography was done to try and determine the amount of muscular activity present mainly in the 'set' position. Regarding the rear foot, it appears that all 3 muscle groups liberated some 20% more electric potential during the standing start. In the case of the front foot all 3 muscle groups were still more active.

Conclusion

Within the limitations of this study, it appears that there are marked differences between the two starts during the first 1.0 sec. Instantaneous greater acceleration is apparent

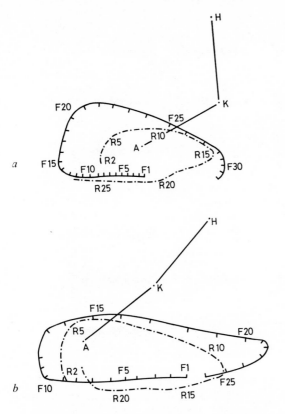

Fig. 2. Rotary motion of ball of foot in relation to hip joint center. *a* Kneeling start: o = foot off block; y = first stride, 0.39 sec, distance 110 cm; A–X = on block; x = off block; z = second stride, 0.19 sec (total 0.58 sec), distance 91 cm. *b* Standing start: o = foot off block; y = first stride, 0.50 sec, distance 153 cm; A–X = on block; x = off block; z = second stride, 0.25 sec (total 0.75 sec), distance 111 cm. 1 = rear foot; 2 = front foot; Scale: only applicable for sequent links and foot spacing.

during the kneeling start. In the upright position, the athlete experiences greater effort in overcoming inertia if his center of gravity is not directly above the front foot.

Considering the muscle activity of the knee and ankle extensors and the opposing flexors it appears that the standing start has a greater burden placed on these muscle groups. This can be tiring if there are false starts.

In conclusion, it seems that the kneeling start holds an advantage over the standing start. Hurdlers, however, benefit by using the standing start in that they arrive at the take-off mark with the body center of gravity in a more favorable position for take-off.

It is suggested, however, that this investigation be extended to cover a distance of 20–30 m from the start. Muscle potential can be transmitted better by radio telemetry under these circumstances. In this way more particulars can be gained.

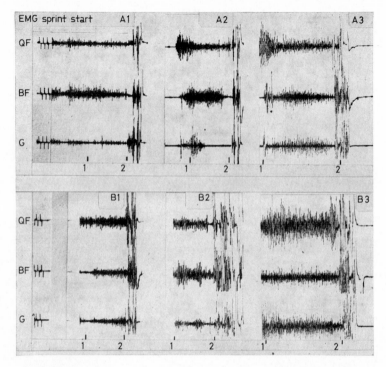

Fig.3. EMG sprint start. A1 = rear foot kneeling; A2 = front foot kneeling; A3 = front foot kneeling; B1 = rear foot standing; B2 = front foot standing; B3 = front foot standing; 1 = command 'set'; 2 = pistol; QF = m.quadriceps femoris; BF = m.biceps femoris; G = m.gastrocnemius. Specifications: Sensitivity, A1, A2, B1, 6 mm = 500 μV; frequency, 700 Hz/0.006 sec; time, 5 mm = 1.0 sec; sensitivity, A3, B3, 6 mm = 500 μV; frequency, 700 Hz/0.006 sec; time, 10 mm = 1.0 sec.

References

BALLREICH, R.: Multivariable Analyse kinematischer Merkmale von Sprintbewegungen. Abstract. 2nd Sem. on Biomechanics, 1969.

BASMAJIAN, J.V.: Muscles alive. Their functions revealed by electromyography (Williams & Wilkins, Baltimore 1962).

BATTYE, C.K. and JOSEPH, J.: An investigation by telemetering of the activity of some muscles in walking. Med. biol. Engng. *4:* 125–135 (1966).

DÄHNE, R.: Theoretische Bestimmungen des Körperschwerpunktes durch Vereinigung der Teilschwerpunkte auf graphischem Wege. Theorie Praxis Körperkultur *9:* pp.836–839 (1966).

DEMPSTER, W.T.: Space requirements of the seated operator. WADC Tech.Rep. 55–159, pp. 254 (Wright Patterson Air Force Base, Ohio 1955).

GROH, H.; THÖS, F. R. und BAUMANN, W.: Die Beinbewegung des Menschen beim Lauf. Int. Z. angew. Physiol. *3:* pp. 231–250 (1966).

HUMPHREY, R. A.: A comparison between the crouch sprint start and an experimental semi-standing start in performance time over a distance of ten yards. Unpublished M. Ed. thesis. Pennsylvania State University (1970).

SHORT, J.: Standing start modernised. Track Techn. *39:* pp. 1227–1228 (1970).

Author's address: Dr. M. DESIPRÉS, Research Laboratory, Department of Physical Education, University of the Orange Free State, P.O. Box 339, *Bloemfontein* (Republic of South Africa)

Medicine and Sport, vol. 8: Biomechanics III, pp. 370–380 (Karger, Basel 1973)

Use of Force Plates for Long-Jump Studies

M. R. RAMEY

University of California, Davis, Calif.

Introduction

The force plate has become a useful tool for the study of many types of human motion. Investigators have utilized it for the study of prosthetic devices [2], the monitoring of human activities in efficiency studies [3], the development of industrial equipment [3], the investigations of the basic methods of human locomotion [7] and, to some degree, for the study of athletic events such as running, jumping and weight-lifting [5, 6, 8]. The force plate yields some fundamental data and substantially assists in the understanding of the motion involved.

In the particular case of the athletic studies, the force plate has been used to identify faults in technique and has led to the development of new ways to perform the event [1].

This article presents the results of a continuing study of the running long-jump, utilizing a specially constructed force plate. Here it is shown that the force plate provides information that can aid in assessing the efficiency of the jumper, assist in the development of the training program, and possibly help in improving a jumper's performance.

Background

The most critical phase of the long-jump generally occurs at the take-off board where the athlete must transfer the horizontal momentum that he has developed on his approach to both horizontal and vertical momentum at take-off. To effect this change in momentum, he exerts force components on the ground that propel him upward and outward to the landing area. These force components, which can be measured by a force plate, are of basic importance in a study of the event since their time-histories determine, to a large extent, the distance that is jumped.

It has been shown that the force-time relationships obtained from a force plate can be used to determine the horizontal and vertical velocities of the jumper's mass center at take-off [6]. These take-off velocities are obtained from Newton's impulse-momentum equations. For example, the equation for the vertical component of the take-off velocity is found to be

$$v_{2y} = \frac{1}{m} \int_0^{t_1} G(t)dt + v_{1y},$$ (1)

where m = the mass of the athlete (his weight divided by the acceleration due to gravity); v_{1y}, v_{2y} = the initial and final vertical velocities respectively of the jumper's mass center while he is in contact with the jumping surface, (here, v_{2y} is the vertical component of the take-off velocity); G(t) = the vertical force component as a function of time t, as measured by the force plate, minus the weight of the athlete; dt = an infinitesimal time increment; t_1 = the time when last contact with the jumping surface ends.

Taking t = 0 when v_{1y} = 0 we obtain

$$v_{2y} = \frac{1}{m} \int_0^{t_1} G(t)\,dt.$$ (2)

Equation (2) shows that the vertical take-off velocity depends on the force function, G(t), its duration, t_1, and the mass of the athlete, m. A similar equation can be written for v_{2x}, the horizontal velocity component of the jumper's mass center at take-off. This equation indicates that the net effect of the horizontal force is to reduce the horizontal velocity developed during the approach on the runway.

The take-off velocities can be substituted into the ballistics equations of physics to yield the distance that the mass center moves. Referring to figure 1 and neglecting air resistance, the horizontal distance is given by the expression

$$x = \frac{v_{2x}[v_{2y} + \sqrt{(v_{2y})^2 + 2gy_0}]}{g},$$ (3)

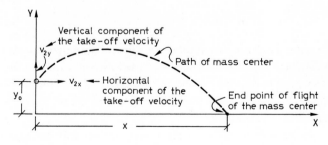

Fig. 1

where x = the horizontal distance that the mass center moves; v_{2x} = the horizontal component of the take-off velocity of the jumper's mass center; y_0 = the initial height of the jumper's mass center; g = the acceleration due to gravity. Equations (2) and (3) provide the necessary relationships that can be used to study the effects of some of the various parameters that influence the distance jumped.

Experimental Method

The force plate used in this study consisted of a 10-in wide by 18-in long laminated plate of aluminum and wood that spanned two force transducers capable of sensing vertical forces and restrained by another force transducer that sensed the horizontal force component. Figure 2 shows a schematic representation of the force plate. Such a force plate is basically an electronic scale which responds to changes in the acceleration of the mass center of the object placed upon it and, hence, senses forces in accordance with Newton's law:

$$F = ma, \tag{4}$$

where F = the force in pounds; m = the mass of the object in slugs; a = the acceleration of the center of mass of the object. The term on the right side of the equation is often called the 'inertia force'.

The tests reported herein were conducted in the laboratory where an elevated runway, terminated by the force plate, was used for the approach while the landing region consisted of foam-rubber mats (fig. 3). This arrangement will permit jumps as long as 22 ft but much shorter jumps are discussed here. [Additional details of the apparatus and jumping area can be found in reference 6.] The athletes were asked to take a short run of approximately 40 ft before jumping from the force plate. A record of the horizontal and vertical force-time relationships was obtained along with the magnitude of the horizontal approach velocity just prior to contact with the jumping surface. This velocity was measured by recording the elapsed time for the athlete to pass two photoelectric sensors spaced 2 ft apart. A 16-mm high-speed camera was positioned normal to the plane of the jump to photograph the take-off. These films were used to obtain the data necessary to locate the mass center of the athlete at specified time intervals [6].

Results

The results obtained for a typical jump are illustrated in figure 4. At the top is the vertical force-time relationship, while below is a sequence of 16-mm stills at 0.02-sec intervals and a plot of the vertical position of the jumper's mass center with respect to time. All three representations are set to the same horizontal scale for ease of reading.

It can be seen from figure 4 that the foot is in contact with the take-off surface for less than 0.20 sec, with the maximum thrusting force occurring

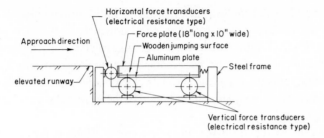

Fig. 2. Schematic representation of the force plate.

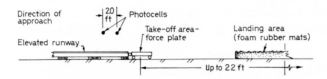

Fig. 3. Side view of the jumping arrangement.

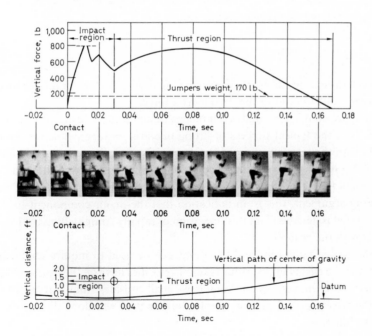

Fig. 4. A typical take-off. *a* Vertical force-time curve. *b* Photographs of the jump. *c* Vertical path of the center of gravity.

near the middle of the total contact time. During the initial contact period, or impact region on the force-time curve, the initial downward velocity of the jumper's center of mass is reduced to zero. This change can be seen in figure 4c. At the end of the impact region the mass center begins to move upward from its lowest position, indicating that the vertical velocity is zero at that instant. Thus, for calculations using equation (2), the end of the impact region was the time designated $t = 0$.

During the thrust region of the force-time curve, which makes up the bulk of the contact time, the vertical force typically rose to a peak of about $4\frac{1}{2}$ times the weight of the athlete. This force, being the inertia force of the mass center of the jumper, consists of the vertical components of the inertia forces of all of the individual moving masses, some of which contribute more to the total inertia force than others. That is, since the inertia force of each moving mass is

$$f_i = m_i a_i, \text{ (i designates a particular moving mass)}, \tag{5}$$

where f_i = the force produced by the acceleration of mass, m_i; a_i = the acceleration of mass, m_i; $m_i a_i$ = the inertia force of mass, m_i; we can write the total force exerted on the mass center as

$$F = \sum_{i=1}^{n} f_i = \sum_{i=1}^{n} m_i a_i. \tag{6}$$

For accelerations, a_i, of nearly the same magnitude, the largest m_i produces the largest contribution to the total inertia force.

In figure 4b it is seen that when the maximum force is being exerted, the jumper is in the typical process of lifting one arm and leg, while the leg in contact with the take-off surface is being straightened. One can quickly note that the leg in contact with the ground is supporting (and accelerating) a far larger mass than the other portions of the body in motion. In light of the preceding argument, this is an indication that the major component to the force exerted is derived from this operation of rapidly straightening the leg in contact with the ground.

The preceding graphs and equations can be used to inquire about the nature and the effects that various take-off velocities have on the distance that the mass center moves. As an example we shall consider what is required to have the athlete propel his mass center farther than was done during some previous jump.

Assuming, for simplicity, that the height of the center of mass is the same at take-off as at landing, i.e. $y_o = 0$, equation (3) simplifies to

$$x = \frac{2}{g} (v_{2x}) (v_{2y}). \tag{7}$$

Here it is observed that x can be increased by either increasing the horizontal take-off velocity, v_{2x}, or the vertical take-off velocity, v_{2y}.

For the jump that produced the force-time curve of figure 4a one can find the vertical take-off velocity. Using equation (2) with $t = 0$ when $v_{1y} = 0$, we find that

$$v_{2y} = \int_0^{t_1} \frac{G(t) \, dt}{m} = 9.7 \, fps.$$

Arbitrarily taking $v_{2x} = 16.1$ fps and using equation (7) to calculate the horizontal distance, x, that the mass center moves, we find that

$$x = \frac{(2)(16.1)(9.7)}{32.2} = 9.7 \, ft.$$

Now suppose we desire to increase the distance jumped from 9.7 ft to some larger value, say 11 ft. Using equation (4) we can determine how much the initial velocities, v_{2x} and v_{2y}, must be changed to produce this result. We shall denote Δv_{2x} as the change in the horizontal take-off velocity, and let Δv_{2y} denote the change in the vertical take-off velocity. Equation (7) can then be written in the form

$$x' = \frac{2}{g} (v_{2x} + \Delta v_{2x}) (v_{2y} + \Delta v_{2y}), \tag{8}$$

where $x' =$ the new horizontal distance that the mass center moves.

We shall determine the change required in only the horizontal take-off velocity necessary to improve the distance jumped from 9.7 to 11 ft. Solving equation (8) for Δv_{2x} with $\Delta v_{2y} = 0$, and upon substitution of the appropriate numbers, we find that

$$\Delta v_{2x} = 2.2 \, fps \, (13.4\% \, increase).$$

Similarly, the required change in only the vertical velocity necessary to accomplish the same improvement of the distance jumped can be determined from equation (8) as

$$\Delta v_{2y} = \frac{gx'}{(2)(v_{2x})} - v_{2y}, (\Delta v_{2x} = 0). \tag{9}$$

Upon substitution of the appropriate numbers we obtain

$\Delta v_{2y} = 1.3$ fps (13.4% increase).

In this example the percentage increase in take-off velocity is the same for both cases, but, because $v_{2x} < v_{2y}$, the required changes in velocity are different.

The preceding example shows that if one take-off velocity is increased without altering the other from the previous jump we naturally increase the distance. However, it is often difficult to increase one velocity without seriously decreasing the other. This situation does occur for jumpers in competition where the athlete makes an effort to achieve more height in the jump but finds the horizontal distance jumped is less than a previous jump because, possibly, the horizontal velocity was reduced in the process.

Using equation (9), we can determine how much the vertical velocity can be reduced while the horizontal velocity is increased from 16.1 to 18.3 fps, as above, but yet keep the distance jumped the same as it originally was – 9.7 ft.

$$\Delta v_{2y} = \frac{gx'}{2(v_{2x} + \Delta v_{2x})} - v_{2y} = \frac{(32.2)(9.7)}{(2)(18.3)} - 9.7 = -1.17 \text{ fps.}$$

That is, if we increase v_{2x} by 2.2 fps but at the same time decrease v_{2y} by 1.17 fps, the distance jumped does not change. A similar situation can be developed by increasing v_{2y} but decreasing v_{2x}. Thus, increasing either of the initial velocities must be done while simultaneously monitoring the other velocity, otherwise the distance jumped can be the same as or even less than a previous jump.

The force-time curve from figure 4a can be used to provide a measure of what the change in force must be to produce a change in the vertical take-off velocity of 1.3 fps as in one of the previous examples.

To illustrate, we shall assume that the required *increase* in force varies linearly from the beginning of the thrust region to the maximum force and linearly down to the end of the curve[1]. (It is assumed that $v_{1y} = 0$ at the beginning of the thrust region and that the duration of the contact time is the same.) From figure 4a the time duration from the beginning of the thrust region to the maximum force is 0.05 sec, while the duration from the maximum force to the end of the contact time is 0.09 sec. Referring to figure 5 we may solve for the maximum force increase, G', by relating the area under the triangle to the net impulse.

1   This assumption is used because it permits an easy solution in these examples. More realistic force increases must be obtained from additional data; however, it will be used in a manner similar to that shown in the example.

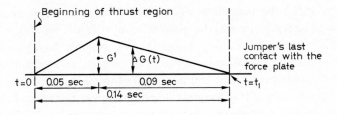

Fig.5. Thrust region.

That is, using equation (2):

$$(\Delta v_{2y})(m) = \int_0^{t_1} \Delta G(t)\,dt = \text{area of the triangle in figure 5,}$$

where $\Delta G(t)$ is the increase in force at a specified time, t. Upon substitution of the appropriate values one obtains.

$$(1.3)(5.3) = \frac{1}{2}(0.14)(G'),$$

or

$$G' = 98.5 \text{ lb,}$$

where $G' =$ maximum force increment necessary to increase the vertical take-off velocity by 1.3 fps. Figure 6 shows how the new force-time curve appears with the above increase in the vertical force component.

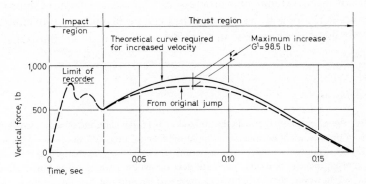

Fig.6. Theoretical force-time relationship required to increase the vertical velocity by 1.3 fps.

In this example, the magnitude of the additional force represents a reasonably large increase and may be quite difficult to expect an athlete to produce without additional training. However, this example does indicate how we can utilize the information from the force plate to ascertain the *magnitude* of the additional forces that are required to make improvements in a jumper's performance. It is recognized here that other factors will also influence the jumper's performance but, all things being equal, the forces exerted will have a vital effect.

Discussion

Figures 4 and 5 and the examples show that to increase the vertical take-off velocity the athlete must increase the upward force, $G(t)$, that accelerates his mass center. Thus one of the functions of his training program should be to develop the strength and timing to provide the necessary acceleration. It is interesting to note that many jumpers use a weight-training technique that provides just this effect. The athletes support weights on their shoulders and do very shallow knee-bends, jumping from leg to leg or onto and off a sturdy chair or bench [4]. This training method effectively works to develop the muscles that can provide increased acceleration to the center of mass. However, this author feels that the one drawback to this method of training is that the athlete does not have any quantitive information as to how his training is progressing since he is in no way monitoring his efforts. If one instrumented the weight-training apparatus and provided a record of the acceleration for the athlete to continuously monitor his effort he would have a much better sense of how his development is progressing. The rate at which the improvement in acceleration is to be accomplished can be planned by providing an analysis of the desired increase in force by using the methods illustrated in this paper.

Conclusions

It has been shown here that with the utilization of a force plate and of some elementary equations of physics, a better understanding of the take-off phase of the long-jump can be obtained. These equations involve various parameters which may be changed to produce a theoretical result that can be used to aid in the development of a training program. It should be recognized, however, that the equations presented here do not contain all of the parameters that influence performance. They do, though, contain some of the more important parameters and illustrate how these can be used to assess performance.

This study is by no means complete for it must yet utilize data obtained from jumps under competition circumstances, consider the horizontal force component in greater detail, and consider various approach techniques at take-off. Future studies will pursue these items along with the effects of measured changes in body mass, the effect of the resistive forces due to air resistance, and the effect that the arms, 'lead' leg and rotation of the mass center have on the horizontal distance that the mass center moves.

Appendix

F = force.

G' = maximum increase in the vertical component of force in the thrust region to increase vertical take-off velocity by a specified amount.

$G(t)$ = the vertical force component minus the weight of the athlete.

$\Delta G(t)$ = increase in the vertical component of force in the thrust region.

a = acceleration of the mass center of an object.

a_i = acceleration of mass, m_i.

f_i = force produced by the acceleration of mass, m_i.

g = the acceleration due to gravity.

m = the mass of the athlete.

m_i = mass of the ith moving object.

t_1 = the time when contact with the jumping surface ends.

v_{1x}, v_{2x} = initial and final horizontal velocities of the jumper's mass center while he is in contact with the jumping surface.

v_{1y}, v_{2y} = initial and final vertical velocities of the jumper's mass center while he is in contact with the jumping surface.

$\Delta v_{2x}, \Delta v_{2y}$ = the changes in the horizontal and vertical take-off velocities respectively.

x = the horizontal distance that the mass center moves.

x' = the horizontal distance that the center of mass moves due to increasing the take-off velocities.

y_0 = the initial height of the jumper's mass center.

References

1   DYATCHKOV, V. M.: The high jump. Track Technique (34): 1059–1075 (1968).

2   EBERHART, H.D. et al.: Fundamental studies of human locomotion and other information relating to the design of artificial limbs. Report to the National Research Council, Committee on Artificial Limbs (University of California, Berkeley 1947).

3   LAURU, L.: Physiological study of motions. Adv. Manag. 22 (3): 17–24 (1957).

4   MURRAY, J. and KARPOVITCH, P.V.: Weight training in athletics. (Prentice Hall, Englewood Cliffs 1956).

5   PAYNE, A.H.; SLATER, W.J., and TELFORD, T.: The use of a force platform in the study of athletic activities. A preliminary investigation. Ergonomics 11 (2): 123–143 (1968).

6   RAMEY, M. R.: Force relationships of the running long jump. Med. Sci. Sports *2* (2): 146–151 (1970).

7   REHMAN, I.; PATEK, P., and GREGSON, G.: Some of the forces exerted in the normal human gait. Arch. phys. Med. *29:* 689–702 (1948).

8   WHITNEY, R. J.: The strength of the lifting action in man. Ergonomics *1* (2): 101 (1958).

Author's address: Prof. MELVIN R. RAMEY, Civil Engineering Department, University of California, *Davis, CA 95616* (USA)

Medicine and Sport, vol. 8: Biomechanics III, pp. 381–386 (Karger, Basel 1973)

Kinesiology of the Long-Jump

J. M. Cooper, R. Ward, P. Taylor and D. Barlow

Biomechanics Laboratory, School of HPER, Indiana University, Bloomington, Ind.

Introduction

Long-jumping is believed to involve a combination of factors including the vertical and horizontal components of speed; the angle of projection; a forceful upward swing with the lead leg, head and trunk erect position at take-off; a vigorous arm action in flight while maintaining a controlled position; and an extended leg position as great as possible at the moment of landing.

The greatest determinator of these factors in the distance of the jump is the horizontal velocity. However, this velocity may be so great that the jumper cannot effectively take off from the board and project himself into the air.

Long-jumping is governed by the physical laws involving force and the laws determining the mechanics of projection [13]. Perhaps the speed of leg muscle contraction as well as the location of the pertinent muscles on the limb involved may also influence performance. Certainly the fast and powerful action of the extensors of the ankle, knee and hip give impetus to the upward thrust of the body from the foot plant against the take-off board.

Theoretically, a projection angle of almost 45° with a constant projection velocity value will enable the long-jumper to attain the longest distance. It must be remembered that the take-off board is slightly higher than that of the landing. In spite of the near 45° theoretical angle, it has proven impossible for humans to long-jump at such an angle. It has been variously reported that the take-off angle ranges from 22 to 31° [1, 2, 6].

The theoretical distance of a jump may be ascertained with the use of the following formula:

$$\text{distance} = \frac{\text{velocity} \times \text{cosine } \theta}{\text{gravity (constant)}} \times \sin \theta \text{ velocity}$$
$$+ \sqrt{\sin^2\theta \text{ velocity}^2 + \text{height gravity (constant)}}.$$

It was believed previously that within certain limits the greater the horizontal velocity and the larger the angle of take-off is toward 45°, the farther the jump. A short time on the board coupled with maximum impulse may enable the take-off velocity and projection energy ($KE = \frac{1}{2} MV^2$) to be harmonized.

Table I. Summary of forces exerted by two long jumpers at foot plant

Jump No.	Horizontal								Moment			
	2 Vertical		× (+) forward		(−) backward		Y-lateral		M (+)		(−)	
	lb-sec	max.	lb-sec	max.	lb-sec	max.	lb-sec	max.	lb-sec	max.	lb-sec	max.
No. 5-Litner-190.08 lb	83.46	696.8	27.05	536.0	2.68	73.44	2.14	80.4	0.64	73.44	00	00
Mass = 5.90 = w/g												
minus (body weight) X	3.66 ×						−Y					
(Contact) for 'Z'	bdy wt						1.61	69.68				
190.08 lb × 0.11 sec =	20.41											
NET	63.05											
Time	0.11 sec		0.10 sec		0.04 sec				0.05 sec	00	0.06 sec	46.2
No. 7-Highbaugh-165	77.04	600.32	21.97	477.04	4.29	85.76		output	00		3.03	
Mass = 5.12 = w/g												
minus (body weight) X	3.63 ×											
(Contact) for 'Z'	bdy wt							48.24				
165 × 0.17 =	28.05											
NET	48.99											
Time	0.17 sec		0.10 sec		0.06 sec				0.03 sec		0.10 sec	
Velocities calculated from No 5 trial	10.68		4.59		0.45							
Data No 7 trial	9.55		4.29		0.83							
Means	10.12		4.44		0.64							

Table II. Length of strides in 2 long-jumpers for 4 to 5 steps before take-off recorded during competition

Subject	Number of steps before take-off	Stride length, ft
Miller	3	6.77
	2	6.66
	1	6.11
	0 (take-off)	6.99
Litner	4	7.10
	3	7.10
	2	7.38
	1	7.49
	0 (take-off)	6.60

Procedures

Basic procedures for this investigation involved the utilization of tri-axial kinematography, precision motion analysis, force-platform, and kinematographic data examination. Three motor-driven moderately high-speed 16-mm cameras, placed in a tri-axial arrangement, were synchronized by a time mark generator which enabled the exact calibration of frame rates to be accomplished. A 3-dimensional force-platform, used in conjunction with a low voltage integrator and 8-channel recorder, was located at the take-off position of the long jump.

The three main subjects were long-jumpers who were members of the Indiana University track team during the spring of 1971. In addition, the action in several meets was filmed during this same period. This involves the use of subjects from other universities.

Summary of Findings

Within the limitations of this study, the following findings seem warranted:

1. The jumper with the longest jump shortened his last stride length at the take-off (table II, III).

2. The jumper with the shortest time on the force platform jumped the farthest (table III).

3. A slight decrease in horizontal velocity occurred in the last few feet of approach of all the jumpers before take-off.

Table III. Horizontal displacement of the center of gravity and time in contact with the force plate from foot-plant to take-off in the long-jump

Subject	Trial No.	Horizontal displacement, ft	Time in contact, sec
Miller	4	3.39	0.173
Litner	5	3.48	0.154
Highbaugh	7	3.56	0.164

Note: Some of the information above was calculated from DEMPSTER's data [5] using high-speed kinematographic procedures.

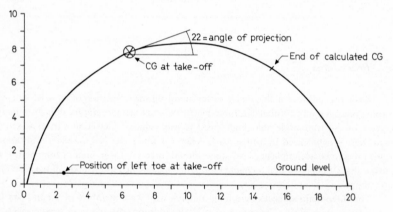

Fig. 1. The parabolic flight and angle of projection of a long-jumper as calculated from film analysis. The solid curved line is an extension of the arc formed by the calculated path of the center of gravity (CG) while the jumpers were in the air.

4. The long-jumpers rotate one foot outward one stride from the take-off board. Possibly this is done in order to be able to position their center of mass so that it will move directly over the take-off foot. This may help to minimize sideward rotational components and to help attain maximum lift.

5. The angle of projection of a long-jumper appears to not be greater than 31° as measured by the center of gravity at take-off (fig. 1).

6. Greater horizontal velocity and a larger angle of projection do not necessarily mean a longer jump. Both horizontal and vertical velocities must

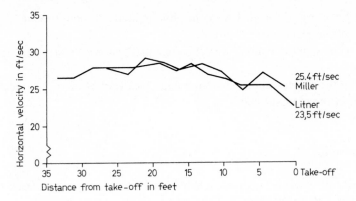

Fig. 2. Velocity displacement curves of 2 long-jumpers' approaches to the take-off during competition.

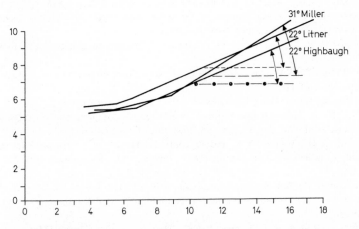

Fig. 3. Path of the center of gravity and angle of projection of 3 jumpers while on the force platform and until $^1/_{110}$ of a second after take-off. The straight extended lines are tangent to the parabolic path of the jumper.

be maximized in order to jump long distances. Increasing the braking force could be essential in order to obtain maximum vertical velocity (fig. 2, 3).

7. A tumbler jumped 18 in. farther with the execution of a front somersault while in the air than when he used the conventional take-off style of jumping. In both instances he used a 1-foot take-off. Further study of this method should be carried out.

8. Vertical impulse was found to be largest for the longest jump (table I).

9. Braking impulse was greatest for the best effort. Conversely, a smaller

braking force was associated with the shorter jump. Braking impulse is a function of force integrated over time.

10. Time spent in contact with the force platform was least for the best jump.

11. In the shortest jump, the horizontal backward velocity was greater than that of the longest jump.

References

1 Bunn, J.W.: Scientific principles of coaching (Prentice-Hall; Englewood Cliffs 1955).
2 Cooper, J.M. and Glassow, R.B.: Kinesiology (Mosby, Saint Louis 1968).
3 Cooper, J.M.; Lavery, J., and Perrin, W.: Track and field for coach and athlete, 2nd ed. (Prentice-Hall, Englewood Cliffs 1970).
4 Cureton, T.K.: Mechanics of the broad jump. Schol. Coach vol. 8, 9, 21 (1935).
5 Dempster, W.T.: Requirements of the space seated operator. WADC Technical Report (US Department of Commerce 1955).
6 Dyson, G.H.G.: The mechanics of athletics (University of London Press, London 1962).
7 Ecker, T.: Track and field dynamics (TAF News Press, Book Division of Track and Field News 1971).
8 Ecker, T.: It could have been greater. Athlet.J., p. 118 (1971).
9 Jokl, E.: A report on Bob Beamon's world record long jump and his subsequent collapse at Mexico City, October 18, 1968. Phys. Educ.: 68–70 (1970).
10 Payne, A.H.; Slater, W.J., and Telford, T.: The use of a force platform in the study of athletic activities. A preliminary investigation. Ergonomics 2(2): 123–143 (1968).
11 Pickering, R.: Bob Beamon's long jump. Athlet.J.: 114–119 (1971).
12 Ramey, M.R.: Force relationships of the long jump, Med.Sci.Sport 2(3): 143–151 (1970).
13 Tricker, R.A.R. and Tricker, B.J.K.: The science of movement (Mills & Boon, London 1967).

Author's address: Dr. John M. Cooper, Biomechanics Laboratory, School of HPER, Indiana University, *Bloomington, Ind.* (USA)

Medicine and Sport, vol. 8: Biomechanics III, pp. 387–393 (Karger, Basel 1973)

A Model for the Setting-up of Representative Motor Skills Explained by using the Results of Research on a Movement Analysis of High-Hurdling

K. Willimczik

Institut für Leibesübungen, Technische Hochschule, Darmstadt

It is obvious and rarely questioned that an analysis of movement in sports in general and the setting-up of standardized motor skills specifically for the theory of physical education is important and necessary. This is valid for the science of training of top athletes as well as for a methodology which is didactically orientated. By this, I mean methodology, which does not orientate itself only on principles such as economy and usefulness, but finds its basis on so-called pedagogical principles [1].

In literature, there are two different ways of setting up fixed motor skills, which are defined 'as an optimal solution of a problem of movement form'. Either one starts from principles of mechanics, especially biomechanics, and one develops a so-called 'Idealkonstruktion', i. e. a 'rationalkonstruiertes Lösungsverfahren' [2], which has a certain validity and is detached from the performing person [3]. This construction is tested then in practice. The other possibility is that one chooses the actual performance; this approach attempts the explanation of phenomenological movement images with the help of biomechanical principles.

Without the intention to lend greater importance to either method, I am choosing the second one. I will give an example for setting up representative motor skills scientifically, and prove and explain this – abstract – model with results of my own research on high-hurdle technique [4].

I started from the following reflections and definitions:

We call a motor skill that solution of a problem that has been proved to be optimal by a competent sample. A sample can be considered competent when it has been made up using a criterion of achievement. Exemplified by the high-hurdle technique, we call that solution 'hurdling' which is demonstrated by a sample that was set up according to the ability of clearing one hurdle as fast as possible.

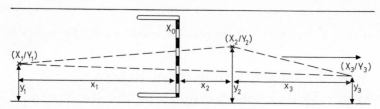

Fig. 1. Distances of the clearance of the hurdle ($L_{H1} = x_1$; $L_{H2} = x_2$; $L_H = L_{H1} + L_{H2}$).

Such an isolated view of the movement of top athletes, however, is of hardly any use in the science of training. It is a disadvantage, because it gives no information whether the intensity of the item, x_i, is relevant for the achievement or not. It is a question whether the intensity of the item varies with the hurdle technique, or whether the subjects do not differ in this item independent of their level of achievement. For the science of training it would mean that only in the first case would an influence be successful.

The above-mentioned problem of relevance of achievement can be decided by an analysis of variance. The classification into groups can be achieved with the help of a criterion of achievement [5]. For my research, 30 hurdlers who ranged from European top athletes to sport students were devided into 3 groups: G_1, G_2, G_3.

In order to clarify the problem of the relevance of achievement I have chosen 3 items:

1. The length of the clearance of the hurdle $L_H = x_1 + x_2$ (fig. 1). The means of the distance from the take-off to the landing are 3.60 m for group G_1; 3.55 m for G_2 and 3.65 m for G_3. Because the standard deviations in relation to the mean differences are relatively high (s = 0.10, 0.13 and 0.22), there are no significant differences between the groups G_1 to G_3. Therefore, good and bad hurdlers do not show significant differences in the length of clearing the hurdle.

2. The opposite is true for the distance take-off-hurdle (L_{H1}) and for the distance hurdle-landing ($L_{H2} = x_2$). Here I have found significant and high significant differences in the analysis of variance between groups G_1, G_2 and G_3. Further statistical treatment indicated that the significant differences were between the pairs G_1 and G_3 and G_2 and G_3, but not of the comparison G_1 and G_2. The values for L_{H2} are: $G_1 = 1.42$ m, $G_2 = 1.45$ m, $G_3 = 1.72$ m.

The above-mentioned differences of the individual items, which are relevant for the achievement, are a special case, because the intensity of the items between the groups changes to a different extent. This permits a differentiated statement concerning the level of achievement on which a change of intensity is possible and necessary.

3. The angle of the 'positive horizontal velocity phase' (ζ_2, fig. 2) belongs to the items which show significant or highly significant differences for each of the pair comparisons. For ζ_2 the mean is 20, 17 and 14°, the standard deviations are 2–4° and the levels of significance are 0.1, 0.1, 0.5 and 0.1.

One often hears reasonable warnings against the research of technique elements by means of empirical methods and the setting-up of norms for practical training. It is said that it is impossible to set up norms of those scores which are dependent on somatometric characteristics or of those scores which are a part of style. It is necessary to examine both these objections.

A generalization of features due to anthropometric and therefore individual characteristics is possible by means of the definition of the so-called indexe. These are defined as movement characteristics, which are related to anthropometric features, if one correlates information obtained about the degree of the dependence of the movement items on anthropometric features.

In hurdling it is said that it is not possible to set up norms concerning the item of clearance of the hurdle, because the clearance of the hurdle depends on the individual features, in this case the length of the leg [6].

A comparison of the clearance of the hurdle and the different phases which make up the clearance with the corresponding indexes (this means the clearance of the hurdles related to the length of the legs) shows relatively high correlation coefficients (0.87 and 0.97). One cannot find norms published for comparing these correlation coefficients. The relatively high level of these coefficients makes it possible to say that a dependence of the characteristics of the clearance to the length of the legs is not all that important.

The question whether the above-discussed features due to anthropometric features are part of the normal technique or whether they must be characterized as style of movement, is answered differently by the movement theorists. MEINEL divides technique in 'rationelle Hauptbestandteile der Technik' and 'zweitrangige Besonderheiten der Ausführung' and places stylistic features as an expression of a 'gereiften Persönlichkeit' [7] in opposition to technique. HOCHMUTH [8] explicitly rejects the expression 'individuelle, zweckmässigste Technik', which is similar to that summed up by MEINEL with the words 'zweitrangige Besonderheiten' because with that ‚die Existenz einer allgemein gültigen zweckmässigen Technik (the existence of one ‚generally valid functional technique) is denied. Not considering those items which MEINEL associates with style HOCHMUTH defines all those 'scores of biomechanical characteristics of individuals' ('biomechanische Eigenschaften der einzelnen Menschen') as style [8].

Like MEINEL [7], I consider a differentiation of features due to anthropometric and stylistic items to be useful and necessary. But contrary to him, I will not define style as a judgement of value but as those individual and personal scores, which do not influence motor behavior.

The stylistic features can also be investigated and statistically treated by the analysis of variance. One can speak of stylistic items, when the achievement groups show relatively high standard deviations when having the same means or relatively small mean differences that not only in relation to the mean differences but also in relation to absolute values. A stylistic feature which was discussed for a long time is the position of the arms while clearing a hurdle. Here the performance varies from double-arm thrust independent of the level of achievement up to an extended arm position to the front and rear.

By means of the prementioned statistical methods it is possible to recognize and work out the essential and general in various concrete forms of the existence of kinetics [9] beyond a mere description of the movements. Results achieved with those methods, however, give no information about the actual change of skills when changing conditions. Have certain movement directions an effect on individual scores and the complex achievement of hurdling or not? The experimental work is a further possibility to test the structure of technique and its variability.

For the statistical analysis of this problem one can use the 2-way analysis of variance. With this method it is possible to test not only the contingent efficiency of the measures concerning methodology of training but to come to the decision whether a noticed change of scores is dependent on the achievement level of the subjects or not.

The number of features which must be analyzed differs from discipline to discipline. It grows with the number of available methods of research for collecting data as well as with the level of complexity of the technical discipline.

I tested 241 kinematic and dynamic items of movement. Such a lot of data needs systematization, in the sense of scientific generalization and the sense of a future use of training.

For the latter, a reduction of movement scores and consequently a reduction of the number of points, which must be corrected, is of great importance. Such a reduction of data can be done with the help of factor analysis. In this way, the large number of movement scores is reduced to a few elementary and independent itemcomplexes, in this case dimensions of technique.

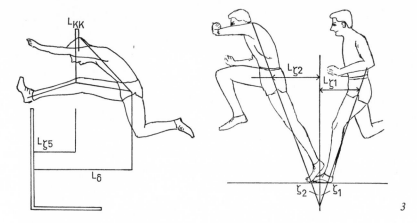

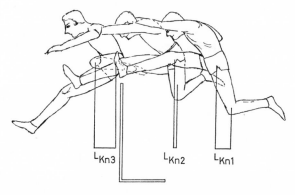

Fig. 2. The 'negative (ζ_1) and positive (ζ_2) horizontal velocity phase' during take-off.
Fig. 3. Angle between upper body and thigh axes (δ_1).

Fig. 4. Horizontal distances of the rear knee to the hip point, $L_{Kn1,2,3}$ when clearing the hurdle.

For hurdling, the factor analysis shows that the analyzed feature com-
plexes are not the same as the traditional categories (front leg, rear leg and
so on). They rather have an overlapping character. The distance hurdle-landing
(L_H), the length of the first step (L_{Ln}) and the position of the foot (of the rear
leg when landing [α_3], belong to such a complex. The angle between the
upper body and thigh axes, when clearing a hurdle, ζ_1 (fig. 3) horizontal
distances of the rear knee to the hip point $L_{Kn1,2,3}$ (fig. 4) when clearing the
hurdle, belong to yet another complex.

The construction and publication of a fixed motor skill by means of th-
suggested methods are not without problems. There is the danger of considere
ing the intensity of the given features to be an absolute infallible norm, but
this should not be so for such a complicated technique as hurdling. A statement
like that can be proved by means of a regressionanalysis. You can start from
the following: the setting of absolute norms of technique varies in the same
direction with the prediction of complex technique on account of specially
chosen variables. The degree of prediction can be defined as the difference
between a predicted and an observed achievement by means of regressionanalysis with regard to the given confidence limits.

In research on hurdle technique one can say that the differences between
the predicted and observed achievements are very small – their mean is
0.05 m/sec when having a clearance velocity of about 7 m/sec – but the confidence limits of the regressors become relatively large, so that a significant
prediction does not seem possible. It is possible to lead this back to the fact
that complex technical disciplines have a certain compensatory possibility.
Which means that small weaknesses of a feature can be compensated by the
performance of other features. The volume of this compensatory factor and
the degree of prediction of a technical achievement must be tested empirically
in each discipline.

Annotations

1. Such pedagogical principles are, among others, natural behavior, vividness, sponta-
neity, entirety, and so on [1].

2. K. MEINEL, Bewegungslehre, Berlin 1969, S. 242.

3. See a.o. T. NETT, Was ist 'Technik' und was ist 'Stil'?, in: Die Lehre der Leicht-
athletik (1955) Nr. 15, S. 8–10.

4. The results of this research with the title 'Leistungsbestimmende Bewegungsmerk-
male der 110-m-Hürdentechnik' have been published by Bartels & Wernitz, West-Berlin,
1972.

5. It is often difficult to discover that achievement which is thought to be a criterion.
This results from difficulties, in general, in measuring an isolated part of a technique of a
discipline. Another component besides technique is the importance of motor abilities. So
the complex performance – that is, the time stopped for the high hurdle race – consists
besides technique of the different forms of speed. Thus, time fails to be a criterion for the
hurdle technique, the hurdle technique factor mentioned in literature, though for other
reasons, is useless as well. It is defined as the difference of the best time of the high-hurdlers
and the 100-meter sprint. What is against the use of this hurdle factor (as a specific criterion
of achievement of the study) is that it only represents the best time (personal record) and not
the actual achievements during the test. If we want to set up a classification of the investigated

specific elements it is absolutely necessary to set them into relation with the achievement performance in this test.

This demand is satisfied by the definition of the mean velocity ($\bar{v}_H$); it is that velocity with which the hurdle is cleared. The almost ideal correlation between the mean velocity ($\bar{v}_H$) and the above-mentioned hurdle technique factor of 0.97 can serve as proof that both criteria can be considered to be equivalent external criteria. That $\bar{v}_H$ is really a criterion of the achievement.

6. T. NETT, Die Technik beim Hürdenlauf und Sprung, Berlin 1954, S. 38.

O. PELZER, Das Trainingsbuch des Leichtathleten, Stuttgart 1926, S. 213.

7. K. MEINEL, Bewegungslehre, S. 244 und 250.

G. HOCHMUTH, Biomechanik sportlicher Bewegungen, Frankfurt 1967, S. 181.

9. K. MEINEL, Bewegungslehre, S. 128.

Author's address: Prof. Dr. KLAUS WILLIMCZIK, Institut für Leibesübungen der Technischen Hochschule, *D-61 Darmstadt* (FRG)

Medicine and Sport, vol. 8: Biomechanics III, pp. 394–402 (Karger, Basel 1973)

An Analysis of Long-Jump

A Model and the Results of a Multivariate Analysis of Kinematic and Dynamic Scores of Jumping Performances

R. BALLREICH

Institut für Sport und Sportwissenschaften,
Johann-Wolfgang-Goethe-Universität, Frankfurt a.M.

*I. Aims and Description
of the Research as well as its Presentation*

In order to optimise the level of physical achievement it cannot only be the aim of biomechanics to give a mere description of the characteristics of motor performance, which is devoid of explanation and which can be re-examined intersubjectively, but its purpose is also to offer information concerning the conditional and causal relations between the described characteristics. This very reason makes it necessary to split up our research into 1 descriptive and 3 analytic approaches, the presentation of which must be differentiated according to the problem, methods and results.

The sample of scores of the present analysis of jumping performances is confined to kinematic and dynamic as well as to anthropometrical significant features. The term 'multivariate analysis' is a result of the procedures of mathematical statistics which were applied in our analysis in order to carry out multidimensional random variables and multivariate distributions which are the analysis of variance, the factor analysis and the regression analysis. With regard to the selection of the sample of kinematic and dynamic features as well as to the procedures of the data extraction and the data processing our model can be used for the analysis of any jumping performance in athletics. This quality of the model and the applied statistic analysis made it obvious to call it 'multivariate analysis of jumping performances'. We shall try to illustrate this model by means of an analysis of the long-jump.

A. The Descriptive Approach

1. The object is the metrical determination of kinematic and dynamic characteristics of jumping performances with reference to various anthropometrical characteristics of the performer.

1. The Applied Method
a) The Sample of Features

The complex achievement of the long-jump was differentiated into the following factors which all attribute to the result of the performance: run-up, take-off, flight and landing. Each of these achievement factors was split up again into a central aspect as for example a kinematic respectively a dynamic one as well as into the spatial-temporal aspect (spatial [1], temporal [2], spatial-temporal [3]), which is a partial one. If the analysed features are not self-defining we will explain them.

1. The run-up. Kinematic features: (1) length of step, length of step index, XL (quotient of the length of step and the length of leg), width of track, direction of the run-up, foot position; (2) quotient of the flight and support phase, frequency of step, frequency of step index, XF (the length of step multiplied by the length of leg) and, (3) number of steps per time unit corresponding to the respective length of steps, to be transformed into mean velocity and mean acceleration.

2. The take-off. Kinematic features: (1) features of the jumper's position (angle of joints, body position) at 3 points of the take-off phase, which can be defined exactly (beginning, the maximum inclination of the knee joint of the take-off foot, ending); (2) the complete time used for the take-off (duration of flexion and extension), and (3) mean angular velocity of the features of the jumper's position, reduction of the horizontal velocity.

Dynamic features: (1) absolute and relative extreme values of the horizontal and vertical forces at the moment of the positive and negative acceleration; (2) the relative temporal position of the extreme forces, the duration of the horizontal and vertical impulses of the negative and positive acceleration, and (3) horizontal and vertical impulses of the negative and positive acceleration as well as the mean forces of negative and positive acceleration, indicator of the distribution of forces (coefficient of variation corresponding to the extreme forces) reactivity index, R(z) (quotient of the mean of the negative vertical and positive vertical acceleration force and the product of the jumper's weight and the time needed for the take-off).

3. The flight. Kinematic features: (1) jumped distance, D, reduced jumped distance, Dred (x-component of the vertical projected trajectory of the jumper's centre of gravity on the level of the take-off, x = parallel to the direction of run-up), economy of the jumping performance with regard to direction and energy, E (E = D − Dred), the angle of projection at take-off, features of the jumper's position during the flight at various phases, which were exactly defined by means of extreme inclination of knee joint; (2) duration of the flight, duration of various phases of the flight, and (3) horizontal and vertical velocity of the projection.

4. The landing. Kinematic features: (1) features of the jumper's position (angle of the joints, body position) at the moment when the landing contact is made, landing economy, EL (EL = [L · H] : [B · Z], L = difference between the x-coordinate of the jumper's centre

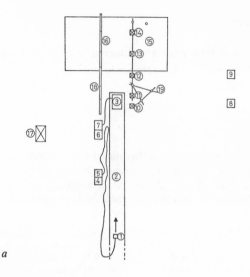

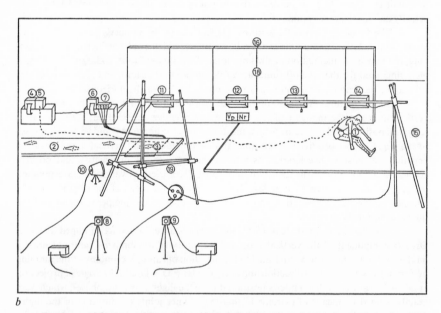

Fig. 1. a and *b:* 1 = subject; 2 = plastic foil (Genotherm); 3 = force plate; 4 = UV-recorder (Helcograph); 5 = amplifier; 6 = UV-recorder (Lumiscript); 7 = amplifier; 8 and 9 = high-speed camera (Hycam); 10–14 = spotlights (Halogen); 15 = jumping pit; 16 = screen; 17 anthropometrical instruments; 18 = lead; 19 = scaffolding for the spotlights. Vp = number of the subject.

of gravity at the moment when the jumper contacts the ground and the x-coordinate of the rear landing mark, H = absolute height of the centre of gravity, B = length of leg, Z = height of landing distance between the jumper's centre of gravity and the ground at the moment when he contacts the ground), and (2) the relative duration of that phase, which immediately precedes the landing.

b) Method of Measurement and Procedures

1. Kinematic features of the run-up. The recording of the kinematic features of the run-up is based on the following data: (x_i, y_i) – which means the coordinates of homologous spike marks of the i-th step; α_i = data of the angles of the foot position during the support phase of the i-th step; Ts_i, Tf_i = the duration of the support and flight phases of the i-th step. In order to be able to measure the xy- and α-data a procedure which registered the marks left by the spikes was developed by means of a plastique foil, which was fixed on the run-up track (Genotherm foil) (measuring error: $\Delta x = \pm 5$ mm; $\Delta y = \pm 3$ mm, $\Delta\alpha = \pm 1°$). In order to measure Ts and Tf we had a timingunit built which was constructed on the strain-gauge principle (measuring error: $\Delta T = \pm 3$ msec). We used a non-telemetric transmission (fig. 1).

2. Kinematic features of the take-off. The phase of the take-off was taken by a 16-mm slow-motion camera. Width of the photographed field, 3 m; frequency, 200 pictures/sec; the spatial resolving power of a certain point of the taken object, ± 5 mm; the temporal resolving power: ± 5 msec.

Dynamic features of the take-off. The registration of the horizontal and the vertical forces was made by means of a 3-component force-plate by Kistler Instrumente AG. Linear dimensions of the force-plate, $600 \times 400 \times 45$ mm; the conditions of the installation: the longitudinal axis of the force-plate was parallel to the direction of the run-up; on the surface of the force-plate we attached a hard material (Mipolam), thickness: 8 mm. It was important that it was not ruined by take-offs with spikes (measuring error $\Delta F: \pm 4\%$; $\Delta T: \pm 2$ msec).

3. and 4. Kinematic features of the flight and the landing. The phases of flight and landing were taken by a second Hycam slow-motion camera. Width of the photographed field, 6 m; frequency, 200 pictures/sec; the spatial resolving power of a certain point of the taken object, ± 5 mm; the temporal resolving power, ± 5 msec.

c) Sample

According to various approaches of the problem we had to divide the sample into 3 groups of performers, G_{1-3}, each of the groups with 20 subjects. They were distinguished by the following variation levels of the jumped distances they achieved: G_1: 7.22–6.49 m; G_2: 6.74–5.98 m, and G_3: 5.95–5.00 m.

2. Some of the Main Descriptive Results

1. With regard to the groups G_{1-3} all features of the run-up, which were related to the anthropometrical characteristics (length of step index, frequency of step index) show tendencies which all go in the same direction. This fact

Table I

	G_1	G_2	G_3	
Length of step index, XL	2.06	1.99	1.93	$(-)$
Frequency of step index, XF	4.4	4.2	4.2	$(\frac{m}{sec})$
Horizontal velocity of the projection V_{ox} / Vertical velocity of the projection V_{oz}	3.5	3.2	3.1	$(-)$
Landing economy, EL	0.8	0.8	0.7	$(-)$

supports the hypothesis that these indices can be reckoned among those features of the motor performance which are relevant for the achievement. They tend to become interindividual independent limit values if the achievement of jumping is progressing (table I).

2. The moment of the maximum flexion of the knee joint of the take-off foot is about 60 msec later than the end of the vertical negative acceleration. This is the moment of the terminal phase of the centre of gravity movement in the vertical direction. This fact has to be interpreted thus: right at the end of the negative vertical acceleration, the negative accelerating forces (in direction of gravity) are not only compensated but also overcompensated by forces which work in the positive direction (swinging movements: swinging leg and arms).

B. The Comparative-Analytical Approach

1. The problem is the registration of differences between the groups of performers, G_{1-3}, with regard to the central tendencies of the kinematic or the dynamic features of the run-up, the take-off, the flight and landing.

1. The Applied Method

H-test of Kruskal/Wallis, level of significance $\alpha = 0.01$; questioning: two-tailed.

2. Some of the Main Results

1. If we consider the moment when the force of the take-off sets in, the highly qualified performer is characterized by the ability to change immediately

from the negative to the positive acceleration phases by acceleration impulses which are medium as well as homogeneously distributed and by a short phase of negative acceleration which is paralleled by a fast increase of the extreme acceleration forces. The reactivity index, which was defined in order to enable us to give a simultaneous characterization of the reactive neuro-muscular activity, variates in the same direction with the measured jumped distance. The effect of these features or the effect of their intensities is the increase of the vertical acceleration impulse; that is, the vertical velocity of the projection.

2. The increment of the jumping performance within the range of achievement which was tested by us (5.0–7.2 m) can be attributed to the simultaneous increase of the horizontal and vertical velocity of projection as well as to the economy of landing. It can be stated that the relative increase of the vertical velocity of the projection is larger than that of the horizontal one. The increase of the landing economy shows that the optimisation of the performance is not only a result of the improvement of the dynamic quality but also a result of the improvement of the coordination (table I).

C. Factor-Analytical Approach

The problem is to identify those elementary and independent kinematic and dynamic feature dimensions (motor dimensions) which make up the jumped distance. In this way we get independent regressors for the functional-analytical approach, which allow to find out their relative influence on the jumping performance.

1. The Applied Method

R-technique of the multiple linear factor analysis, factor extraction: method of principal axis by Hotelling, factor transformation: Varimax criterium by Kaiser; factor loading of the 'marker variable', 0.8.

2. Some of the Main Results

1. By means of the factors length of step index, frequency of step index and foot position we were able to reproduce the variance of the mean velocity of the run-up nearly up to 97%. 56% of the variance of the mean run-up velocity can be explained by differences within the length of step index and 30% by differences within the frequency of step index.

2. With regard to the horizontal velocity the following statement can be made: the variance of the reduction of the horizontal velocity is by 72% a

result of inter-individual differences during the phase of the negative horizontal acceleration. The reduction of velocity depends primarily on the duration of the impulse of the horizontal negative acceleration and not on the mean horizontal force of the negative acceleration.

3. The dimension of the negative vertical acceleration phase is indicated by the quotient of the vertical impulse of the negative and positive acceleration. Jumpers who have high loading on this factor are characterised by the poor ability to switch over from plyometric to myometric muscle activity. By this the horizontal velocity of projection is influenced negatively. The determined part of the variance is 30%.

4. The vertical explosive strength index, that means the temporal increase of the maximum vertical force of acceleration can be accepted as the nominal 'marker variable' for the dimension of the vertical acceleration phase. From the point of the relevance of the factor achievement the high loading of the vertical velocity of projection has to be considered. 59% of the variance are a result of the explosive strength index differences.

D. The Functional-Analytical Approach

The problem is the registration of the mathematical-functional relation between the distance achieved by the subject and the intensities of the selected kinematic and dynamic features of the subject's jumping performance. By means of this relation we try to find out quantitatively the relative influence (weight) of the biomechanical features on the jumping performance.

1. The Applied Method

Polynomial regression; number of regressor, 8 (all phases of performance are represented); model of function: parabolic (table II); confidence limits of the regression coefficients, 95%.

The relative weight of the i-th regressor, x_i, set into relation to the jumped distance, D, was defined by the alteration, ΔD, of the jumped distance in relation to the alteration, Δx_i, of the regressor, x_i, when x_k is kept constant ($k = 1, ..., 8, k \neq i$). The definition of the relative weight is as follows:

$$\Delta D \, (\bar{x}_i, \Delta x_i) = \frac{\partial D}{\partial x_i} \cdot \Delta x_i \, (x_i = \bar{x}_i, \Delta x = s[\bar{x}_i]).$$

By means of the confidence limits of the regression coefficients we determined that region of the relative weight ΔD (ΔD_{min}, ΔD_{max}) within which the true relative weight can be located with 95% of probability (table II).

Table II. Relative weights, D, of the regressors XL, ..., EL. The confidence limits of 95% D_{min} and D_{max} are given

	G_1		G_2		G_3	
	ΔD_{min}	ΔD_{max}	ΔD_{min}	ΔD_{max}	ΔD_{min}	ΔD_{max}
XL	0.13[1]	0.13	−0.02	0.07	0.13	0.30
XT	0.22	0.23	0.02	0.07	0.17	0.23
QLI 32	0.03	0.04	−0.02	0.04	0	0.12
Q (Ix)	−0.06	−0.01	−0.04	0.03	−0.84	0.23
Q (T6)	−0.11	−0.06	−0.21	0.31	−0.66	0.45
Q (Iz)	0	0.05	−0.52	0.18	−0.79	−0.28
Rz	0	0.02	−0.11	0.32	−0.17	0.42
EL	0.05	0.07	−0.10	0.10	−0.12	0.33

$D = a_0 + a_1XL + ... + a_8EL + a_9XL^2 + ... + a_{16}EL^2$

1 The dimension of the numbers is meter (m)

XL = length of step index; XF = frequency of step index; QLI = length of step quotient, quotient of the length of step 3 and the length of step 2 (step 1: take-off step); Q (Ix) = quotient of the horizontal impulse of the negative and positive acceleration; Q (T6) = quotient of the duration of the vertical impulse of the negative and positive acceleration; Q (Iz) = quotient of the vertical impulse of the negative and positive acceleration; R (z) = reactivity index; EL = landing economy.

2. Some of the Main Results

1. In correspondence with the results found out by the analysis of variance and the factor-analysis the regressors: length of step index, frequency of step index, length of step quotient and landing economy, show a positive influence; the regressors: the quotient of the horizontal impulse of the negative and positive acceleration, the quotient of the duration of the vertical impulse of the negative and positive acceleration and the quotient of the vertical impulse of the negative and positive acceleration show a negative influence on the jumped distance, if one arranges the observed alteration D of the jumped distance according to its direction. Extent and intensity of these influences are specific for the 3 groups.

2. In regard to G_1, the frequency of step index and the length of step index have the highest influence on the jumped distance. If the standard deviation of these variables whose independence was guaranteed factor-analytically is progressing the distance can be improved at an average of a least 0.22 or 0.13 m, if only the constancy of the other regressors is kept. The influence of the landing economy and the length of step quotient is far more

insignificant for the jumping achievement if the conditions remain the same, namely about 0.05 m, respectively 0.03 m. The influences of the quotient of the duration of the vertical impulse of the negative and positive acceleration of −0.06 m and the quotient of the horizontal impulse of the negative and positive acceleration of − 0.01 m are about the same. In G_3 the quotient of the vertical impulse of the negative and positive acceleration has a minimum influence of − 0.28 m. That is about double the contribution of the frequency of step index of 0.17 m or the length of step index of 0.13 m. To summarise, it can be stated that the jumped distance of G_1 is mainly influenced by the run-up and is less influenced by the take-off and the landing. In G_3 the take-off dominates the run-up.

Reference

BALLREICH, R.: Weitsprung-Analyse (Sportverlag, Berlin 1970).

Author's address: Prof. Dr. R. BALLREICH, Institut für Sport und Sportwissenschaften, Ginnheimer Landstrasse 39, *D-6 Frankfurt a. M.* (FRG)

Medicine and Sport, vol. 8: Biomechanics III, pp. 403–408 (Karger, Basel 1973)

A Comparative Analysis of Dynamic Take-Off Features of Flop and Straddle

Angela Kuhlow

Institute of Physical Education, J.W. Goethe University,
Frankfurt a. M.

The here-presented comparison of selected dynamic take-off features of flop and straddle is hopefully a contribution to the discussion about the dynamic pattern of the take-off movement of both these techniques.

Object of our comparative analysis is the identification of those dynamic take-off features which show a techno-specific evident relation to flop and straddle, i.e. those distinct characteristics which differentiate significantly jumpers of both the techniques (applied method: analysis of variance).

The dynamic pattern of the take-off movement was characterized by: (1) the time needed for the take-off, T_t; (2) by a quotient of 2 vertical forces, $Q(F_z)$; (3) by the temporal position of the positive vertical acceleration force, $Q(T)$, set into relation to the time of the take-off; (4) by the indicator of the distribution of the vertical forces, $D(F_z)$; (5) by the vertical impulse, J_z; (6) by the indicator of the take-off economy, $E\Delta v_z$, as well as by (7) the reduction of the horizontal velocity, Δv_R.

1. The beginning and ending of the time needed for the take-off are defined by the first and last contact between the take-off foot and ground.

2. The quotient of 2 vertical forces as defined by us is the quotient of the absolute maximum and the relative maximum vertical force.

3. The quotient, $Q(T)$, determines the temporal position of the positive vertical acceleration force set into relation to the time of the take-off.

4. The indicator of the distribution of the vertical forces describes the distribution of the 3 extreme vertical forces: the absolute maximum, the relative minimum and the relative maximum force; the lower the indicator score the more homogeneous are the scores of the respective extreme forces.

5. As we did not carry out a kinematographic movement analysis, we could not calculate the time when the jumper's center of gravity changed its direction from the downward to the upward movement so that it was impossible to differentiate the vertical impulse into a negative and a positive vertical acceleration impulse. Therefore, we confined ourselves to the registration of the total alteration of the vertical velocity during the take-off.

6. The indicator of the take-off economy is defined as the quotient of the alteration of the vertical velocity during the take-off and the height of jump, i.e. the alteration of the downward and upward velocity of the jumper's center of gravity.

7. The reduction of the horizontal velocity gives information about the velocity reduction of the jumper's center of gravity induced by the negative and positive horizontal acceleration impulse during the take-off.

For the registration of the dynamic take-off features we used a Kistler force-plate which works on the principle of piezo-electric crystals; it measures the 3 orthogonal planes of the forces (F_x, F_y, F_z) of any directed dynamic force. The dynamograph was so arranged that the run-up and the take-off were on the same level of height. Via the charge amplifier the signals of the transmitter reached the 3-channel recorder. Here the signals were recorded as intensity-time diagrams. By means of a time-, force-, and plane-scale the corresponding dynamic measurable features were calculated. The apparatus measuring the force had an error band of $\pm 4\%$, the apparatus for the temporal registration of the forces had an error band of $\pm 0.1\%$.

Our research was done with 40 women high jumpers: 20 straddlers and 20 floppers. Because of the variance of achievement (1.30–1.65 m) we expected, for both techniques, a large spread of distribution concerning the intensities of the dynamic take-off features. The classification of the 40 jumpers was divided into 4 groups (G_{1-4}).

G_1: 10 good straddlers with a height of 1.55–1.65 m.

G_2: 10 poor straddlers with a height of 1.30–1.40 m.

G_3: 10 good floppers with a height of 1.55–1.65 m.

G_4: 10 poor floppers with a height of 1.30–1.40 m.

The subjects had to jump as high as possible while the bar was continuously raised in stages of 5 cm. The position of the bar with respect to the dynamograph was arranged individually for each jumper.

The object of the following chapter is the presentation, interpretation and discussion of our results.

1. Presentation:

Table I. Scores of the dynamic take-off features

V_i	G_1		G_2		G_3		G_4		D
	M_1	s_1	M_2	s_2	M_3	s_3	M_4	s_4	
T_t	0.26	0.04	0.20	0.02	0.17	0.01	0.19	0.02	sec
$Q(F_z)$	1.5	0.3	2.3	0.4	1.8	0.4	1.9	0.2	–
$Q(T)$	0.53	0.04	0.49	0.05	0.47	0.07	0.49	0.07	–
$D(F_z)$	0.42	0.07	0.57	0.09	0.44	0.06	0.44	0.10	–
J_z	24.2	2.7	20.0	2.4	21.8	4.2	21.4	2.7	kpsec
$E\Delta v_z$	0.15	0.02	0.19	0.01	0.15	0.02	0.19	0.03	sec³/m
Δv_R	1.9	0.2	1.4	0.3	1.9	0.4	1.3	0.4	m/sec

V_i = dynamic variables; G_i = groups of jumpers (G_{1-4}); M_i = group-specific (G_{1-4}) arithmetic means (M_{1-4}) of the dynamic features; s_i = standard deviation (s_{1-4}) of the group-specific arithmetic means; D = physical dimension of the dynamic features; for other abbreviations see text.

2. Interpretation and discussion. Our results underline the fact that the time of the take-off is a techno-specific feature. This can be deduced from the different spatial-temporal movement pattern during the phase immediately preceding the take-off (straddle: relatively low run-up velocity; flop: relatively high run-up velocity) as well as from the take-off movement itself (straddle: relatively extreme lay-out, stiff-legged kick of the swinging leg, simultaneous upward swinging of both arms; flop: relatively median lay-out, bent-legged kick of the swinging leg, alternate swinging of both arms).

We regard it as an important finding of our research that good straddlers and floppers are distinguished by a higher positive acceleration force if set in relation to the negative vertical acceleration force. The above-mentioned fact can be seen as a result of the relatively higher positive vertical acceleration force of the good straddlers and floppers compared with the poor jumpers of both techniques while the negative vertical acceleration force is indifferent for both groups. Whether the increase of the vertical projection velocity and the positive vertical acceleration impulse besides the increase of the positive vertical acceleration force can be attributed to the homogeneity of the force distribution and to the time of the positive acceleration impulse could not be solved as our instrumental equipment did not differentiate negative and positive vertical acceleration impulses. The temporal position of the positive vertical acceleration force set into relation to the time of the take-off is a feature which attributes to the achievement of the straddle technique and which is – with respect to the comparison of both techniques – techno-specific. The relatively late moment when the maximum positive vertical acceleration force of the good straddlers sets in can be explained by the stifflegged kick of the swinging leg contrary to a bent one of the poor straddlers. According to HESS there is the following relation between the swinging and stretching movement and, with that, the moment of the maximum positive vertical acceleration force: the intensive stretching of the take-off foot which coinduces the maximum force is not possible until the maximum swinging acceleration has been concluded. As the moment of the maximum acceleration of the stiff-legged kick of the swinging leg is later than that of the bent swinging leg, the intensive stretching of the take-off leg and with it the maximum acceleration is carried out later in this swinging technique. If we observe the kick of the swinging leg, the flop differs from the straddle by a bent swinging leg. This means that the phenomenon – the relatively early moment of the maximum positive vertical acceleration force – as well as its conditions correspond to the above-mentioned information.

If an analysis of the dynamic characteristics of the take-off based on the vertical impulse, the homogeneity of the vertical force distribution and the indicator of the take-off economy as well as the reduction of the horizontal velocity is carried out, then the indicator of the take-off economy and the reduction of the horizontal velocity are techno-indifferent features, whereas the indicator of the force distribution is relevant for the technique only in the case of straddle. Good jumpers of both techniques – straddle and flop – can be classified by high economy and a high reduction of the horizontal velocity. The indicator of the take-off economy as defined by us takes into consideration the ratio of the alteration of the vertical velocity (in or against the direction of gravity) and the height of jump. This indicator enables us to make predictions about the economy of the vertical impulse without being compelled to consider the differentiation of the vertical impulse into a positive and a negative vertical acceleration impulse. The low indicator scores of the good jumpers show that their good heights were achieved by a relatively small alteration of the vertical velocity. The question whether straddle or flop are characterized by a higher economy must be rejected on the basis of our indicator.

Another unexpected finding is the intensity of the reduction of the horizontal velocity which is independent from the respective technique. For the good jumpers of flop and straddle it is 1.9 m/sec and for the poor jumpers it is 1.4 m/sec. The higher reduction of the horizontal velocity of the good jumpers results from a relatively high negative horizontal acceleration impulse and from a small positive horizontal acceleration impulse which can be neglected, however, because of its insignificance. The conditions of the identity of the reduction of the horizontal velocity for both the good straddlers and floppers, on the one hand, and the poor jumpers on the other can be explained only by means of a morphological analysis of the take-off movement.

In order to evaluate the relevance of the positive and negative vertical acceleration impulse for the achievement of the flop we started from the proved identity of the vertical impulse of both groups. As the jumped height and the positive vertical acceleration impulse correlate almost ideally and, in the same direction, the positive vertical acceleration impulse of the good floppers has to be higher than that of the poor ones. The identity of the vertical impulse of both groups and a higher positive vertical acceleration impulse which is necessary for better achievements makes it obvious that the good group of floppers, contrary to the bad one, can be classified by a smaller negative vertical acceleration impulse.

Contrary to the flop technique, the vertical impulse is a criterion which influences the achievement of the straddle-technique; i.e. a good straddler is distinguished from a bad one by: (1) a higher vertical impulse, and (2) a higher positive vertical acceleration impulse which is necessary for better achievements.

From these conditions, however, no prediction about the greater-smaller relation of the negative vertical acceleration impulse of both groups can be deduced, because an equal, a larger or a smaller negative vertical acceleration impulse is compatible with the two above-mentioned conditions. But this does not mean that our results exclude that the scores of the negative vertical acceleration impulse of good and poor straddlers do not differ.

The object of the last paragraph is the analysis of the quantitative relation between the height of jump and the dynamic features of the take-off with regard to the specific straddle and flop technique; for this, a nonlinear regression analysis was applied. Besides the velocity of the run-up as a kinematic parameter, the following regressors were selected: the time needed for the take-off and the temporal position of the vertical acceleration force in relation to the time of take-off. With regard to the analyzed sample of regressors and subjects the time of the take-off in straddle has the greatest influence on height of jump. In the flop, however, the velocity of the run-up dominates. If one increases these variables by only 1 SD, the height of jump is at last 4 cm better for the straddle technique and 6 cm for the flop technique, provided that the two other regressors are kept constant. Under equal conditions the temporal position of the positive vertical acceleration force in relation to the time needed for the take-off influences the straddler's height by about 2 cm; considering the straddle, the influence of the velocity of the run-up on the height is about the same size, namely 1 cm.

Different from the straddlers is the achievement for the floppers, which is influenced by neither the relative temporal position of the positive vertical acceleration force nor the time needed for the take-off if one increases these variables by only 1 SD. Summing up these results, we can say that the analyzed regressors have a techno-specific primary influence: whilst the straddle technique is mainly influenced by the time taken for the take-off, flop is influenced by the velocity of the run-up.

Summary

The present comparison of selected dynamic features of (1) the time of the take-off; (2) a quotient of 2 vertical forces defined by us; (3) the temporal position of the positive vertical acceleration force in relation to the time needed for the take-off; (4) the indicator of the distribution of the vertical forces; (5) the vertical impulse; (6) the indicator of the take-off economy, and (7) the reduction of the horizontal velocity of both straddle and flop is a contribution to the discussion about the dynamic pattern of the take-off movement of both techniques. The sample comprised 40 women high-jumpers (20 straddlers and 20 floppers) whose achievements varied from 1.30 to 1.65 m. By means of the analysis of variance we arrived at the following results: while the indicator of the take-off economy, the reduction of the horizontal velocity and the ratio of two vertical forces are relevant for the achievement in both techniques, the time of take-off is a techno-specific parameter. The conditions for these and the other results were discussed. The analysis of the quantitative relation of the height of jump and the 3 regressors (1) time of the take-off; (2) the relative temporal position of the positive vertical acceleration force, and (3) the velocity of this part of the run-up which immediately precedes the take-off was done by means of a nonlinear regression analysis. We discovered that the analyzed regressors have techno-specific primary influences: while the straddle technique is mainly influenced by the time of the take-off, flop is influenced by the velocity of the run-up.

Author's address: Dr. ANGELA KUHLOW, Institut für Sport und Sportwissenschaften der Johann-Wolfgang-Goethe-Universität, Ginnheimer Landstrasse 39, *D-6 Frankfurt a. M.* (FRG)

Medicine and Sport, vol. 8: Biomechanics III, pp. 409–416 (Karger, Basel 1973)

Kinematics and Kinetics of the Standing Long-Jump in 7-, 10-, 13- and 16-Year-Old Boys

B. Roy, Y. Youm and Elizabeth M. Roberts

Département d'Education Physique, Université Laval, Québec, and Department of Engineering Mechanics and Department of Physical Education for Women, University of Wisconsin, Madison, Wisc.

Purpose of the Study

This study focuses specifically on the vertical and horizontal components of force, impulse and power generated during the propulsive phase of the jump.

Previous investigators [HALVERSON, 1958; GLASSOW *et al.*, 1965; ROBERTS, 1971] suggest that the kinematic patterns of some basic skills like throwing, running, kicking and jumping are already established by school age. One purpose of the present work was to investigate these suggestions further, using a force platform and kinematographic records.

In order to get kinematic and kinetic data from the kinematographic records, a suitable mathematical approach was investigated. The method of least square curve fitting was used. Under the assumption that human motion is relatively smooth, the fifth degree of polynomial was used.

Procedures

50 male students, aged 7, 10 and 13 years were selected for this investigation. The middle 30% on the American norms [American Association for Health, Physical Education and Recreation, 1965] for the distance jumped, was used as a criterion of selection of the subjects in each group. As the American norms do not include the 7-year-old category, the Canadian norms [Canadian Association of Health, Physical Education and Recreation, 1966] were used to compute the equation of the regression line for this group. Each group was finally made up as follows: 7- and 10-year-old groups, 15 subjects each; 13-year-old group, 20 subjects. At a first testing session, the horizontal

and vertical forces were recorded as the subjects jumped from a force platform[1].

At a second session, kinematographic records of the standing long-jump were obtained from 5 subjects randomly selected from each group. Five subjects were selected in the same manner from a 16-year-old group and included in the kinematographic part of the study.

An X-Y coordinate reader was used to measure and card-punch the horizontal and vertical displacement of the segment center of mass on the film.

A computer program originally developed by Garrett et al. [1968] and subsequently adapted by Robert W. Schutz, of the University of Wisconsin, was used to compute the location and linear displacement of the segment and total body center of mass.

Results and Discussion

For illustrative purposes, a force platform record of a 13-year-old boy is presented in figure 1. It should be noted that the total propulsive phase lasts less than 1 sec. The upper graph is the vertical force, Fy, measured in pounds. Body weight is set at zero. The middle graph shows the horizontal force, Fx, also calibrated in pounds; the bottom graph shows the sagittal moment, M_z, in in/lb. The maximal negative vertical force was recorded as the subject started to drop into a crouch. Prior to the drop, the horizontal force reflected the forward shift of the body. The vertical force turned positive as the drop into the crouch was slowed down. It can be noticed that the maximal vertical force occurred at the time when the center of mass was changing its vertical motion from downward to upward.

The kinematographic records were used to obtain the vertical and horizontal components of displacement, velocity, acceleration and power. The similarity between the force measures obtained with the force platform and the same parameter derived through differentiation of the displacement of the body center of mass was studied. It was found that the mean differences between the two measures were 6% for the horizontal force and 4.2% for the vertical force [Roy, 1971].

The mean, standard deviation and coefficient of variability of the vertical and horizontal components of velocity, acceleration and power are incorporated in table I.

---

1   This platform was designed by Prof. Ali H. Seireg from the Department of Mechanical Engineering of the University of Wisconsin, Madison, USA.

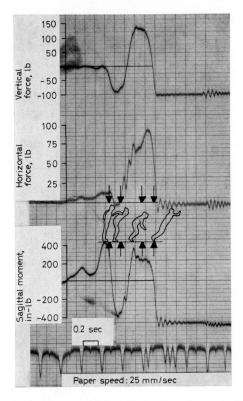

Fig. 1. Record (13-year-old boy) of the vertical and horizontal forces and sagittal moment.

Figures 2 and 3 present sample curves of the vertical and horizontal components of displacement, velocity, force and power during the propulsive phase of the jump of a 16-year-old boy.

The statistical analysis seemed to confirm a definite trend in maximal horizontal velocity and in both components of maximal power. In terms of maximal horizontal acceleration, only the 7-year-old group was found significantly different from the 16-year-old group. This last fact leads us to conclude that this parameter tended to remain very much constant from 10 years of age on. The other comparisons in maximal vertical velocity and acceleration did not reveal any signifiant differences between groups.

The resultant velocity at take-off was also computed. Generally it increased from 7 through 16 years of ages. A trend seemed to exist in this parameter as the 7-year-old group showed a resultant velocity of 8.9 ft/sec, the

Table I. Mean, standard deviation and coefficient of variation of the maximal vertical and horizontal components of velocity, acceleration and power

Age	7 (N = 5)			10 (N = 5)			13 (N = 5)			16 (N = 5)		
	M	σ	σ/M	M	σ	σ/M	M	σ	σ/M	M	σ	σ/M
Maximal vertical velocity, ft/sec	5.1	0.4	7.8	5.1	0.5	9.8	5.2	0.8	15.3	6.1	1.5	24.5
Maximal horizontal velocity, ft/sec	7.4	0.4	5.4	8.7	1.2	13.7	9.2	0.1	10.0	11.4	1.1	9.6
Maximal vertical acceleration, ft/sec²	43.8	2.9	2.0	42.8	7.9	18.4	43.3	14.3	33.0	37.9	9.5	25.0
Maximal horizontal acceleration, ft/sec²	22.6	3.6	16.0	31.4	7.8	24.8	33.1	9.6	29.0	33.4	4.8	14.3
Maximal vertical power, ft-lb/sec	494.6	61.7	12.4	603.4	122.9	20.0	960.6	412.9	63.3	1428.4	498.7	34.9
Maximal horizontal power, ft-lb/sec	241.0	94.7	39.2	481.0	144.5	30.0	815.0	200.8	24.6	1258.0	196.5	15.5

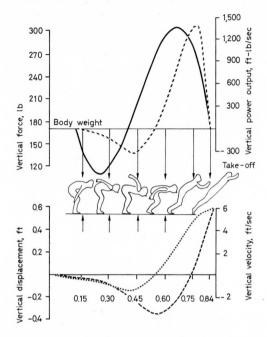

Fig. 2. Sample (16-year-old boy) of the center of gravity vertical displacement (- - -), velocity (.....), force (—) and power (- - -) during the propulsive phase of the standing long-jump.

10-year-old group, 9.5 ft/sec, the 13-year-old group, 10.6 ft/sec and the 16-year-old group, 12.4 ft/sec.

The increase in resultant velocity was mostly due to the increase in the horizontal component of velocity, whereas the vertical component tended to remain constant from 7 through 16 years of age.

The present investigation seems, therefore, to suggest a definite trend, at least in the horizontal component of velocity. The resultant velocity calculated by previous workers [HALVERSON, 1958; ECKERT, 1961] was in some instances 1–2 ft/sec less than that obtained in this study, even for skilled women performers [FELTON, 1960].

The acceleration measures strongly suggest that no trend existed in the vertical component of acceleration. Studies by DAVIES and RENNIE [1968] and PAYNE *et al.* [1968] also suggest that the vertical acceleration tends to remain quite constant in jumping events; mass would be, therefore, the only factor which differentiates between groups.

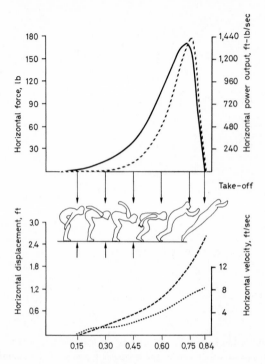

Fig. 3. Sample (16-year-old boy) of the center of gravity horizontal displacement (- - -), velocity (.........), force (—) and power (- - -) during the propulsive phase of the standing long-jump.

The maximal vertical force recorded for any age group of our study was about twice the subject's body weight (2 G); the horizontal force was equivalent to the subject's weight (1 G). Previous investigators [MURRAY *et al.*, 1967; PAYNE *et al.*, 1968; RAMEY, 1970] have reported maximal vertical thrusts up to 5 G in running and jumping events performed by skilled adult performers. According to RAMEY [1970] force per mass and impulse per mass would be the primary variables in the execution of a jump.

GERRISH [1934] and DAVIES and RENNIE [1968] reported vertical output of 5.2–5.9 hp for male subjects and 3.1 hp for the female jumpers. The 16-year-old boys of the present study had a power output of roughly half the values presented by these investigators for their male subjects. The present findings stress once more the influence of mass as a determining factor in power output since the vertical velocity and acceleration reported by these investigators were very similar to those obtained in the present study.

Previous investigators [ZIMMERMAN, 1951; HALVERSON, 1958; FELTON, 1960] had pointed out that the angle of take-off had a tendency to decrease as the performance improved. The results of our study tend to support the constancy of this parameter from 7 through 16 years of age. The 4 groups had a take-off angle (with the horizontal) of 26–29° (computed from horizontal and vertical velocity at take-off).

Although the statistical level of significance was reached in certain parameters, the magnitude of changes in the kinematics of jumping is small. It is suggested that the increase in force and power components from one group to the next could be attributed partly to the increase in body mass as the acceleration tended to remain relatively constant. Therefore, the neuromuscular patterns in terms of muscle action potential and their occurrence in time are well established by 7 years of age in the standing long-jump and probably in other basic skills.

References

American Association for Health, Physical Education and Recreation: Youth fitness test manual, 3th ed. (Washington 1965).

Canadian Association of Health, Physical Education and Recreation: Fitness performance test manual for girls and boys, 7–17 (Ottawa 1966).

DAVIES, C. Y. and RENNIE, R.: Human power output. Nature, Lond. *216:* 770–771 (1968).

DEMPSTER, W. T.: Space requirements of the seated operator. Kinematic and mechanical aspects of the body with special reference to the limbs. Wright Air Development Center technical report, 55–159 (1965).

ECKERT, H. M.: Linear relationship of isometric strength to propulsive force, angular velocity and angular acceleration in the standing broad jump. Doct. diss., University of Wisconsin, Madison (1961).

FELTON, E. A.: A kinesiological comparison of good and poor jumpers in the standing broad jump. Master's thesis, University of Wisconsin, Madison (1960).

GARRETT, R. E.; WIDULE, C. J., and GARRETT, G. E.: Computed aided analysis of human motion. Kinesiol. Rev. (1968).

GERRISH, P. H.: A dynamic analysis of the standing vertical jump. Doct. diss., Columbia University, New-York (1934).

GLASSOW, R. B.; HALVERSON, L. E., and RARICK, G. L.: Improvement of motor development and physical fitness in elementary school children. Cooperation research project, No. 696 (University of Wisconsin, Madison 1965).

HALVERSON, L. E.: A comparison of performance of kindergarden children in the take-off phase of the standing broad jump. Doct. diss., University of Wisconsin, Madison (1958).

MURRAY, P. M.; SEIREG, A., and SCHOLZ, R. C.: Center of gravity, center of pressure, and supportive forces during human activities. J. appl. Physiol. *23:* 831–838 (1967).

Payne, A.H.; Slater, W.J., and Telford, T.: The use of a force platform in the study of athletic activities. A preliminary investigation. Ergonomics *11:* 123–143 (1968).

Ramey, M.R.: Force relationships of the running long jump. Med. Scie. Sports *2:* 146–151 (1970).

Roberts, E.M.: Mechanics; in Larson Encyclopedia of sport sciences and medicine. American College of Sport Medicine, pp. 1154–1157 (MacMillan, New York 1971).

Roy, B.G.: Contribution relative du membre supérieur aux forces verticales et horizontales dans le saut en longueur sans élan chez des garçons de 7, 10, 13 et 16 ans. Kinanthropologie (in press, 1971).

Zimmerman, H.: Characteristic likeness and differences between skilled and non-skilled performance in the standing broad jump. Doc. diss., University of Wisconsin, Madison (1951).

Author's address: Prof. B. Roy, Département d'Education Physique, Université Laval, *Quebec* (Canada)

Medicine and Sport, vol. 8: Biomechanics III, pp. 417–425 (Karger, Basel 1973)

Relations between Performance in High-Jump and Graph of Impulsion

Experiment Realized by Means of a Force Platform

Ginette Hunebelle and J. Damoiseau

University of Liège, Liège

Introduction

In the recording of impulsions, it is well-known that impulsion strength is represented by the whole area situated above the line corresponding to the body weight. The coincidence between the periods of impulsion and position or movement of body segments is studied during normal and pathological gait [7, 8, 19, 24] in running [6] and in jumping [3, 11, 12, 17, 23, 26, 30].

In this study, we are interested in researching relations between ability in high-jump and the form of impulsion. Similar modifications were observed on the impulsion pattern of the start of the run [22] and of high-jump [9] with training. They consist of an increase of the impulsion area characterized by a faster up-motion and by the appearance of a plateau replacing the peak at the apex of the curve. Three characteristics of high-jump impulsion will be defined and their relation with performance will be proved.

Experimental

24 students, 18–21 years old, performed a standing high-jump on a force platform. Although this jump is simple, each subject had the possibility to become familiar with the apparatus, repeating the jump 3–4 times before its recording.

The physical fitness of the students was very varied. None of them had a particular training in high-jump.

Performances in front-running high-jump were also noted. The height was increased in 5-cm stages; 3 trials were allowed at each height.

Since Marey and Demeny's paper [27], the recording systems for impulsions have improved greatly. At first they were mechanical [1, 3, 14]; but they soon became electrical, utilizing either strain gauges [2, 16, 19, 21, 24, 28] or piezoelectric quartz [10, 13, 16].

We recorded impulsions by means of a force platform provided with 3 strain gauges and connected to an amplifier and a recorder. The system is fully described in another paper [15]. The apparatus only records vertical strengths, though there exists more refined appar-

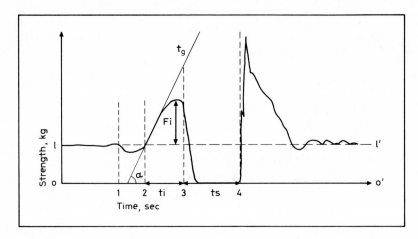

Fig.1. Four recorded periods in any vertical jump. 1–2 = preparatory period; 2–3 = impulsion; 3–4 = suspension period; from 4 upwards = reception and equilibration. α = slope of the curve. For abbreviations, see text.

atus recording simultaneously horizontal, vertical and torsion [19,30]. Such a level of accuracy was not necessary in this study. The apparatus shows a high sensibility and a low inertia. It is precise enough not to exceed a maximal deviation of 5% [15]. The speed of the paper is stable (25 mm/sec).

Analysis of Graphs

Four well-known periods can be found in any impulsion graph (fig. 1): a preparatory period (1–2), an over-pressure which corresponds to the impulsion (2–3), a suspension period (3–4), and the reception period or search for equilibrium (4 upwards).

The area representing the quantity of movement of the impulsion increases while (1) the duration, or length of the graph; (2) the instantaneous maximal force (Fi), or height of the graph, and (3) the ascension velocity of the curve, or slope of the graph, of the impulsion increase. In order to study the influence of these 3 factors on the performance, the following variables are measured (fig. 1).

1. The angle (α) limited by the basal line of the graph (00′) and the tangent of the rising part of the curve (tg).

2. The value of the Fi expressed in percent of the body weight (FIP = Fi · 100/body weight).

Table I. Mean values ($\overline{X}$), standard deviation (SD) and variation coefficient (V) of the variables studied to determine fitness factors in high-jump (N = 24)

	$\overline{X}$	SD	V
α = slope curve, degrees	58.4	24.8	0.253
Log. α	1.7547	0.1140	0.0650
FIP = maximal force in percent of body weight	150.2	44.8	0.298
Ti = duration of impulsion, sec	38.9	8.06	0.245
High-jumps			
Elevation center of gravity, cm	42.8	4.47	0.105
Running high-jump, cm	133.3	13.2	0.0993

3. The duration of impulsion (Ti), from the moment when the pressure overcomes the body weight to the point where the feet are leaving the platform.

4. The measurement of the suspension period (ts) is used to calculate the elevation of the center of gravity.

Results and Discussion

Table I represents mean values and dispersion characteristics of the slope of the curve (α) and its logarithm, the Fi expressed in percent of body weight (FIP), Ti and the performances realized in standing and running high-jumps. The range of dispersion of the results shows the absence of selection among the subjects. The variation coefficient of the variables relative to the impulsion are about twice as high as that concerning the performances.

Table II shows the correlations between performances in high-jumps and log. α, FIP and Ti.

The highly significant correlation coefficients indicate a strong relation between: (1) the performances in standing and running jumps ($r = 0.7024$); (2) the variables relative to impulsion: α or log. α, FIP and Ti (columns 1–4), indicating that these 4 variables are representative of impulsion and (3) the variables relative to impulsion, except Ti, and the performances in jump (columns 5 and 6), indicating that α or log. α and FIP are important factors in jumping.

While α and FIP increase, Ti decreases significantly (column 4); nevertheless, the relations between Ti and the two performances are not significant. We think that, among nonselected jumpers, the weak subjects reveal a tri-

Table II. Intercorrelations between the different variables studied

α	Log. α	FIP	Ti	HG	HJ
	0.9886	0.6311	−0.6152	0.6774	0.6838
		0.6117	−0.6406	0.6947	0.6983
			−0.5546	0.5482	0.4384
				−0.3444	−0.3849
					0.7024

α = slope angle; FIP = instantaneous maximal force; Ti = duration of impulsion; HG = elevation of center of gravity in the vertical jump; HJ = performance in high-jump. N = 24. r = 0.4001 (p = 0.05); r = 0.5169 (p = 0.01); r = 0.6304 (p = 0.001) [19].

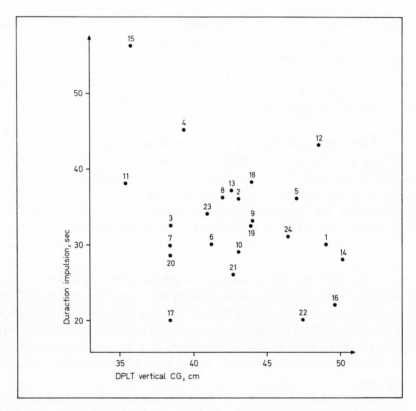

Fig. 2. Evolution of the slope of the curve (log. α) with the vertical elevation of the center of gravity.

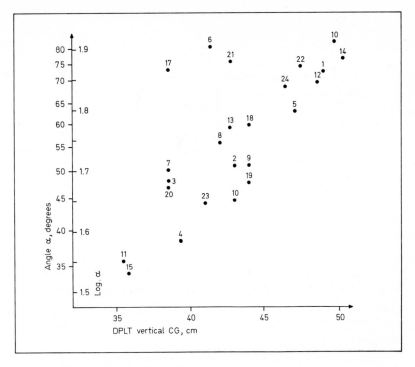

Fig.3. Evolution of impulsion duration with the elevation of the center of gravity.

angular impulsion with a very slow slope which corresponds to a long and slow impulsion and explains the negative correlation observed in table II. Figure 2 shows a large range of dispersion in duration of impulsion among the good jumpers (i.e., those whose center of gravity rises higher than 45 cm) (No. 1, 5, 12, 14, 16, 22, 24). We also found these subjects to have the fastest slope of impulsion ($\alpha > 60°$ for all of them). These facts prevented us to take Ti into consideration in this study for determining factors in high-jumps. But we think that among selected jumpers a long duration of impulsion could be favorable to the performance provided it is not associated with the decrease of angle α.

The most accurate relation appears between α and the performances (table II). An anamorphosis, the replacement of α by its logarithm, transforms the parabolic relation into a linear regression (see fig. 3 for vertical jump).

The regressions are:

Log. $\alpha = 0.998 + 0.018$ CG.   (1)
Log. $\alpha = 0.953 + 0.0066$ HJ.   (2)

Table III. Estimation of the level of significance of the multiple regressions $y_i = a + bx_i + cx_i'$

y_i	*F1*	*F2*	*b*-CI	*c*-CI
Elevation center of gravity	*6.59*	3.26	3.35	0
	(p = 0.01)	NS	(p = 0.01)	
High-jump	5.46	0.102	1.2	0
	(p = 0.02)		(p = 0.05)	

CI = confidence interval.
NS = not significant.

Where CG = elevation of center of gravity in the vertical jump,
   HJ = performance in high-jump.
Both angular coefficients are significantly greater than 0 (p $\leqslant$ 0.001): for equation (1), $0.018 - t.s_b = 0.018 - 3.8 \times 0.004 = 0.003$; for equation (2), $0.0066 - t.s_b' = 0.0066 - 3.8 \times 0.00015 = 0.0054$. t = the Student t-test (p = 0.001) [19], and s_b and s_b' = estimated standard error on the corresponding angular coefficient.

The relations between FIP and the performances are not so accurate, although significant. They are also linear:

$$FIP = -84.5^1 + 5.489 \ CG. \qquad (3)$$
$$FIP = -47.6^1 + 1.483 \ HJ. \qquad (4)$$

Both angular coefficients are also significantly greater than 0: for equation (3), $5.489 - t.s_c = 5.489 - 2.5 \times 1.79 = 5.489 - 4.475 = 1.014 \ (p = 0.01)$; for equation (4), $1.483 - t.s_c' = 1.483 - 1.71 \times 0.648 = 1.483 - 1.108 = 0.375$ (p = 0.05).

A multiple regression equation including log. α and FIP is calculated, utilizing the method of orthogonalization of the independant variables. The following values are found [4, 5, 31].

$$CG = \quad 6.76^1 + 18.85 \ \log. \alpha + 0.020 \ FIP. \qquad (5)$$
$$HJ = -4.5^1 \quad + 78.70 \ \log. \alpha + 0.0052 \ FIP. \qquad (6)$$

The level of significance of these regressions is estimated by comparing the pattern corresponding to the regression $y_i = a + bx + cx_i'$ to the one corresponding to $y_i = a$ (*F1* in table III). While utilizing a similar method, *F2* estimates whether patterns corresponding to $y_i = a + bx_i + cx_i'$ and to

1   Values nonsignificantly different from 0.

$y_i = a' + bx_i$ are significantly different. *F1* is highly significant for standing and running high-jumps: a multiple linear regression can thus be admitted. But the introduction of FIP does not bring any supplementary advantages (*F2* nonsignificant). The confidence interval calculated for *b* and *c* confirms the conclusion showing that *b* is significantly different from 0, but not *c*.

The results are highly reliable (r = 0.978 for α and 0.899 for FIP) provided the boy does not fail his jump. The 'accident' in the jump is characterized by a decrease of one or both of the variables, α and FIP (15–25% decrease).

Conclusion

While looking for relations between the shape of the impulsion and the performances in high-jump, we defined 3 variables: (1) the slope of the curve (α); (2) the maximal instantaneous strength (expressed in percent of body weight = FIP), and (3) Ti.

The measurement of these variables and of jump performances on a sample of 24 boys and the comparison of results allow one to conclude that: (1) the weak jumpers show a triangular impulsion characterized by a slow and low increase of the over-pressure, and a long duration of impulsion, and (2) the good jumpers show a fast and important increase of the curve. The duration of impulsion is very dispersed but shorter than among the weak subjects.

There exists a relation between log. α, FIP and the performances. The estimation of simple and multiple regressions (with log. α and FIP as independant variables) makes it obvious that the relations are linear, and that it is not advantageous to consider the multiple regression, as good results are obtained if only log. α is taken into consideration.

We proved the observations of HOWELL [22] and CAVAGNA *et al.* [9] to be correct. The recording of impulsion from time to time during training may provide useful information about the development of the jumper's ability and about the means to improve it. Even at the beginning of training, we can say that the boy who draws a triangular impulsion will never be a good jumper.

It would be interesting to undertake a similar study on a selected exercise, because we think that when α is already very high, other factors, in particular Ti (increasing of the plateau) may gain more influence.

References

1   AMAR, J.: Le moteur humain (Dunod, Paris 1923).
2   BARANY, J. W. and WHETSEL, R. G.: Construction of a portable platform for measuring bodily movements (Purdue University Press, Purdue 1962).
3   BOIGEY, M.: Manuel scientifique d'éducation physique (Masson, Paris 1923).
4   BRENY, H.: Les modèles linéaires en analyse statistique. Enseign. math. *6:* 51–76 (1960).

5 Breny, H.: Les modèles linéaires en analyse statistique. Enseign. math. 6: 220–243 (1960).

6 Cavagna, G. A.; Saibene, F. P., and Margaria, R.: External work in walking. J. appl. Physiol. 18: 1–9 (1963).

7 Cavagna, G. A.; Saibene, F. P., and Margaria, R.: Mechanical work in running. J. appl. Physiol. 19: 249–256 (1964).

8 Cavagna, G. A.: Mechanics of walking. J. appl. Physiol. 21: 271–278 (1966).

9 Cavagna, G. A.; Komarek, L.; Citterio, G., and Margaria, R.: Power output of the previously stretched muscle. Medicine and Sport. Biomechanics II, vol. 6, pp. 159–167 (Karger, Basel 1971).

10 Chailley-Bert, P.: Thèse sur la physiologie de la marche. Paris (1921).

11 Demeny, G.: Evolution de l'éducation physique. Ecole Française (L. Fournier, Paris 1909).

12 Dufour, W.: Le mouvement humain et la locomotion bipédale. Gymnast. éduc. 4: 128–140 (1959).

13 Elftman, H.: The measurement of the external force in walking. Science 88: 152–153 (1938).

14 Elftman, H.: The force exerted by the ground in walking. Int. Z. angew. Physiol. 10: 485–489 (1939).

15 Falize, J. L.; Lucassen, J. P. et Hunebelle, G.: Analyse de l'impulsion dans le saut en hauteur sans élan. Kinanthropologie 1: 25–44 (1969).

16 Fayt, V.: Le piézodynamographe du centre scientifique de l'homme à Paris peut-il devenir une base scientifique nouvelle par la méthode de l'éducation physique. Mémoire de licence, Université de Bruxelles, non publié (1957).

17 Fayt, V.: Au sujet de l'enseignement de sauts en gymnastique scolaire. Thèse, Université de Bruxelles (1956).

18 Fischer, R. A. and Yates, F.: Statistical tables for biological agricultural and medical research (Oliver & Boyd, London 1963).

19 Hearn, N. K. and Konz, S.: Equipment note. An improved design for a force platform. Ergonomics 11: 383–389 (1968).

20 Herron, R. E. and Ramsden, R. W.: A telepedometer for the remote measurement of human locomotor activity. Psychophysiology 4: 112–115 (1967).

21 Hoes, M. J.; Binkhorst, R. A.; Smeekes-Kuyl, A. E., and Vissers, A. C.: Measurement of forces exerted on pedal and crank during work on a bicycle ergometer at different loads. Int. Z. angew. Physiol. 26: 33–42 (1968).

22 Howell, M. L.: Use of force time graphs for performance analysis in facilitating motor learning, Res. Quart. amer. Ass. Hlth phys. Educ. 27: 12–22 (1956).

23 Hunebelle, G.: Dans la chute des sauts, l'homme ne se comporte pas comme un corps inanimé, élastique ou non. Thèse annexe du doctorat en Education Physique, Université de Liège (1968).

24 Ismail, A. H.; Barany, J. W., and Manning, K. R.: Assessment and evaluation of haemiplegic gait (Purdue University Press, Purdue 1965).

25 Lapeyrie, M.; Rabischong, P.; Avril, S.; Pous, J. G. et Perruchon, E.: l'Electropodographie. Son intérêt en orthopédie Montpellier chir. 13: 303–309 (1967).

26 Lauru, L.: Physiological study of motion. Adv. Manag. 22: 17–25 (1957).

27 MAREY, M. et DEMENY, M.G.: Locomotion humaine. Mécanisme du saut. C.R. Acad.
    Sci. *101:* 489–494 (1885).
28 PAYNE, A.H.; SLATER, W.J., and TELFORD, T.: The use of a force platform in the
    study of athletic activities. Ergonomics *11:* 123–143 (1968).
29 RABISCHONG, P.; AVRIL, S.; AUBRIOT, E.; GUIBAL, C. et BLUCHE, H.: L'électropodo-
    graphie. Une méthode nouvelle de mesure des forces de pression plantaire. Rev. Pédi-
    curie *44:* 9–15 (1967).
30 ROY, B.G.: Kinematics and kinetics of the standing long jump in seven, ten, thirteen
    and sixteen years old boys. Unpublished doctoral thesis, University of Wisconsin
    (1971).
31 WILLIAMS, E.J.: Regression analysis (Wiley, New York 1959).

Author's address: Dr. GINETTE HUNEBELLE, Institut Supérieur d'Education Physique,
Université de Liège au Sart Tilman, *B-4000 Liège* (Belgium)

Medicine and Sport, vol. 8: Biomechanics III, pp. 426–428 (Karger, Basel 1973)

The Biomechanical Studies on the Superposition of Angular Speeds in Joints of Lower Extremities of Sportsmen

M. Koniar

Faculty of Physical Education and Sport, Bratislava

At our Department of Biomechanics at the Faculty of Physical Education and Sport in Bratislava, we are interested in the question of angular speeds in joints and their influence upon the effectiveness of the movements of sportsmen. One of the reasons for work on this problem is that the question of the effectiveness of movements as well as the factors which influence this effectiveness have been investigated and explained in different ways.

In physical training, the effectiveness of movements of sportsmen is often judged by the achieved performance; effectiveness is very often judged by somatic, physiological and psychological abilities of sportsmen. But this is not sufficient in spite of the fact that even though the sportsman might possess all functional abilities to reach the highest performance, he might be unable to reach it. One of the reasons might be a bad sports technique, which means that a good sports technique is reflected in the high effectiveness of movements of sportsmen and also in the high sports performance.

But, as our researches show, neither good functional abilities nor sports techniques are able to secure a high performance if the socalled 'principle of superposition of angular speeds in joints' is not respected in movements. This means that if the sportsman wants to reach the highest sports performance he has to move his body segments in such a way that the maximum values of angular speeds in joints may be reached in *one moment*. This problem became the subject of our experimental researches. Our experiments were carried out on a group of students of our faculty (N = 20).

Electrogoniography and chronography were the methods used for the registration of angular speeds in joints and the initial speed of the sportsman's movement.

We elaborated the data by using the methods of graphic derivation and

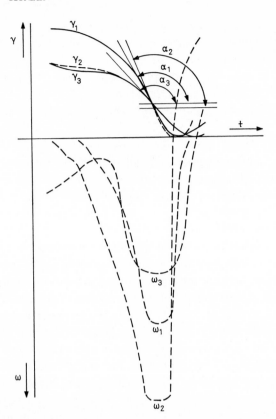

Fig. 1. The superposition of angular speeds in joints of lower extremities on vertical jump: γ_1 = angular changes in the hip joint; γ_2 = angular changes in the knee joint; γ_3 = angular changes in the ankle joint; ω_1 = angular speed in the hip joint; ω_2 = angular speed in the knee joint; ω_3 = angular speed in the ankle joint; $\alpha_1, \alpha_2, \alpha_3$ = values of angular changes in joints of lower extremities in the moment of superposition.

mathematical statistics. The method of graphic derivation was used to determine the angular speeds in joints and the method of mathematical statistics to determine the relations between the angular speeds in joints and the effectiveness of the resulting movement. Two variants of experiments were used: (1) flexion and extension in joints of the lower extremities, and (2) the vertical jump.

The angular speeds of the hip, knee and ankle joints were measured in both types of experiments and the initial speed of the sportsman in vertical jump.

The purpose of our experimental research was to discover the maximum values of angular speeds in joints of lower extremities and their influence upon the effectiveness of the resulting movement of sportsmen. The main purpose was, of course, to prove the existence of 'the principle of superposition of angular speeds in joints' in the movements of sportsmen.

On the basis of our experimental researches we have arrived at the following conclusions:

1. As for the maximum values of angular speeds in joints of lower extremities: (a) we have found out high angular speeds in all joints of lower extremities at flexion and extension. The angular speed in the hip joint was greater than $1,500° \cdot sec^{-1}$, in the knee joint it was greater than $1,700° \cdot sec^{-1}$ and in the ankle joint it was greater than $1,000° \cdot sec^{-1}$, and (b) these high angular speeds in joints of lower extremities reflect good functional abilities of sportsmen and they also represent the high effectiveness of sportsmen's movements.

2. As for 'the principle of superposition of angular speeds in joints', we have investigated the relations between the angular speeds in joints of lower extremities, the initial speed of sportsmen in vertical jump and the height of this jump, thereby using the method of mathematical statistics (correlations). From the analysis of these relations it follows that the height of the vertical jump not only depends on the height of angular speeds in joints but also on the summation of these speeds. The vertical jump was highest when the maximum values of angular speeds in joints of lower extremities were reached in *one moment* (fig. 1).

We have named this special summation of movements 'the principle of superposition of angular speeds in joints'.

As our further experimental researches show, the knowledge of this principle is a very important one in the evaluation and further improvement of the technique of sportsmen's movements.

Author's address: Dr. MILOSLAV KONIAR, CSc., Faculty of Physical Education and Sport, *Bratislava-Lafranconi* (Czechoslovakia)

Medicine and Sport, vol. 8: Biomechanics III, pp. 429–433 (Karger, Basel 1973)

Biomechanical Analysis of the Full Twist Back Somersault

J. BORMS, W. DUQUET and M. HEBBELINCK

Navorsingslaboratorium HILO, Vrije Universiteit Brussel

The present study was concerned with a biomechanical analysis of a typical twisting movement by means of a projection method. An attempt was made to gather information regarding our theory of the twisting movements, which is based on physical principles [4]. According to this theory there are 4 basic techniques [1] which may be used singly or in combination. These are identified as: (1) the contact twist (twist taken while in contact with the apparatus; (2) the action-reaction twist; (3) the twist effected by means of well coordinated hip movements, and (4) the twist using the gyroscopic effect. The projection method itself is explained in a previous communication.

Methods

All data included in this paper were taken from 16-mm movie film. Overhead and side views of each performance were taken simultaneously. Both cameras were set at 64 frames/sec and placed at 6.5 and 13 m, respectively, from the subject. An electronic counter to establish true frame time was shown in each shot. A 1-meter standard and a timer to synchronize the two films were also included in the two views.

The subject used for the experiment was a well trained all-round gymnast, especially skilled in tumbling events, who performed 10 vaults. Six experts in gymnastics evaluated the vaults, and on the basis of this judgment one performance was selected as being excellently performed from the technical point of view. In order to have reliable measurements of the center of gravity, according to the method of BRAUNE and FISCHER [3], check points were marked on the segmental centers of gravity from upper arm, for arm and hand, trunk and head, upper leg, under leg and foot [2].

Besides the usual corrections, some other important corrections were made while drawing the overhead views. Indeed, while the gymnast performs the twist somersault, his body first approaches the overhead camera, then withdraws again.

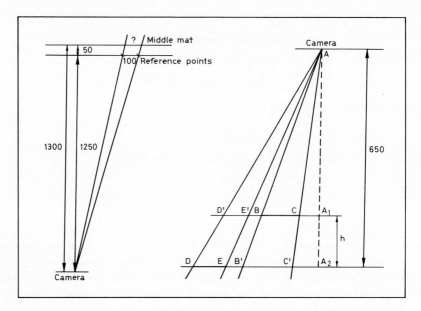

Fig. 1. Scheme of the correction that had to be made because of the approach of the performer's body to the overhead camera.

To reduce B′C′ to a scale of 1/10, the reference distance DE (= 100 mm) had to be reduced to the same extent (fig. 1). Thus, if one wants to measure a distance BC, with a scale of 1/10, D′E′ has to be reduced to 1/10.

But

$$\frac{D'E'}{DE} = \frac{BC}{B'C'} = \frac{AA_1}{AA_2}$$

Thus:

$$D'E' = \frac{AA_1}{AA_2} \cdot DE$$

$$DE = 100$$

$$AA_2 = 650$$

$$AA_1 = AA_2 - h = 650 - h$$

$$D'E' = \frac{100 \cdot (650 - h)}{650}$$

$$x = \frac{10 \cdot (650 - h)}{650}$$

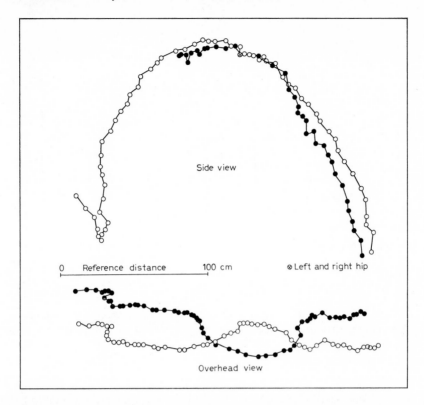

Fig. 2. Curves of the movement pattern of left (○) and right hip (●) constructed for overhead and side views.

D′E′ = the size, to be given to the reference points, in order to obtain a drawing with a scale of 1/1.

x = the size, to be given to the reference points, to obtain a drawing with a scale of 1/10.

The height, h, of the body segments was, for practical reasons, replaced by the height of the center of gravity of the body at that moment.

Results

Curves of the movement patterns of hands, shoulders, hips and feet were constructed for both overhead and side views (fig. 2). Combining the corresponding images and applying the projection method provided the curves of real paths and velocities of these body parts (fig. 3). The variations of the

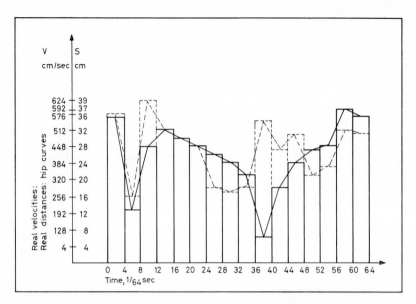

Fig.3. Curves of the real magnitude of velocities and distances of left and right hand. These curves have been obtained by combining overhead and side views. —— = left hip; ––– = right hip.

size of the trunk-legs angle were calculated by applying the same method (fig. 4).

α stands for the trunk-legs angle as measured on the front side of the body; β stands for the smallest angle between trunk and legs at any phase of the movement.

From these movemental traces the following results were obtained: (1) the twisting direction was to the right; (2) no twisting impulsion was taken while pushing off the floor; (3) the arms started the twisting action, using the gyroscopical effect; (4) the right arm provided a supplementary impulse, by applying the action-reaction effect; (5) the left arm, brought upwards, rotated in the direction of the twist and used an action-reaction twist towards the end of the movement; (6) the position of the shoulders at the landing showed a twisting motion of less than 360°, which could not be seen by the experts, and the hip curve showed the importance of the gyroscopical effect.

The graphs of the real paths and velocities accentuate the differences in action – in time and space – between the respective body segments and also between left and right segments. The variations of the trunk-legs angle are not

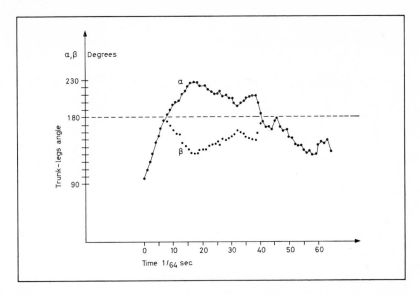

Fig. 4. Example of a curve representing the real magnitude of the trunk-legs angle during the course of the studied movement. α = trunk-legs angle on the front side of the body; β = smallest trunk-legs angle.

only indicators of the importance of these trunk movements; they also show a pattern of the twisting motion which might well be typical and form the basis for a theoretical model for similar movements in other sport situations.

References

1   BATTERMAN, C.: The techniques of springboard diving (MIT Press, Massachussetts 1968).
2   BEVAN, R. and CORSER, T.: The biomechanical study of gymnastic movements. Gymnast 7: 23–26 (1969).
3   BRAUNE, W. and FISCHER, O.: Über den Schwerpunkt des menschlichen Körpers. Sächs. Akad. Wiss., Leipzig 15: 559–672 (1889).
4   DUQUET, W.: De Schroefbeweging; Licentiate thesis, Vrije Universiteit, Brussels (1970).

Author's address: Dr. J. BORMS, Navorsingslaboratorium van het HILO, Vrije Universiteit Brussel, Hegerlaan 28, *B-1050 Brussels* (Belgium)

Medicine and Sport, vol. 8: Biomechanics III, pp. 434–439 (Karger, Basel 1973)

Biomechanical Study of Competitive Cycling

The Forces Exercised on the Pedals

A. Dal Monte, A. Manoni and S. Fucci

Institute of Sports Medicine, Italian Olympic Committee, Rome

I. Purpose of the Work

The present research work aims to establish a method capable of analysing the muscular tension exercised on the pedals by a cyclist during specific sport activity.

In order to measure the particular technical aspect of pedalling, a study was carried out on the direction and intensity of the forces exercised on the pedals. The present study emphasises the main techniques used by the cyclist during the phases where most effort is required by the pedalling action. Maximum effort is required on the part of the cyclist in the following cases, namely: (1) cyclist sitting on bicycle at horizontal level; (2) cyclist sitting on bicycle at an angle to horizontal level; (3) cyclist standing while pedalling on bicycle moving at horizontal level, and (4) cyclist standing while pedalling on bicycle in position described in (2).

II. Material and Method

A. Equipment – L. normal

1. Pedals equipped with tension transducers with linear recording.
2. Oscilloscope.
3. Polaroid photographic camera.
4. Roller cyclo-ergometer.
5. Bicycle.
6. Mobile support to determine the dynamic dead centre of pedalling.
7. Arriflex cine-camera.
8. Reflection moviola.

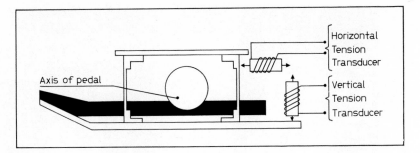

Fig. 1. Scheme of the dynamometric pedal.

1. *Pedals Equipped with Tension Transducers with Linear Recording*

Each of the 2 pedals records the tension exercised, according to 2 orthogonal axes: one coinciding with the foot plane and the other perpendicular to it. This was obtained by using 2 potentio-metric linear transducers in each pedal, whose signals after being amplified by 4 twin amplifiers were transmitted to a cathode ray oscilloscope.

2. *Oscilloscope*

This is a 2-channel cathode ray oscilloscope.

3. *Polaroid Photographic Camera*

This is used for the photographic recording of the scan in the oscilloscope screen.

4. *Roller Cyclo-Ergometer*

This special roller cyclo-ergometer uses the actual bicycles adopted by cylists for the races. This instrument is equipped with a hydraulic brake, calibrated for progressive work charges between 0 and 3,000 kg-m/min.

5. *Bicycle*

Road-race bicycles, equipped with proper dynamometric pedals like those previously described, were used in the experiment.

6. *Mobile Support to Determine the Dynamic Dead Centre of Pedalling*

The dynamic dead centre of pedalling is the point determined by the alignment between the 3 following axes; the axis of the hip rotation, pedal rotation centre and pedal crank centre. The above-mentioned centre moves backward and forward according to the position assumed by the cyclist.

A system which causes the interruption of the scan correspondent to the dead centre was studied for the identification of the dynamic dead centre.

This was possible by using a micro-switch applied to the body by an oscillating pantograph.

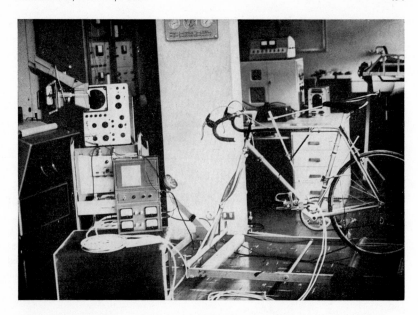

Fig. 2. Assembly of the equipment for obtaining data from dynamometric pedal.

7. *Arriflex Cine-Camera*

For the kinematographic record of the pedalling an Arriflex 16-mm cine-camera was used at the following speed: 80 photogrammes/sec, equipped with a zoom lens of 12–120 mm.

B. Method

The following tests were recorded.

Table I. Trial on flat ground

Cyclist's position	Weight, kg-m/min	Pedalling frequency, pedals/min
seated	2,400	90
standing	2,400	120

Table II. Trial on slope

Cyclist's position	Weight, kg-m/min	Pedalling frequency, pedals/min
seated	2,400	60
standing	2,400	90

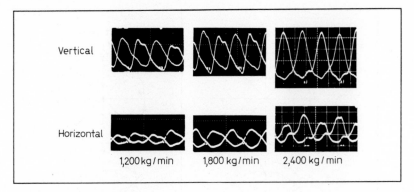

Fig. 3. The curves obtained at different work loads on the dynamometer. Bicycle horizontal, cyclists sitting. The upper graph shows the tension curves on the vertical axis of the pedal. The lower graph shows the tension curves on the horizontal axis of the pedal. Work frequency, 90 pedals/min.

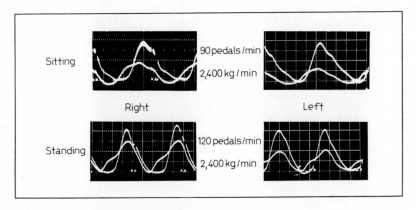

Fig. 4. Two different curves at different positions of the cyclist. Bicycle horizontal. The upper graph shows curves in sitting position. The lower graph shows curves obtained pedalling in standing position. In every graph the taller curve refers to the vertical tensions and the smaller curve to the horizontal tension. The left side refers to the tension of the right feet and *vice versa.*

In the slope trails the simulated gradient was 10%. The oscilloscope was experimentally calibrated in order to obtain a signal amplitude equal to 1 cm for every 10 kg of tension recorded.

The scan sliding speed was chosen so as to include the complete pedalling within the screen limits of the oscilloscope. The scan thus obtained was photographed so that the data could be recorded by the oscilloscope. The photographic scan interruption corresponds to the dynamic dead centre of the pedalling analysed.

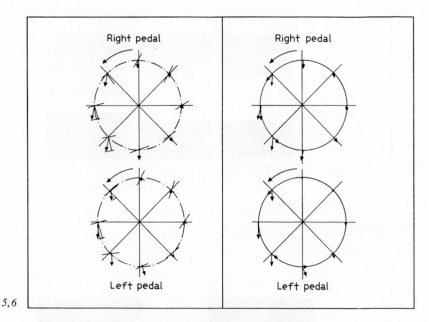

5,6

Fig. 5. Method to obtain total tension applied on the pedals.
Fig. 6. Tangential tensions on both pedals.

1. Method Used in the Analysis of the Data

After working out the parallel and orthogonal tension values exercised on the pedal and recording the positions taken at chosen points by the pedal with a kinematographic analyser, a graph of the tangential forces and the results of the points themselves was drawn with the purpose of comparing the actual phases with the exercised tensions.

III. Conclusion

The method may be developed according to the degree of the inquiry obtained through the elaboration of the data. In fact, the photographs showing the scans already allow a first interpretation of the mistakes made by the cyclist: such as asymmetry in the form of the curve obtained by both pedals, different intensity in the extension phases, i.e. of the thrust exercised by the limbs. This is useful for the correction of the pedalling technique.

A more thorough analysis, i.e. one which employs kinematographic inquiry, leads to the distribution of the forces during the whole pedalling action. Through analysis of both the forces exercised on the 2 pedals and the points on which the forces are applied it is possible to evaluate exactly the efficiency of pedalling. This inquiry will be the more accurate, the more numerous the points analysed during the whole pedalling action, up to an optimum total integration through a computer.

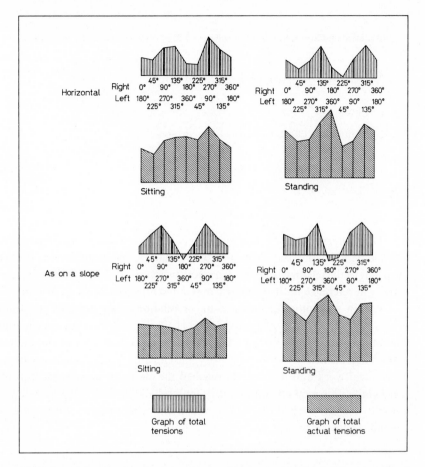

Fig. 7 Comparing graphs of total and tangential tensions in the 4 different explored positions.

From analysis of the forces obtained with the method already described, one can arrive at a comparison between the forces which can be usefully employed (tangential forces), the total tension applied and the dispersed tensions. From this, an idea can be had of the mechanical rendering of the alternating movement of the transformations, in rotary movement which takes place in the bicycle.

It can be deduced by this comparison that the high efficiency of the man-bicycle system can be attributed solely to the mechanic of the bicycle because, in the energy applied by the man to the machine, dispersion is great.

Author's address: Prof. Dr. ANTONIO DAL MONTE, Istituto di Medicina dello Sport, Via dei Campi Sportivi 46, *I-00197 Rome* (Italy)

Medicine and Sport, vol. 8: Biomechanics III, pp. 440–445 (Karger, Basel 1973)

Muscular Activity in Swimming

L. Lewillie

Laboratoire de l'Effort, Université Libre, Brussels

The problem of appraisal of the activity of muscles during free movement is fundamental in the study of human biomechanics. The swimmer was studied by means of the telemetric method previously described [2, 3] and the muscular activity in reference to the isometric maximum [4] was defined quantitatively.

Within physiological limits, a linear relation exists between the static muscular force and the integrated EMG [1, 4]. Some difficulties appear only when the load is below 10% of the isometric maximum, and at the highest loads when the activity of the accessory muscles allows a slight increase of the mechanical work and has not yet reached the electrical maximum. But, for these limits, this method shows a very high reliability even if the question of linearity may arise when the values measured during isotonic contraction go beyond the isometric 100%.

Calibration

The calibration is linear for the two types of integration [4]. It is realized that an isometric contraction (± 6 sec) is short enough to avoid fatigue [5]. It gives identical results when repeated without displacement of the electrodes.

The recordings presented have been made on the same subject, a swimmer of national class swimming the 4 strokes (crawl, breast stroke, back stroke, dolphin) at 3 different speeds (slow, normal, sprint). It may be observed that the movement cycle of the swimmer is even. Taking into account the cyclic differences due to the stroke (e.g. respiration in crawl), the length of one cycle, the duration of the muscular contraction and its quantitative importance are all highly reproducible.

We have measured in each case: (1) the total duration of one cycle, the muscular activity and the interval of relative rest, and (2) the maximum intensity reached and the total electrical activity developed by the muscle during the movement. These results are represented in table I.

It may be observed that the duration of the arm and leg movements is shortened with the increase of swimming speed and that the improvement in sprint is due exclusively to a reduction of the interval time. The intensity of muscular activity increases for the triceps brachii as well as for the rectus femoris.

Crawl

The shortening of interval time at sprint speed is very important: 0.96–0.32 sec for the arm and 0.32–0.18 sec for the leg. The duration of the triceps activity is shortened regularly when one rectus femoris stays perfectly equal. In this case, the lowering of total time for the leg kick is due to a change in rhythm (3 kicks for one arm movement at slow and normal speed, 2 at sprint speed).

The two spikes of the triceps recording, previously described [2], appear clearly. The activity accelerates rapidly and the total activity shows this by following closely the maximum values: 80% at slow speed, 86% when the movement speeds up.

A high level of activity persists during the interval even for a gliding stroke. The subject is a qualified swimmer and one can expect an almost complete relaxation during the non-active periods. This is obviously not the case when the EMG shows an activity of 20% of the isometric maximum at slow speed and of 50% when sprinting. These values are the highest observed for all strokes for that muscle.

The activity of the rectus femoris describes clearly the crawl kick. Leg kicks are perfectly equal at sprint speeds, but the respiratory rhythm increases 1 of every 3 kicks at slow speed, however rolling is not of consideration here.

The activity is not high (47–56% isometric) when movement is used only to stabilize the body. On the contrary, at sprint speed the shoulders rising out of the water cause an effective propulsion from the legs and the activity of the rectus femoris increases to 90% isometric. The activity during the interval also stays high (28–40%).

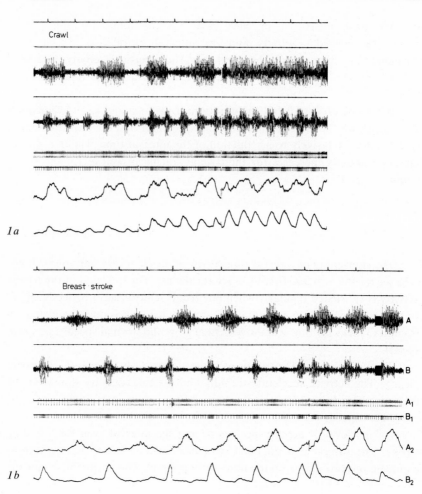

Fig. 1. a–d Recording of the EMG during 4 different swimming strokes performed at different speeds. A = EMG of the triceps brachii; A1 and A2 = integrated EMG; B = EMG of the rectus femoris, and B1 and B2 = integrated EMG.

Breast Stroke

In the case of the arm, the different times follow the same pattern as in the crawl stroke. The activity of the rectus femoris appears to be, on the contrary, very short when compared to the duration of one movement. The improvement in performance is due, at normal speed, to an effective increase

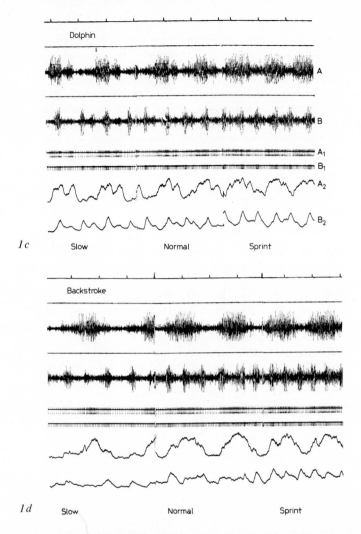

1c

of the duration of the leg thrust, when at sprint speed it is due to an increase of the arm pull and to a shortening of the interval, no change being observed in the rectus femoris.

At slow speed, the activity of the rectus femoris is at a high level (84%) also, of all strokes, the activity of the triceps is at its lowest level (20%). The rise of activity is slower than in the other strokes, as it appears in the important difference between the maximum (56%) and the total activity (20%). The glide of the breast stroke shows a lack of activity during intervals.

Table I

	Triceps						Quadriceps					
	duration in $^1/_{100}$th of a second			intensity in percent of the isometric maximum			duration in $^1/_{100}$th of a second			intensity in percent of the isometric maximum		
	cycle	contraction	interval	maximum	total	interval	cycle	contraction	interval	maximum	total	interval
Crawl												
Slow	218	90	118	108	86	20	68	32	36	47	50	28
Normal	178	82	96	114	96	25	64	32	32	56	51	28
Sprint	102	70	32	128	110	50	50	32	18	90	90	40
Breast stroke												
Slow	216	108	112	56	20	0	216	40	176	84	50	15
Normal	172	84	88	96	60	4	172	52	120	96	68	15
Sprint	120	80	40	120	80	10	120	36	80	96	72	20
Back stroke												
Slow	254	128	116	108	72	10	84	40	44	56	44	28
Normal	200	124	76	114	80	20	60	36	24	76	66	45
Sprint	153	128	25	142	96	22	52	32	20	96	90	52
Dolphin												
Slow	180	100	80	102	80	14	84	64	20	72	52	22
Normal	131	107	24	114	80	20	66	48	18	80	70	36
Sprint	120	90	24	160	104	30	58	32	26	100	96	45

Back Stroke

The general pattern is close to the crawl stroke, but slightly slower. The duration of the triceps activity is important: 80% of the cycle. The maximum intensity is higher than in the crawl when the total activity is lower, showing a slower rise of the contraction, corresponding to a slower arm stroke.

The dorsal position changes the action of the foot; therefore, it may be assumed that the activity of the whole lower limb could be different from the prone position. The activity of the rectus femoris during the leg kick is nearly the same in the crawl and back stroke. In the latter, the activity is slightly higher and ofter more irregular.

Dolphin

At sprint speed, the maximum intensity is the highest observed of all strokes. The interval is very short and the activity of the triceps very low at this time: almost half the values recorded in the crawl. The aspect of the curves seems comparable for the two strokes, but the difference between maximum and total intensity is 3 times higher in the dolphin stroke, showing that the rise of muscular activity is sharper. The rectus femoris also shows a high level of intensity at slow speed, nearing crawl values at high speed.

The telemetrical appraisal of muscular activity by means of integrated EMG and the quantitative estimation by reference to the maximum isometric activity appears highly reliable. This method allows a comparison of strokes, as well as muscles of different subjects. Its accuracy is very high and allows one to follow smallest modifications of activity.

Summary

A quantitative telemetrical analysis of the integrad EMG has been realized during 4 swimming strokes at different speeds by reference to the maximum isometric contraction.

This method appears highly reliable and allows one to study quantitatively the activity of the muscles, their importance in each stroke and the modifications due to training.

References

1   BIGLAND, B. and LIPPOLD, O. C. J.: The relation between force, velocity and integrated electrical activity in human muscles. J. Physiol., Lond. *123:* 214–224 (1954).

2   LEWILLIE, L.: Telemetrical analysis of the electromyogram. Biomechanics, vol. 1, pp. 147–149 (Karger, Basel 1968).

3   LEWILLIE, L.: Graphic and electromyographic analysis of various styles of swimming. Biomechanics, vol. 2, pp. 253–257 (Karger, Basel 1971).

4   LEWILLIE, L.: Quantitative comparison of the electromyogram of the swimmer. Biomechanics in swimming, vol. 1, pp. 155–159 (Brussels 1971).

5   SCHERRER, J. et MONOD, H.: Le travail musculaire local et la fatigue chez l'homme. J. Physiol., Paris *52:* 443–446 (1960).

Author's address: Prof. LÉON LEWILLIE, Laboratoire de l'Effort, Université Libre de Bruxelles, Avenue Paul Heger 28, *B-1050 Brussels* (Belgium)

Medicine and Sport, vol. 8: Biomechanics III, pp. 446–452 (Karger, Basel 1973)

A Kinematographical, Electromyographical, and Resistance Study of Water-Polo and Competition Front Crawl

J.P. CLARYS, J. JISKOOT and L. LEWILLIE

Vrije Universiteit Brussel, Instituut voor Morfologie; Université Libre de Bruxelles, Laboratoire de l'Effort; and Akademie voor Lichamelijke Opvoeding, Amsterdam

There is a tendency in Europe, among competition swimmers, not to take part in water-polo training or matches, because one might have a negative influence on the performance of the other. The study of both styles with different parameters might lead to the acceptance or rejection of this opinion.

Purpose

The purpose of this research is (1) the study of biomechanical, anatomical and resistance phenomena of competition and water-polo front crawl; (2) to describe similarities and differences of both competition and water-polo front crawl, putting the stress on the trajectory of elbow and wrist, using kinematography and a light-trace technique, and (3) to compare the absolute resistance of water on the body positions and movements for both swimming styles.

Methods

All subjects involved were competent swimmers and polo-players of national division level. For the analysis of movement, 16-mm and super-8 kinematography was used for general comparison, while the arm movement of both styles was studied with a light-trace technique, previously described for swimming movements by LEWILLIE [6] and COUNSILMAN [2], and on water-polo shooting by CLARYS [1].

The characteristics of the tele-electromyographical registration has been described by LEWILLIE [5]. Surface electrodes were used to investigate the action of the m. biceps brachii, the m. triceps brachii, the m. flexor carpi ulnaris and the m. brachio radialis. The tele-electromyographical analyses were synchronized with 16-mm kinematography of the

Fig. 1. Test set-up for tele electromyography (swimmer plus antennae, receiver and recorder, 16-mm film and sound-meter).

swimmer and with a sound-meter in order to locate the contraction in the arm movement as shown in figure 1.

The resistance tests were carried out in a towing tank (200 × 4 × 4 m) of the Netherlands Ship Model Basin. This technique has been applied to swimmers by OOSTERVELD and RIJKEN [7], VAN MANEN and RIJKEN [8] and by DE GOEDE et al. [3]. The basin walls are fitted with rails on which a towing carriage can be driven electrically at different, accurately adjustable speeds. The speed of the cariage is measured by a photo-electric cell system, and recorded on direct developing photo-strip chart. Figure 2 shows the test set-up for the measurement of forces acting on the moving body.

Results

The kinematographical analyses show important differences in the body position. The polo stroke is characterized by a lordosis in the lumbar region and an elevation of the shoulders. The light-trace analyses show a longer glid-

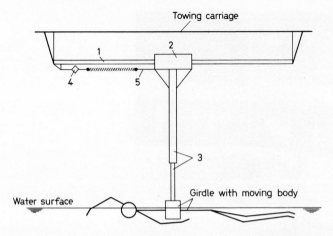

Fig.2. Test set-up for resistance measurement. 1 = horizontal steel shaft; 2 = tube; 3 = vertical telescopic rod; 4 = force transducer; 5 = horizontal connecting rod with spring system.

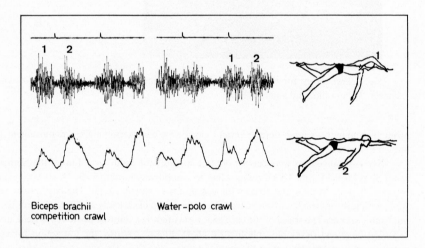

Fig.3. EMG of the biceps brachii.

ing period and consequently a longer arm cycle in the competition crawl. Biomechanically speaking, the arm movement patterns are very similar in both strokes. Generally, the brachial biceps works in the same way and at the same moment in both crawl strokes. The exact moment of the contractions

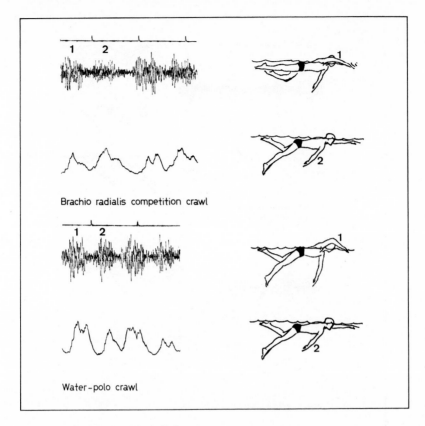

Brachio radialis competition crawl

Water-polo crawl

Fig.4. EMG of the brachio radialis.

is shown in figure 3. This is similar to the findings of IKAI *et al.* [4]. The m. brachio radialis action differs from the biceps in its regularity and its contraction level, but the general pattern is the same in both crawl strokes. Figure 4 shows the timing and the number of contraction peaks. The m. flexor carpi ulnaris action bears a great resemblance to the brachial biceps in the competition crawl movement. The agreement is less in the water-polo stroke due to irregularities in alternative flexion of the hand and wrist, combined with an ulnar abduction (fig. 5). Upon investigating the brachial triceps muscle it was striking that no differences in amplitude were found either between both crawl strokes or between training and sprinting (rhythm). The competition crawl pattern is similar to the findings of LEWILLIE [6], while in water-polo the continuous contraction plateau is dominant, as shown in figure 6.

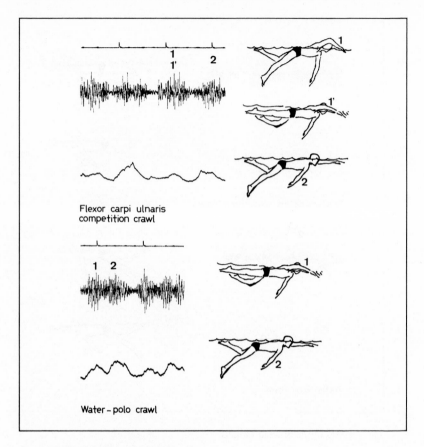

Fig.5. EMG of the flexor carpi Ulnaris.

The results of the resistance measurements are interpreted on their hydro-mechanical meaning[1]. The differences in attitudes for both types of crawl strokes give way to the hypothesis that the water-polo attitude may result in a higher resistance than the competition attitudes. The water-polo stroke with the same energy consequently may result in a lower free-swimming speed. This was indeed measured for 7 of 10 subjects where the competition stroke resulted in higher speed (average 6%); 3 subjects obtained a higher speed with the polo stroke. If the resistances to be overcome at the free-swimming speeds are compared, however, it appears that the resistances for water-polo

1   Report No. 71–276–1–HST; Netherlands Ship Model Basin; H. RIJKEN, August 1971.

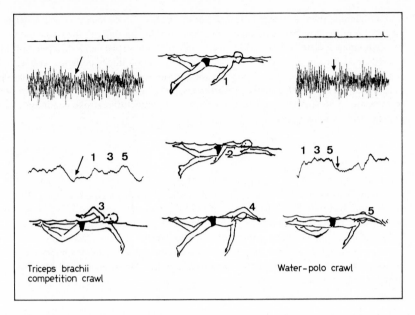

Fig.6. EMG of the triceps brachii.

attitude are between 6 and 45% higher than for the competition attitude at 6% lower speeds. In terms of power it appears from the test results that for subjects using the water-polo stroke (between 15 and 50%) more power output has been developed. This means that under water-polo stroke conditions the total energy output of the subjects is, in general, larger than under competition stroke condition, despite the fact that the free-swimming speeds under competition stroke conditions are higher.

General Conclusions

While studying the crawl movement of swimming and water-polo in 3 different manners with mutual relationships, it has been possible to obtain a fairly complete image of the biomechanical, anatomical and resistance parameters of both competition and water-polo front crawl.

With kinematography, the main difference was found to be in position. Arm and leg movements show great similarities. In relation to kinematography, apart from the contraction level and duration, the electromyographical investigation of the 4 arm muscles did not show any important differences.

Originally, the differences in resistances of the body in competition and water-polo attitude confirmed the hypothesis that both styles would also differ in their execution. The free-swimming speeds and the resistance values of both strokes and their relationships towards the towed position values throws great doubt upon this hypothesis. Although the sample is too small, one can conclude that water-polo front crawl does not have a negative influence on the competition stroke. Nevertheless, there remains the question why the output of the water-polo stroke cannot be delivered under competition stroke condition and so improve free swimming speeds.

Acknowledgements

This investigation has been sponsored by the Ministry of Dutch Culture (BLOSO) and by the 'Instituut voor Morfologie' of the Vrije Universiteit Brussel. The Netherlands Ship Model Basin of Wageningen and the Physical Education Academy of Amsterdam are gratefully acknowledged for their respective apparatus and pool facilities.

We are indebted to Prof. BROUWER for his advice and to all the swimmers and water-polo players for their assistance. We are also grateful to Mr. ROBEAUX and Mr. H. RIJKEN for their technical aid.

References

1   CLARYS, J.P.: Emploi des traces lumineuses lors d'une analyse de trajet au cours de plusieurs shots de water-polo. Kinanthropologie *3/1:* 27–42 (1971).

2   COUNSILMAN, J.E.: The application of Bernoulli's principle to human propulsion in water. 1st Int. Symp. on Biomechanics in Swimming, pp. 59–71 (Lewillie & Clarys, Brussels 1971).

3   GOEDE, H. DE; JISKOOT, J. en Sluis, A. VAN DER: Over de stuwkracht bij zwemmers. Zwemkroniek *48* (2): 78–90 (1971).

4   IKAI, M.; ISHII, K., and MIYASHITA, M.: An electromyographic study of swimming. Jap. Res. J. phys. Educ. *7:* 55–87 (1964).

5   LEWILLIE, L.: Telemetrical analysis of the electromyogram. Biomechanics, vol. 1, 1st Int. Seminar, Zürich 1967, pp. 147–149 (Karger, Basel 1968).

6   LEWILLIE,L.: Quantitative comparison of the electromyogram of the swimmer. 1st Int. Symp. on Biomechanics in Swimming, pp. 155–159 (Lewillie & Clarys, Brussels 1971).

7   OOSTERVELD, M.C.W. en RIJKEN, H.: Analyse van het aandeel van de door de benen, resp. door de armen geleverde stuwkracht bij zwemmers. Rapport n° 70-039-SP (Nederlands Scheepsbouwkundig Proefstation, Wageningen 1970).

8   MAANEN, J.D. VAN en RIJKEN, H.: Voortzetting van het onderzoek naar het aandeel van de door de benen geleverde stuwkracht bij zwemmers. Rapport n° 70-097-SP (Nederlands Scheepsbouwkundig Proefstation, Wageningen 1970).

Author's address: Dr. JAN P. CLARYS, Vrije Universiteit Brussel, Instituut voor Morfologie, Eversstraat 2, *B-1000 Brussels* (Belgium)

Medicine and Sport, vol. 8: Biomechanics III, pp. 453–459 (Karger, Basel 1973)

The Influence of Mechanical Factors on Speed in Tobogganing

W. Baumann

Institut für Biomechanik, Deutsche Sporthochschule, Köln

Based on experimental studies during the Toboggan World Championships in 1970 in Berchtesgaden, Germany, a mechanical model of the descent has been developed. With regard to the geometry of the course and the physical properties of the running toboggan, the velocity displacement function and the time displacement function have been computed. Comparison of calculated values and record performance data results in conformity of a high degree. Time displacement function is approximated with 1% accuracy, for subjective estimated 'faultless' runs with 2% accuracy. Hence, the model is a useful starting point for a numerical description of the toboggan motion under the influence of mechanical factors. Corresponding results are presented in this paper.

Speed and competition performance in tobogganing are determined to a high degree by the geometry of the toboggan course and the physical characteristics of the running toboggan. Based on experimental studies at the Toboggan World Championships in 1970 in Berchtesgaden, Germany, a mechanical model of the descent has been developed. For this purpose the following geometrical and physical quantities enter into the mathematical description of the mechanics of descent: lengths, slopes and averaged radii of curves of the sectionized course, initial speed, total weight of toboggan and tobogganer, coefficient of friction and air resistance. Figure 1 shows the forces applied to the toboggan on the inclined plane.

From an equation of motion for straight-line motion we have:

$$m \frac{d^2 x}{dt^2} = m\,g \sin \alpha - \mu m\,g \cos \alpha - \frac{c_\omega \tau A}{2} \cdot \left(\frac{dx}{dt} \right)^2. \qquad (1)$$

In approximate regard to centrifugal force in curves, which increases

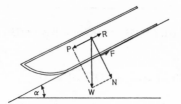

Fig. 1. Forces applied to toboggan on inclined plane. P = propelling force; W = weight; N = normal component of weight; F = frictional resistance; R = air resistance; α = angle of inclination.

restraining friction forces, we get the additional term:

$$-\frac{m\mu}{r} \cdot \left(\frac{dx}{dt}\right)^2.$$

Substitution in equation (1), dividing both sides by m and rearranging, produces finally:

$$\frac{d^2x}{dt^2} = g(\sin\alpha - \mu\cos\alpha) - \left(\frac{c_\omega \tau A}{2m} + \frac{\mu}{r}\right)\left(\frac{dx}{dt}\right)^2, \tag{2}$$

where t = time, x = co-ordinate in direction of motion, α = angle of slope, r = averaged radius of curve, g = gravitational acceleration, m = mass of toboggan and tobogganer, A = shadow area, τ = density of air, c_ω = coefficient of air resistance, and μ = coefficient of sliding friction.
Integration of the second-order differential equation (2) yields the solutions:

$$v(x) = \sqrt{v_1^2 + (v_0^2 - v_1^2)\exp(-2bx)}, \tag{3}$$

and

$$t[v(x)] = \frac{1}{b} \cdot \tanh^{-1}\frac{v_1(v - v_0)}{v_1^2 - v v_0}, \text{ for } |v| < v_1, \tag{4}$$

where

$$b = \frac{c_\omega \tau A}{2m} + \frac{\mu}{r},$$

v_0 = initial velocity and v_1 = limiting velocity.
Limiting velocity v_1 results from the condition:

$$\frac{d^2x}{dt^2} = 0$$

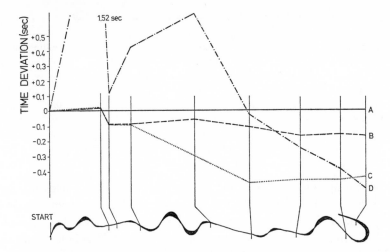

Fig. 2. Time deviations between calculated intermediate and total times and record performance data plotted against course length. A = record performance; B–D = calculated times for different degrees of approximation to the course; B = course sectionized in 15 parts with different lengths, slopes and radii of curves; C = course sectionized in 8 parts with different lengths and slopes but no curves; D = course taken as a straight with averaged slope. Below: tobogganing course of Berchtesgaden, Germany; artificial ice; length, 1,080 m; level difference, 116.7 m.

and is given by the expression:

$$v_1^2 = \frac{g\,(\sin \alpha - \mu \cos \alpha)}{\dfrac{c_\omega \tau A}{2\,m} + \dfrac{\mu}{r}}$$

The solutions (3) and (4) describe speed and time, respectively, as a function of distance covered.

In order to check conformity between calculated values and competition performance data the tobogganing course in Berchtesgaden has been surveyed. Furthermore, we measured initial velocity, intermediate and total times of nearly all World Championship male participants in 8 runs, to some extent also body weight, including sports dress. We then calculated from equation (4) intermediate and total times and compared them with the corresponding values of the best performances. Figure 2 represents the time deviations between record performance (line A) and different approximations to course characteristics.

Table I. Change of total time due to variations of mechanical factors

Quantity	Magnitude of variation	Resulting change of total time, sec
Weight	+ 10 kg	− 0.25 to − 0.35
Initial velocity	+ 1 m/sec	−0.6
Coefficient of friction	+ 10 % (fig. 3 b)	+ 0.25 to + 0.30
Air resistance	+ 10 % (fig. 3 c)	+ 0.45 to + 0.70

Obviously, subdivision of the course into 15 sections yields the best approach (line B): the model describes the motion of the toboggan with good approximation, for the record performance with an accuracy of 1%, generally with a 2-percent accuracy. Since we have a model which holds good for 'faultless' performances (determined by excellent total times and subjective estimation), we own a useful tool for detecting and localizing of the tobogganers faults as well as of critical points of the course. Furthermore, we can derivate more general relations between mechanical factors and velocity displacement respectively time displacement functions at tobogganing. The latter has been carried out for the toboggan course of Berchtesgaden. The following factors have been varied: total weight of toboggan plus tobogganer, initial velocity, coefficient of friction and air resistance.[1] Figure 3a–c, shows the results, weight vs. total time, for different parameter variations.

Common tendency of all graphs is the decrease of total time with increasing weight. Obviously parameter variations are more efficient at low weights. The generally derivable interrelations between physical quantities and connected changes of total times are summarized in table I.

Estimating these values, we must take into account the additional data of best performances: total weight, 95–110 kg; initial speed, 7–8 m/sec; total time 45–45.5 sec; for friction forces and air resistance see figure 3b and c.

Beyond that we tried to clear another problem, namely the question whether the obvious gain in time on the straight due to higher weights is compensated by an equivalent loss of time in the curved sections of the course, or not. From equation (3) we calculated speed changes with respect to weight in different course sections: on a straight, 100 m long, in 90° circular curves with the radii 10 and 40 m, and in a double S-curve. Common slope for all

---

1 Air resistance data are taken from unpublished information about wind-tunnel measurements by other authors.

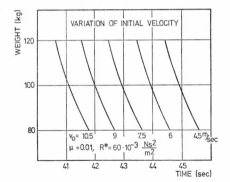

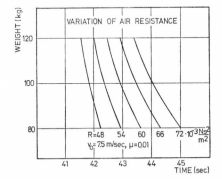

Fig. 3 a–c. Time vs. weight for different parameter variations. The quantities actually kept constant are indicated below the curves. μ = coefficient of friction; v_0 = initial velocity; R^* = air resistance factor (in addition, air resistance factor R^* is varied with total weight according to ± 1 kg = $\pm 3.10^{-3}$ Ns²/m²).

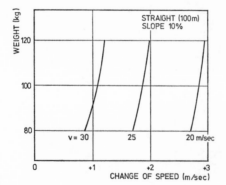

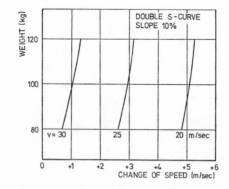

Fig. 4. a–c. Change of speed vs. weight for different course elements. Lengths: straight = 100 m; curve (r = 10 m) = 15.7 m; curve (r = 40 m) = 62.8 m. Double S-curve = 215 m. R = 6.10⁻² Ns²/m² (for weight = 100 kg), μ = 0.01.

sections was assumed to be 10%. The results are graphed in figure 4a–c. Corresponding to different possible allocations on the course, initial speed for the sections is varied from 20 to 30 m/sec. Clearly on a straight we have increased speed at the end of the section proportional to the weight. At the end of the curves we get the same dependence of changes of speed on weight: greater gain and smaller loss of speed with increasing weight for r = 40 m and r = 10 m, respectively. However, calculation has been carried out without consideration of the fact that the steering of the toboggan – accompanied by considerable transfer of body weight – may be made more difficult by centrifugal accelerations in curves up to 5 times gravitational acceleration.

The nearly complete determination of the toboggan descent by mechanical laws suggested the set-up of a differential equation. The resulting functions give us some insight into the interrelationships between mechanical factors and performance. The active intervention of the tobogganer resulting in a superposition of those functional relationships by stochastic variables implies further problems, which must be solved by extended theoretical and experimental studies.

Author's Address: Dipl. Phys. W. BAUMANN, Institut für Biomechanik der Deutschen Sporthochschule, Carl-Diem-Weg 5, *D-5000 Köln 41* (FRG)

V. Applied Biomechanics in Rehabilitation

Medicine and Sport, vol. 8: Biomechanics III, pp. 462–471 (Karger, Basel 1973)

Biomechanics and Medical Rehabilitation: Introductory Remarks

S. Boccardi

Istituto di Terapia Fisica e Rieducazione Motoria, Ospedale S. Carlo Borromeo, Milan

I am very pleased to address this 3rd International Seminar on Biomechanics, and I thank the Organizing Committee for asking me to participate as an official speaker.

Medical rehabilitation and biomechanics are both new branches of science which begin to co-ordinate the enormous mass of data and experience gathered in the course of centuries. The most technically advanced countries, with their highest economical levels have taken the initiative and have formed the first working teams. Italy is at a disadvantage, due to the lack of means and directives and to the shortage of specialized doctors and of technical training of staff. In spite of all this something has been done, but our position is not stable enough to allow a wide field for the operators of the two sectors to meet.

I myself am a physiatrist: I work in general hospitals and in rehabilitation centers, dealing with everyday problems of diagnostics and therapy. Little time, few collaborators and a shortage of financial aid limit our ambition in research, so that we must turn to medical literature for the information we need. Because of all these conditions, I cannot give any new contribution to an assembly of high-ranking specialists. Therefore, my paper will be truly introductory to a very interesting session. I think that my participation may be useful if I limit myself to a series of questions, to be answered by the experts in biomechanics to the medical rehabilitation operators. There are, of course, many unanswered questions. Medical rehabilitation aims at re-establishing altered functions: and most cases of illness involve alterations of movement. There are three main ways of reaching this result.

1. Use of postures and movements to modify one or the other of the altered elements (kinesitherapy).

2. Learning new postural or movement patterns to ensure the functional activity which can no longer return to the original form (functional re-education).

3. Use of external devices for functional substitution (prothesis and orthesis).

In other words, the aim of medical rehabilitation is the recovery of altered function of motion through motional activities or mechanical devices. I think that in no other sector of applied medicine is biomechanics more necessary than in this one. Biomechanics can and must clarify: (1) details of normal motion; (2) alterations which have taken place after the sickness, and (3) characteristics of rehabilitation techniques to be used.

Biomechanical research has already given a huge mass of results in the first of these three fields; that is, the study of normal movement. Much remains to be done concerning the description of alterations in movement patterns, which are still so often worded with the approximate terms of the clinical terminology of the 19th century. And much more remains to be done concerning the analysis of the techniques of motor re-education and of their effects. Very often new treatments are suggested, built on a somewhat vague as well as fascinating theoretical basis, and their results, described as excellent, are displayed without a serious documentation. A serious and thorough evaluation of the true effects of these techniques is very important if medical rehabilitation is to be given the serious consideration it deserves. As examples of our need for a sound kinesiological knowledge I will now present you with a small series of problems which are particularly open to biomechanical analysis. For some of them I will offer the result of some of our non-sophisticated observations.

As to the analysis of normal motion, one of the most interesting aspects is that of the individual variability of many of the morphological and functional elements of which it is made up. It is very important to know the abstract synthesis of the human gait, but it is also important to know how differently men walk, and even the different ways an individual man can walk. One of these variable factors to be precised is the range of motion of the joints. These are (table I), for example, the normal limits of the movements of the hip, following different textbooks of kinesiology. On these grounds, it is very difficult for a physiatrist or a physiotherapist to decide if the hip joint of his patient is normal and must be respected, or if it is limited and must be treated.

We have carried out some observations on spine, hip, knee, shoulder and ankle joints in homogeneous groups of subjects. As an example you can see data on the variability of the range of motion of the hip joint in a group of

Table I. Range of motion of the hip joint

	Lapierre	G. Scott	Cappellini	Rocher
Flexion, degrees	120	115–125	120	130
Extension, degrees	10–15	–	13–15	10–35
Adduction, degrees	30	–	40	10–15
Abduction, degrees	–	45–50	60	45
Inward rotation, degrees	37	–	–	45
Outward rotation, degrees	13	–	50–60	60
Rotations, degrees	50	90–100	50–60	105

Table II. Range of motion of the hip joint

	Mean, degrees	SD, degrees	Range, degrees
Flexion, straight knee	90	±8	82–98
Flexion, bent knee	128	±9	119–137
Extension	12	±5	7–27
Adduction	20	±6	14–26
Abduction	36	±7	29–43
Inward rotation	43	±5	39–47
Outward rotation	40	±7	35–45

Table III. Lateral bending of the trunk

	Mean, degrees	SD, degrees	Range, degrees
Left bending	44	± 7.8	21–67
Right bending	41	± 6.9	20–54
Total bending	85	±12.3	43–115

Right bending/Total bending 48.2%. Difference between means: left-right bending, 3°, significant at 0.01 level.

young women, homogeneous of age and physical fitness (table II, fig. 1). For what concerns the lateral mobility of the spine, the variability is perhaps greater (table III). It is interesting to point out that the difference between the average values of left-bending and right-bending is statistically significant, this may depend on the presence of the so-called physiological scoliosis.

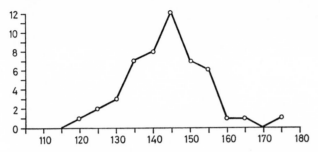

Fig. 1. Distribution of the range of motion of the hip on the sagittal plane.

Another interesting aspect is that of the limits of the normal variety of some relations between different body segments in the habitual postures. We must remember, on this subject, that findings of 80% of scapula alata and 50% of postural scoliosis at school age have been reported in these last years. On these grounds, a very expensive campaign of depistage and corrective treatment has been undertaken in Italy. I think that the approach to the problem will be wrong if we do not have some homogeneous and efficient technique of evaluation at our disposal and a set of parameters representing the 'normality' for comparison. We have dedicated particular attention to the 'normal' erect position. The term 'normal' must be considered in its statistical meaning, as the average of a great number of individual postures, each one of them normal for a given subject. Unfortunately, the term 'normal' is often interpreted as 'ideal', a standardized model of postural alignment to be reached by everybody through years of tedious and seldom effective exercises. A series of observations carried out by a group of students of the Istituto Superiore di Educazione Fisica of Milan gave the large range of individual variations listed in table IV.

A third field, concerning medical rehabilitation, is on muscular strength. A lot of time is dedicated, in physical therapy departments, to rebuilding the power of directly or indirectly damaged muscles. The research should aim at defining the normal limits of the strength of the different muscular groups, in both sexes, in different ages and conditions. The technique of evaluation should be specified with great accuracy. As an example of the usefulness of this kind of work I will mention the problem of the 'dépistage' of the carriers of Duchenne-type muscular dystrophy. The knowledge of the normal range of the strength of the girdle muscles in young women can help in a first screening of the potential carriers of the disease.

Another worthwhile research, for practical purposes, is that concerning

Table IV. Body alignment in standing

	Mean, degrees	SD, degrees	Range, degrees
Head posture	54	±6	38–70
Dorsal kyphosis	159	±6.8	142–178
Lumbar lordosis	150	±7.3	131–166
Trunk forward shift	86	±2.8	73–92
Pelvis sagittal tilt	12	±5	2–28
Hip forward shift	96	±2	92–102
Abdomen protrusion	8	±2.8	3–18
Knee sagittal alignment	174	±5	165–190

Table V. Grip strength in different elbow positions

	Mean	SD
Straight elbow	43.3	±10.2
Elbow at 90°	41	±9.7
Elbow at 90°, supported	41.6	±9.4
Elbow at 70°	41.4	±9.8

Differences between means are *non-significant.*
Correlation coefficients:

	B	C	D
A	0.895	0.925	0.864
B	–	0.966	0.905
C	–	–	0.950

Table VI. Strength of the abductor muscles in different positions of the shoulder

			Difference between means	t	p
A Inward rotation at 90°	36.14	±9.2			
B Inward rotation at 45°	37.39	±8.7			
C Anatomical position	43.14	±12.8			
D Outward rotation at 90°	49.16	±14.9			
A–B			−1.25	0.68	NS
B–C			−7	3.07	<.01
C–D			−6.02	2.82	<.05

NS = not significant.

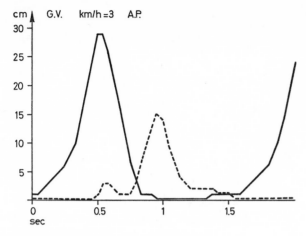

Fig. 2. Heel and toe trajectory in normal gait.

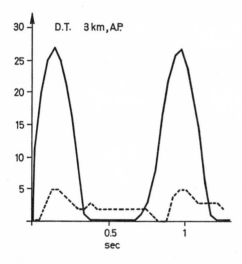

Fig. 3. Heel and toe trajectory of a below-knee prosthesis wearer.

the variations of the isometric force in various positions of body segments. It has been suggested that isometric exercise is particularly effective if the muscle is in a position of highest mechanical efficiency; that is, in the position in which electrogenesis is greatest. To this purpose, a series of observations on grasp muscles and on abductors of the shoulder has been made (table V, VI).

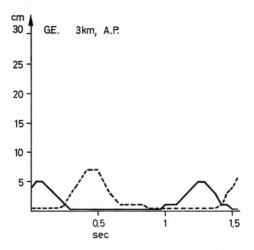

Fig.4. Heel and toe trajectory of a long-leg brace wearer.

Table VII. Lateral flexibility in 51 polio scolioses

	Mean[1]	SD
At first examination	7.38	±2.77
At last examination	7.69	±2.70
Normal range of motion	8	±2.18

1 In degrees pro vertebra.

Differences between means are *non-significant*.
Correlation coefficients:

	Last control	Increase, %
First control	0.805	−0.001
Last control	–	0.016

As to the analysis of pathological movement, I think it is superficial and useless to try to correct a motion trouble without knowing its characteristics. This field of research is immense and almost unexplored. For example, we still do not have an effective and sure method for evaluating muscular spasticity, even for rehabilitation purposes.

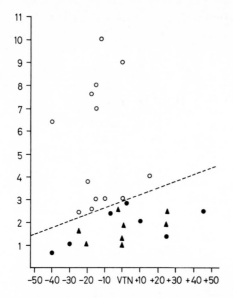

Fig. 5. Evolution of polio scolioses related to gravity and to mobility. x = range of motion at first examination; y = degrees pro vertebra at first examination; VTN = normal range of motion; + = increased; ▲ = improved; ● = unchanged.

A very interesting sector is that of pathological gait. We have studied the vertical displacement of the foot in walking, as its normal pattern is clearly affected by pathological alterations as the rigidity of a joint, the weakness of a muscular group, the use of a prothesis. Through a kinematographic analysis of the gait against a measured background, we obtained graphs of the normal trajectory and its modifications (fig. 2–4). We can see the graphs of a below-knee prothesis wearer and of a long brace wearer. I think that even these findings can help in making the choice between different kinds of prosthesis or braces. We have also studied the mobility of polio scoliosis on more than 50 polio patients who were under our care from a very early age. We have thus confirmed that polio scoliosis remains relatively mobile on a frontal plane, at least for the first 6–8 years (table VII). The prognostic utility of an early evaluation of the severity and the mobility of the curve has also been confirmed. We can see from the graph (fig. 5) that if the X-rays show a curve of more than 3° for each vertebra involved at the first examination such cases will deteriorate, while reduced mobility does not imply a negative prognosis.

Table VIII. Effects of brief isometric exercises, following different authors

Authors	Muscles	Isometric contraction, duration sec	Number of Contractions	Joint position °	Average gain %	Contro-lateral effects	Hypertrophy
1954 HETTINGER	–	5	1–7	–	–	–	yes
1955 BAER	wrist flexor	?	10	0	35	–	–
1957 RASCH	elbow flexor	15	1	90	0	–	–
1957 ROSE	quadriceps	5	1	0	30	yes	no
1959 LIBERSON	abd. 5° dig.	6	20	0	203	no	yes
			1		114		
1960 WALTERS	elbow flexor	15	1	90	36	yes	–
1960 LAWRENCE	quadriceps	30–45	10	0	50	–	–
1961 GERSTEN	triceps			70	15		–
	quadriceps[1]	5	5	90		?	–
	hamstrings[1]			0	40		
1962 LAWRENCE	quadriceps	30	10	0	62	yes	–
1962 SCHWEID	quadriceps children	5[2]	1	0	77	no	no
1962 SUTTON	deltoid			90	90		
	gluteus med.	5[2]	1	45		–	–
	triceps			0			
1962 SUTTON	dors. interos.	5[2]		0	30	–	–

1 Abnormal muscles.
2 After isotonic contraction.

But, as I said, it is especially in the field of the analysis of re-education techniques that detailed research and precise data are needed. To prove this we show, on this chart (table VIII), contrasting evaluations of the effects of brief isometric exercises reported by several authors. We have carried out tests involving 200 measurements on 10 normal people on the effects of manœuvres usually employed as neuromuscular facilitation techniques. We wanted to know how the maximal isometric force of shoulder abductor muscles is modified by the strong simultaneous contraction of other muscular groups or by a different position of the head.

To reassume: the execution of a whole mass pattern, according to KABAT, determines a surprising reduction of the isometric force, while it should facilitate the recruitment of motor units of all the muscles that belong to the synergy. This probably depends on the experimental conditions, but in any

Table IX. Influence of facilitation techniques on the isometric strength of the abductor muscles of the shoulder

	Right	Left	Mean	Difference between means	t	p
1 Standard test	61.8	54.6	58.2			
2 Bilateral contraction	62.3	58.4	60.3			
3 Contralateral grip	62.5	59.6	61.0			
4 Homolateral head rotation	62.5	53.5	59.4			
5 Contralateral head rotation	60.3	54.8	57.5			
6 Mass pattern	45.4	44.0	44.7			
1–2				−2.1	1.70	NS
1–3				−2.8	2.18	0.05
1–4				−1.2	0.98	NS
1–5				0.7	0.66	NS
1–6				13.5	7.64	0.001
4–5				1.9	2.08	0.05

case it confirms the need of a very accurate execution of the facilitatory technique. The use of the contralateral grasping muscles has a considerably positive influence on the isometric force of the abductor muscles (table IX).

I hope that my brief exposition of a few interesting and so different questions may have shown the wide range of problems open to exploration and the benefits which biomechanics can bring to a rational use of rehabilitation techniques.

Author's address: Prof. SILVANO BOCCARDI, Ospedale San Carlo Borromeo, *Milan* (Italy)

Medicine and Sport, vol. 8: Biomechanics III, pp. 472–481 (Karger, Basel 1973)

Bio-Mechanics and Bio-Electronics in the Application of Protheses to the Upper Limbs

V. Guardascione and H. Schmidl

Instituto Nationale per l'Assicurazione contro gli Infortuni sul Lavoro, Rome

One of the most problematic areas in prosthetic research relates to the upper limbs. The problem has existed since man began to use tools and machinery to increase his functional possibilities. However, in the design of artificial limbs, the main purpose is build a mechanism that will enable a functioning that is as natural as possible, with the greatest range of functions.

The expresion 'as natural as possible' is the key to the problem, since it implies that the device should appear like and function as the limb that it replaces without the control of the artificial device interfering with any of the other physiological movements. Whereas it is possible for a person to be entirely absorbed in the operation of a machine – for example, an automobile – it is quite impossible that a person should be obviously and totally absorbed in making a prosthesis work. For this reason, the more naturally the limb can be made to operate without engaging the possessor's attention, the better it is for the disabled. A research of the possibilities to construct an artificial limb can be divided into two aspects: one concerns the artificial replacement of the lost articulations, the other concerns the control of these articulations. At the beginning, when electronics did not offer the possibility of such perfect control through the use of muscle potential, it was thought that the greatest obstacle was the impossibility of exercising control over movements. Today, on the contrary, the greatest obstacle is considered to be the natural substitution of the various movements of the arm, a substitution that is almost impossible – with the means at present available – above all, because at this moment we are not even in a position to define precisely the measurements of the movements of the joint.

Many centres of research have investigated, and are still investigating, interphase couplings – particularly with regard to persons with a high level of amputation, in order to determine whether these methods can be used, improved or rendered multifunctional.

In particular, the investigation consists of studies of motion in order to determine what movements are required, how many degrees of liberty and the range of movement in each degree for the performance of the maximum possible number of actions in daily living, how much benefit can be drawn from each degree of liberty and the optimum range for each degree. All this might be summed up in the expression 'efficiency of motion'. The ultimate purpose is that of simplifying prosthetic design and reducing the requirements of the control areas.

LEONARDO DA VINCI (1452–1519) carried out investigations on human motion, but scientific investigations, properly so-called, began in the 19th century with German and Swiss researchers, i.e. ALBERT (1876), STRASSER, GASSMAN, BRAUNE and FISCHER (1887).

Generally, these studies were concerned with the investigation of the lower limbs. It is interesting to note that up to 1947, nobody had made any attempt to measure the co-ordinated movements of the upper limbs in daily life. Most of the investigations dealt with terminology or were based on enquiries to determine the range of rotation of the joint, each acting independently.

At the beginning, measurements were made with a goniometer to determine the ranges of movement of a joint. The measurement techniques have gradually changed until now research workers use photography, photogrammetry and cine-radiography and, more recently, computers, a technique that has made possible what promises to be the quickest method for the study of movements of the joint. According to the reports of some research workers, the method makes use of a television camera directly linked to a computer. The parts of the body to be examined are identified by means of very small sources of light or special marks. The desired movements are then photographed with the television camera. The television signal is then fed directly into the computer, where it is immediately analysed and the results plotted, projected or printed. The method promises to be the best collector of data on combined movements but calls for considerable quantities of equipment and advanced technology.

These studies have specially interested research workers in connection with the construction of artificial limbs and, above all, in order to have precise information on the possible partial or total elimination of articulation without worsening the essential daily movements.

As I have elaborated before, our present technological resources do not permit us to construct an artificial upper extremity that perfectly imitates the natural limb.

The investigations carried out at the Inail Prostheses Centre, where a group of investigators has been working since 1964, have shown that with the methods available to us at the present time, an artificial limb with combined movements creates confusion in the patient.

These combined movements can be controlled electronically by means of a programmer or they can be coupled by means of micro-switches but they do not permit, in this case, the control – even singly – of the various movements.

It has, moreover, been observed, and this has been noted by other groups working in this field, that the construction of a prosthesis for the upper limbs cannot be based on the movements of a natural limb; but consideration must also be given, above all else, to the mental and physical capabilities of the amputee in order that he may be given the means of controlling in a natural manner every movement of the artificial limb.

Many years of investigations have shown that the good functioning of prostheses depends, especially, on the control of the movements and above all on the amputee-prosthesis contact; the man-machine relationship.

With the possibility of utilising muscular potential there has been created, for the first time, a direct contact between the stump and the prosthesis so that it is possible to affirm that we have attained a physiological control of the artificial limb. Since 1964, the research department of the Inail Prostheses Centre at Vigorso di Budrio has dedicated its efforts to the utilisation of muscle potential as a control signal for an artificial limb for an upper extremity amputation. As early as the same year, it was found possible to apply a prosthesis to a forearm amputee with myo-electric control for the regulation of hand movements.

If, at the beginning, the control signal was simply an on-off (that is, all or nothing) signal and it was possible to utilise for each muscle only one signal, it has subsequently been possible to create electronic components that have made possible the utilisation of the muscle potential as a proportional control. The achievement of an electronic device for the utilisation of several signals from one muscle opened the way to the achievement of an entire artificial arm with up to 8 operative control signals.

At the present time, it is possible to construct an artificial limb for the upper extremity as far as the disarticulation of the shoulder. The limb itself is equipped with a precise control of the hand, wrist and elbow, with proportional control of the grip and control of the speed of movements. Sensitive control of the hand is also possible, and this control is incorporated into the

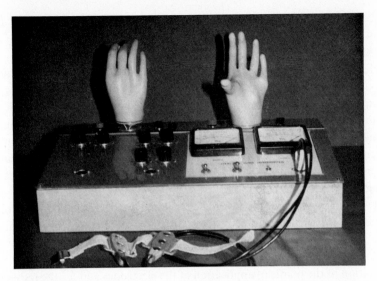

Fig. 1. Myometer and detector of myo-electrical signals for multi-channel myo-electrical control.

prosthesis itself. The preliminary condition for all this was, as I have just mentioned above, the achievement of an electronic device that draws at least two signals from a single muscle and which, at the same time, makes possible a proportional control.

The most important step in this research was the production of an amplifier which would make it possible to utilise the muscle potential at various levels that can be sensed and controlled by the amputee himself. Consequently, as the ultimate result of the research, the amplifier is made up of two elements or parts, one of which consists of an integrated circuit differential amplifier followed by a transistor for further signal amplification, with the possibility of output level adjustment through a potentiometer. The other part of the amplifier has, instead, the function of transforming the output signal coming from the first part into a 2-level signal. The low level operates a power transistor directly connected to the motor power supply for the movement; thus, a control is achieved which is proportional to the contraction of the muscle itself. At the high level, as soon as the power transistor is saturated, a special relay arranges to invert the motor polarity. The motor continues functioning as long as the muscle remains contracted. On release of the muscle contraction, the motor returns for an instant to the previous condition, a braking action thus being achieved. The two movements are independent of each other and do not interfere with one another.

After the construction of the amplifier, which enabled us to utilise one muscle for a 2-signal myo-electric control, we realised, on the basis of practical experiments on the amputee, that the instruments normally used for myo-electric prosthetic application are quite insufficient. We observed, indeed, that the amputee needs special preparation for the application of the multi-channel control. To carry out this preparation it was necessary to develop an instrument capable of recording the necessary values and, at the same time, of showing the amputee how to operate the muscles used as controls, this through light signals.

Our long experience in this field, an experience derived from the practical application of myo-electric prostheses with normal, but not proportional, multi-channel control has helped us greatly in developing this work.

The instrument mentioned above incorporates a myo-meter for the localisation of the points of application of the electrodes and for the measurement of the muscle potential. In addition, the instrument is provided with leads for practical tests of the proportional multi-channel control to be made by the amputee. These tests give the technician the necessary information for a correct adjustment of the two levels of the muscle signal according to the peculiarities of the amputee himself.

Several warning lamps of different colours are also inserted in the instrument, the various colours corresponding to the movements for opening, closing and rotating the wrist and artificial hand are mounted on the instrument. From the variation in intensity of the lamps the amputee can observe the proportionality of his own signal. The construction of this instrument, which is absolutely necessary, was motivated by the need to offer the amputee the possibility of adequate practice, so that he might be able to achieve the maximum possible functionality from his myo-electric artificial limb with multi-channel control.

At this point, still in the research phase, it became possible to apply a myo-electrically controlled prosthesis to an above-elbow amputee, taking into consideration the utilisation of the pectoral, biceps, deltoid and triceps muscles which together give an output of 8 signals.

It is interesting to note that the first amputee experiencing this new type of prosthesis (at present 4 are trying it) learned in a few days how to control the various movements, independently from each another.

Practically speaking the prosthesis in question is entirely myo-electrically operated with functions of the hand, wrist and elbow joint. Six signals are

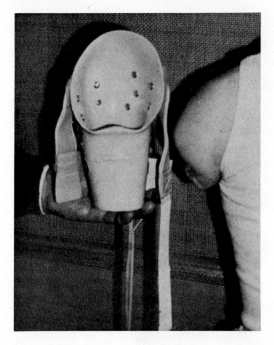

Fig. 2. Socket and electrodes for multi-channel control.

needed to execute these movements, and two more signals are available for a further movement that can be chosen among extra-intra rotation of the arm, flexion-extension of the wrist or, in the shoulder disarticulation cases, for the articulation of an active shoulder.

As can be seen, through the realisation of this amplifier the present needs for movements have been met, but it can be foreseen that these control signals will soon be utilised for further movements that may be deemed necessary.

At present, we can put at the disposal of the prosthesis such a large range of movements as has never before been possible.

At this stage of research we came to realise that a prosthesis can have all the desired possibilities of good operation but it can never be expected to replace the natural limb if it fails to give the amputee sensitive information.

If one consults books dealing with technical orthopedics or reads the proceedings of congresses, one notices that everybody discusses mechanical function but never sensitive function. It is logical to think that an amputee has lost not only mechanical but also sensitive function. For this reason, the

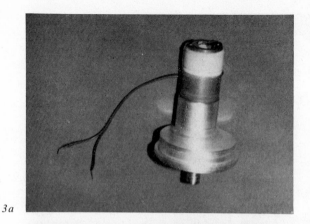

3a

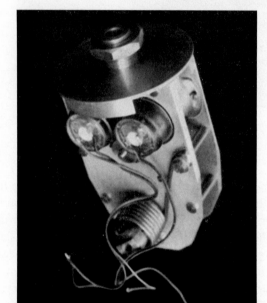

3b

Fig. 3a–b. Wrist joint and elbow joint for myo-electrically control prosthesis with proportional control.

research work of the Inail Prosthetic Center in Budrio has been intensified with the purpose of giving the artificial hand also sensitive control.

Much information can be transmitted to the human body if eyes and ears can be used as receivers. We well know how essential these organs are in everyday life, so that if it is impossible to utilise these for the purpose of prosthetic application, then it will be reasonable to transmit this information through the skin by using either mechanical or electrical stimulation.

Physiological experiments have shown that the skin is more sensitive to vibrations of about 200 cps, but any vibration over 150 cps is perceptible and diverts. For this reason sinusoidal vibrations of about 100 cps whose amplitude is modulated by the signal have been adopted.

At the beginning of the experiment, a skin vibrator was considered for the sensitive information of an artificial hand but, later, the vibrator was replaced by a skin stimulator. The necessary component for sensitive information consists of a receiver of the pressure sense (strain gauge or pressure indicator). This receiver is in contact with the skin vibrator or with an electrode for skin stimulation by means of an electronic device. The finger position receiver, which arranges to give the amputee the information concerning the position of the fingers, consists of a potentiometer of some ohms, whose signals are proportional to the opening or closing of the fingers. In this case, too, the information arrives through an electronic device with either a skin vibrator or a semi-conductor electrode. With this system of sensitive information, a blind amputee, for instance, can sense with the artificial hand a weight greater than 20 oz and distinguishes an object of about 1 mm thickness.

As has been said above, a skin vibrator was utilised at the beginning, but during the research the program was modified and direct skin stimulation through semi-conductor electrodes was used. Practical trials have shown that skin stimulation, as information for the amputee, is very similar to the sense of touch and this is why it gave the best result.

From the standpoint of prosthetic application, too, fewer difficulties are encountered, as the device requires less space inside the prosthesis than skin vibrators, because we were able to make use of one electrode for skin stimulation.

As a result of the realization of a sensitive device for the artificial hand, the application was made possible of a myo-electrically controlled prosthesis for arm amputation with the movements of the hand, with proportional and speed control, movement of wrist articulation and speed control, control of the elbow joint and its speed, and sensitive information of the hand. The practical application of this prosthesis on an amputee has revealed a new aspect of the prosthesis. The amputee to whom the prosthesis was supplied – the amputee in question was already provided with a multi-channel myo-electric prosthesis with 4 controls which, however, had no sensitive function – stated that he immediately perceived a considerable degree of sensibility in the use

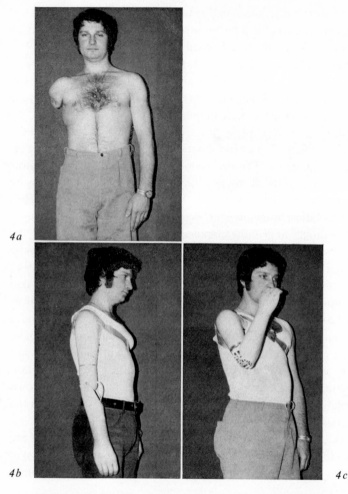

4a

4b   4c

Fig.4a–c. Above-elbow amputee with 8-control myo-electrical prosthesis.

of the artificial limb. Even movements with the prosthesis are safer and firmer. We have noted different behaviour in the amputee, we have seen him much freer and less conditioned in his movements.

From the studies carried out on the function of the prosthesis in its ordinary use, the excellent function of the artificial limb has become clear. At the same time, movements are now found to be smoother and well integrated in the complex of physiological movements that the person carries out,

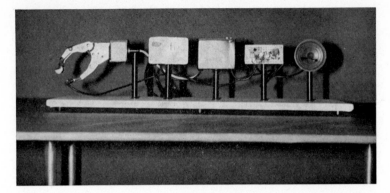

Fig.5. Components for sensitive control to be inserted in the prosthesis.

even if the movements of the limb, as compared with the possibilities of a natural limb, are – quantitatively – more limited. From the mechanical point of view, these limitations are compensated so that the amputee makes use of prostheses with excellent functional qualities.

It is clear that with advances in technology we shall arrive at the possibility of approaching the natural function of the arm but this depends not only on mechanical and electronic factors but also on the invention of means that can transmit impulses picked up directly from the nerve and utilised for the control of the single movements of the limb. Our research group has begun investigations in this direction but, given the special complexity and difficulty of the subject, any forecast would be premature.

If the results of the research find practical application, a first step will have been made towards the realisation of an artificial limb with natural movements.

Authors' address: Dr. V. GUARDASCIONE and Dr. H. SCHMIDL, Instituto per l'Assicurazione contro gli Infortuni sul Lavoro, *Rome* (Italy)

Medicine and Sport, vol. 8: Biomechanics III, pp. 482–488 (Karger, Basel 1973)

Electromyography of the Lower Limb Amputee

D. N. CONDIE

Dundee Limb-Fitting Centre, Dundee

The adoption of the team approach to amputation surgery and prosthetic replacement has brought about an increased understanding of the biomechanics of walking which has been directly responsible for the development of sophisticated socket design philosophies for below and above knee amputation stumps. Post-operative care also has developed with particular attention being devoted to the control of stump volume. One of the effects of these major developments has been to challenge conventional amputation surgery and to place increasing demands on the amputating surgeon to produce a stump which is more than the ineffective flabby prop with which we are still regrettably familiar.

Surprisingly, electromyography has not been applied to the study of the amputee with much success, a notable exception being the work of WEISS [1960] and co-workers, although it should be noted that this work was principally concerned with the physiology of muscle contraction and the subject of proprioceptive feed-back rather than the function of the muscles of the stump in gait.

Experimental Studies

Amputee EMG tests are performed utilizing specially prepared plastic sockets mounted in a modular assembly limb. Identification of stump musculature is performed prior to socket production and markers incorporated in the cast to be carried through the socket production process to provide a precise indication of the best monitoring position for subsequent dynamic studies. Just prior to testing, the electrodes are positioned. Since all the sockets produced are of the total contact variety a disc of material is removed from the socket, to allow positioning of the electrodes, which is replaced by a plug on which are mounted the electrode connectors.

Contrary to the expressed opinions of most experimental kinesiologists surface electrodes have been employed for the initial series of tests. These studies are intended to serve as a basis for more detailed study using intra-muscular electrodes. Records were made of the phasic activity of direct EMG signals. No standard amplifier gain was adopted nor were the signals cali-brated. For these reasons the amplitude of signals from different muscles may not be used to compare the level of muscle activity. Amplitude variations of particular muscle signals may be used as a general indication of the level of muscular activity, but not of the level of muscular tension.

Electrode placement was decided on a basis of functional-anatomical grouping, viz. hip flexors, extensors, abductors and adductors. Standard traces of normal subjects were produced using electrode placements similar to those to be used with amputees.

Results

Above-Knee Amputees – Muscles of the Hip (fig. 1)

The muscles of the thigh stump which remain unaltered by the ampu-tation surgery, i.e. gluteus maximus and the abductor muscles, demonstrate fairly normal myograms of activity. It should, however, be noted that all the amputees who were tested had achieved a rather good gait with no sign of an abductor limp. This would appear to be confirmed by the EMG. Other patients studied on an *ad hoc* basis have demonstrated a greater degree of variability of muscular activity which could be visually correlated with their gait abnormalities, e.g. the abductor muscles appear to be a very sensitive indicator of pelvic stability.

Records of the activity of flexors, long extensors and adductors of the hip, all of which are sectioned as a result of amputation, show surprisingly few unusual features. Certainly there are minor variations in precise muscle phasing and there would appear to be a general prolongation of activity. Regrettably it is not possible to make comparisons of the magnitude of muscle activity, and perhaps this would be a more revealing parameter.

No results from above-knee myoplastic amputations are available as yet; however, it may be of value to consider experience from parallel situations

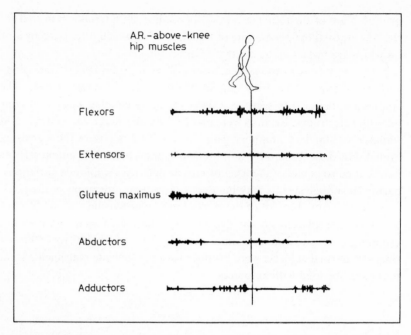

Fig. 1. Typical EMG record of thigh muscles of above-knee amputee with conventional surgery.

such as muscle or tendon transfers. In this field a wealth of experimental EMG studies exists to give some guide-lines on the likely result of myoplastic surgery of the amputation stump. SUTHERLAND *et al.* [1960] in their studies of the treatment of quadriceps paralyses in poliomyelitis by transplanting the hamstrings to the patellar tendon showed an almost 100% successful conversion of hamstring to stance phase activity, nor was there any tendency for the muscles to revert to their old phase under conditions of fatigue or lack of concentration. Particularly important is the belief postulated by these authors that too much emphasis is placed upon the likely effect of reflex pathways and of the need for muscles to re-learn their new function. On the contrary, it is suggested on the basis of clinical results that a far greater degree of voluntary control of the musculo-skeletal system exists than is normally acknowledged, and that it is quite wrong to use reflex behaviour to explain the complexities of locomotion in man.

Conversely, CLOSE and TODD [1959] concluded from their massive studies of the results of tendon transfer that: 'Training in the form of re-education directed towards the use of the muscle during the proper phase of walking appears essential in certain transfers.' Regarding the likelihood of muscles altering their pre-operative phase it is stated that 'Phasic transfer is in general superior to nonphasic transfer. Results of phasic transfer are predictable. In some instances however nonphasic transfers can undergo phasic conversion.' Only further studies of above-knee stumps after myoplasty will reveal how amputation surgery relates to this experience.

Below-Knee Amputees – Muscles of the Hip (fig. 2)

Alterations in phasic activity of these muscles can only be ascribed to changes in the gait pattern as a result of either phasic variations or weakness or inaction of other muscles. Once again, precedents for this type of phasic alteration exist in the study of quadriceps paralysis due to poliomyelitis where EMG recordings have demonstrated complete conversion of the hamstrings to stance phase activity to compensate for gait deviations.

Hip flexor and extensor activity, while exhibiting some irregularities, does not demonstrate any significant phase alterations; the one significant exception being the extensor muscles of the amputee with an osteomyoplastic stump which showed a completely abnormal period of activity (fig. 3). The possible explanation for this conversion will be discussed in conjunction with the results of stump EMG. Hip abductor activity is largely normal; however, the adductor musculature action does tend to be rather erratic and generally prolonged.

Below-Knee Stump Muscles

The muscles sectioned in below-knee amputation soon atrophy if they remain unattached. For these reasons it was considered unnecessary to consider the muscles of the conventional below-knee amputees. Stumps of several amputees who had received myoplastic surgery were examined (fig. 4).

As can be seen, the anterior tibial muscles despite their isometric condition appear to have retained their normal phasic activity. On the contrary, the calf muscles which continue to exert some control at the knee joint have performed a complete conversion to swing phase activity (fig. 5). Two theories may be postulated to explain this occurrence. Possibly the contraction of the calf muscles is a purely reflex action designed to prevent socket slippage during the swing phase. Alternatively, and more intriguingly, could the conversion

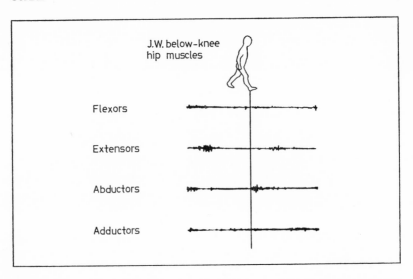

Fig. 2. EMG record of thigh muscles of below-knee amputee with osteomyoplastic amputation surgery. OMP = osteomyoplastic.

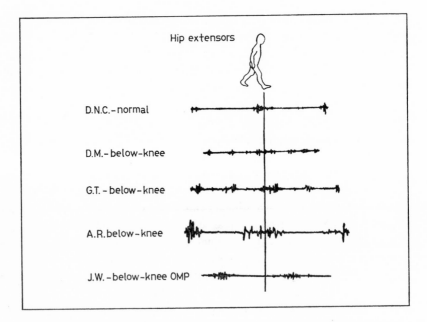

Fig. 3. Comparison of activity of hip extensors of below-knee amputee with that of normal subject. OMP = osteomyoplastic.

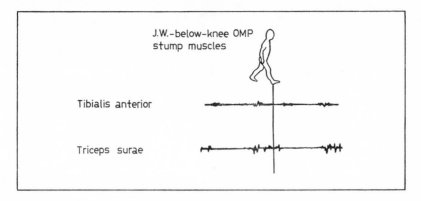

Fig. 4. EMG record of stump muscles of below-knee amputee with osteomyoplastic amputation surgery. OMP = osteomyoplastic.

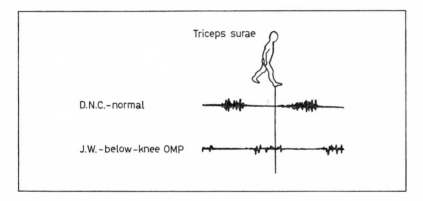

Fig. 5. Comparison of activity of calf muscles of below-knee amputee with that of normal subject. OMP = osteomyoplastic.

of the calf muscles to the apparent role of swing phase terminal decelerators be related to the abnormality noted in the activity of the hip extensors, notably the hamstrings? Certainly the situation deserves further study.

Conclusions

These studies are of a preliminary nature and, as has clearly been demonstrated by work in related fields, a large degree of variability is to be expected in post-operative EMG studies. The following points do, however, deserve some emphasis.

1. EMG studies provide a useful means of studying and/or charting the progress of muscles sectioned during amputation surgery. The importance of this type of follow-up is greatly increased by the development of myoplastic surgical techniques.

2. Studies of muscle phasic activity alone are unlikely to be sufficient. If the full benefit is to be derived quantitative techniques should be exploited.

3. Studies of gait deviations should provide a more secure basis for explanation than current theories.

References

CLOSE, J.R. and TODD, F.: Phasic activity of the muscles of the lower extremity and the effect of tendon transfer. J. Bone Jt Surg. *41:* 189–208 (1959).

SUTHERLAND, D.H.; SCHOTTSTAADT, E.R., and BOST, F.C.: Electromyographic study of transplanted muscles about the knee in poliomyelitis patients. J. Bone Jt Surg. *42:* 919–938 (1960).

WEISS, M.: The results of electromyographic tests carried out on patients after amputations. Prosthet. int. Proc. of the 2nd int. Prosth. Course, Copenhagen, *35:* p. 39 (1960).

Author's address: Mr. D. CONDIE, Dundee Limb Fitting Centre, 133 Queen Street, Broughty Ferry, *Dundee DD5 1AG* (Scotland)

Medicine and Sport, vol. 8: Biomechanics III, pp. 489–491 (Karger, Basel 1973)

A New Method in the Treatment of Gait Disorders

F. VAN FAASSEN and N. H. MOLEN

Faculteit der Geneeskunde, Vrije Universiteit, Amsterdam

In the rehabilitation of the temporarily or permanently handicapped, the evaluation of the gait of the patient under treatment is of utmost importance. It is the base upon which to decide whether a therapy should be extended or in what way it should be altered or, even worse, discontinued when progress is not evident.

The trouble is that the judgement and the evaluation of the gait during therapy depend upon subjective elements. In the Institute of Biomechanics and Experimental Rehabilitation of the Free University in Amsterdam, a couple of measuring methods have been developed in the last few years which made it possible to set down objectively the parameters that determine the gait.

In order to arrive at a description of the gait as accurately as possible in numerical measures, the following measurements have been carried out: the step time measurement, the single support time, the swing time and the double support time of the right and the left legs were recorded and compared with each other so as to eliminate the possibility of asymmetry in the gait. This measurement was completed with an objective observation of the step lengths of the left and the right legs. In recording the momentary speed of the centre of gravity of the body in a forward direction, information is obtained about the effect of propulsive functions of the lower limbs in the entire course of movement. In fact, from this measurement the accelerations and the displacement of the body's centre of gravity are being recorded.

By means of force plates, the external forces of the foot are recorded in 3 directions. The foreward and backward forces render information on the propulsive power generated by the limb. During walking, the angles of deflection of the hip joint, the knee joint and the ankle joint, and the eversion and inversion of the foot, are measured in relation to time by means of potentiometers.

The last 3 measurements create a fairly accurate picture of the heel strike, mid-stance and push-off sequence. The measurements mentioned may or may not be combined with an EMG registration of the muscle functions. Furthermore, measurement of the oxygen consumption at varying speeds is feasible. Preferably, all observations should be recorded simultaneously in order to facilitate comparison to one another.

The determination of the parameters of the normal as well as of the pathological gait will provide a means to draw up precisely any departure from the rule. When the measurements are repeated regularly in the course of the therapeutical treatment, the progress and the final results of the therapy may well be objectified.

Up to now, the psychological rules lying at the root of every course of instruction, have hardly ever been taken into account in conventional walking therapy, such as the continuous feedback information on the patient's walking pattern while under treatment and, in connection with same, a reward afterwards, according to merit.

Experiments are now under way at the Institute for Biomechanics and Experimental Rehabilitation, with a new procedure of walking training, in which the measurements earlier mentioned give the patient an insight into his actual performance. In this method of training a treadmill is used; it has the advantage that the patient is walking without displacement within the room of action.

Moreover, the repetitive and often tiring about-turns at the end of a walkway are thus eliminated. The treadmill, designed by technicians of our institute, differs from those now generally in use in such a way that the patient is linked to a control circuit through which the speed of the treadmill can be automatically adapted to the handicap and the propulsive capabilities of the patient. For safety's sake, the pulse rate is being recorded continuously by means of a cardiotachometer while another device switches off the treadmill automatically whenever the actual pulse rate rises above a limit set in advance. For the elderly patients with cardiovascular diseases, such a precaution is no more than appropriate.

In order to keep the patient informed about the qualities of his walking pattern during training, a number of parameters are recorded and the results displayed for his sake. The measurement of the step times is computerized in such a way that the patient is informed of his time-asymmetry factor after each sequence of 5 steps. In the same way, the patient can take note of differences in the step lengths. The angle of deflection of the joints of the lower limbs in relation to time, speed of walking and frequency, i.e. the number of steps per minute and the distance covered, are recorded as well.

Finally, the patient can visually study his gait by means of 2 monitors, on which he can watch himself from the front and from the back. By these observations the patient is enabled to make corrections in his gait in the course of training.

Although the patient is in a position to observe simultaneously all registered data, it stands to reason that in practice and out of didactical motivations, part of the data will not be conveyed to the patient, so as to stress the correction of one single deviation in the walking pattern. Whenever the patient himself succeeds in making real corrections concerning his gait this can be considered as reward enough.

During training in walking on the treadmill, certain developments by the NASA for use in the space-flight program were applied in selected cases in which either partial or total relief of load on the lower limbs was required. In order to obtain this alleviation the patient, wearing a tight-fitting dress, is attached to a suspension gear overhead. Half a dozen isotonic springs take over the weight of the body, either partially or in full. The isotonic nature of the springs warrants a constant relief of the extremities by a steady force, irrespective of the amount of extension of the springs. Among others, this method is being applied in those cases where it is contraindicated to impose a full load on the lower limbs, as is the case, for instance, with fractures not fully consolidated. This method has proved advantageous for patients who have undergone recent orthopaedic surgery.

It has been noticed that in case of motor diseases in neurology, the training of the extremities concerned is substantially enhanced when they carry only a partial load. The method described has not only the advantage of joints and muscles being exercised while a full load is still undesirable, but it also enables the therapist to start the rehabilitation procedure at an earlier stage.

Author's address: Dr. F. van Faassen, M.D., Ph.D., and N.H. Molen, M.D., M.Sc., Vrije Universiteit, Faculteit der Geneeskunde, van der Boechorststraat 7, *Amsterdam* (The Netherlands)

Medicine and Sport, vol. 8: Biomechanics III, pp. 492–496 (Karger, Basel 1973)

Changes of Electrical Activity, of Actual Circumference of Thigh during Contraction and of Strength-Endurance Product during Isometric Training in Functional and Organic Muscle Atrophy

H. Stoboy and G. Friedebold

Orthopädische Klinik und Poliklinik der Freien Universität, and Institut für Leistungsmedizin, Berlin

It is well-known that voluntary muscle strength of normal subjects and patients with muscular atrophy due to disuse can be enhanced markedly by progressive isometric training [1, 2]. But strength alone is not a sufficient criteria of the effect of isometric training. Additional parameters have to be measured to detect basic differences between normal and atrophied muscles. In previous investigations it was shown that the gain in strength during training is combined with a decrease in static endurance [2]. In atrophied muscles the static endurance is always less than in normal. For this reason one of the most reliable criteria for the effectiveness of isometric training, the strength-endurance product, is also less in atrophied than in normal muscles during the whole course of training. There is also a marked difference in electrical activity between normal and functional atrophied muscles. In the normal group, peak electrical activity is reached at the beginning of training, but then declines far below the starting value. In the atrophied group, however, electrical activity continues to rise until the fifth week of training after which it slowly decreases to a level near the starting point. This decreased electrical activity in normal muscle could be correlated to a decrease in synchronously contracting motorunits [3]. The aim of this investigation was to determine whether there are also special patterns of static endurance, strength endurance and electrical activity in patients with a residual organic paralysis and tabes dorsalis.

Materials and Methods

Four groups of subjects were investigated.
1. Healthy female students of physical therapy who served as controls.

2. Subjects confined to bed for at least 4 months with a variety of disabilities; none, however, due to neuromuscular disorders.

3. Patients with residual paralysis after poliomyelitis.

4. Patients with confirmed tabes dorsalis and a complete absence of reflexes in the lower extremities.

The 4 groups underwent progressive isometric training for 10 weeks. The subjects performed one maximal contraction of the femoris quadriceps muscle for 10 sec daily. Once a week the following values were measured.

1. Maximum strength in kilograms.

2. Static endurance in seconds.

3. Strength endurance (product of strength and static endurance).

4. Integrated electrical activity (the action potentials during contraction were picked up with surface electrodes and the total area was integrated).

5. Average change of circumference of the thigh during contraction.

In order to measure the change in thigh circumference a metal band was wrapped one and a half turns around the thigh. One end of the band was fixed while the other end was attached to a lever which moved the axis of a potentiometer. During muscle relaxation a Wheatstone bridge was balanced to read 0. The contraction of a muscle caused an increase in thigh circumference. This results in a proportional change of resistance in the potentiometer, causing the flow of an alternating current. This current, in turn, is proportional to the resistance and consequently to the change in thigh circumference.

Results

At the beginning of the training, maximal strength in *atrophied muscles* was significantly less than the starting value of normal muscle ($p < 0.005$). The final values of these two groups did not essentially differ so that the gain in strength was steeper in atrophied muscles. Even in the polio group the increase in strength was steeper than in healthy subjects ($p < 0.01$). Only patients with tabes dorsalis showed no increase in strength during training (fig. 1).

The normal as well as the atrophied muscle shows a steady diminution in static endurance. However, during the whole course of training the values were distinctly lower in the atrophied group (p of final values < 0.05). While the tabetic group demonstrated irregular variations, the static endurance in polio patients was so short that its decrease could be considered insignificant.

In the normal group the strength endurance reached a plateau in about the fifth week. In contrast, strength endurance of the atrophied group increased continuously until the ninth week of training ($r + 0.96$), but the strength endurance was significantly less in the atrophied as compared to the normal group. The values in the tabetic group showed an irregular pattern (fig. 2).

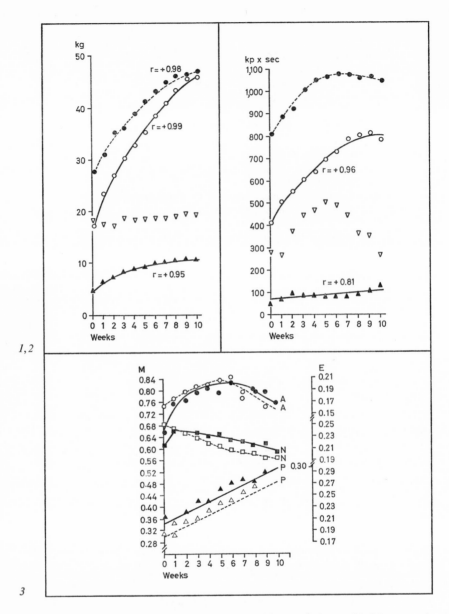

Fig. 1. Changes of maximum strength in normal (●-- ●-- ●), atrophied (○—○—○), poliomyelitic (▲—▲—▲) and tabetic groups (▽ ▽ ▽) due to isometric training.

Fig. 2. Changes of strength endurance in normal (●-- ●-- ●), atrophied (○—○—○), poliomyelitic (▲—▲—▲) and tabetic groups (▽ ▽ ▽) due to isometric training.

Fig. 3. Changes of electrical activity (- - - - - - E) and average circumference of thigh (——————— M) during contraction in normal (□ ■ N), atrophied (○ ● A) and poliomyelitic (△ ▲ P) groups due to isometric training.

In healthy subjects the electrical activity decreased distinctly during training ($p < 0.05$). In the atrophied group it increased until the sixth week of training to fall subsequently to a final value not significantly different from the starting value. In the polio group a steady increase could be observed. Only in the tabetic group were the values scattered over the whole range and no special pattern could be detected (fig. 3).

The pattern of average change of circumference during contraction nearly paralleled the electrical activity. In the atrophied group this value rose until the sixth week of training, then gradually fell to a value approaching the original one. In contrast, the peak value in the normal group was reached after the first week of training, following which the curve fell below the starting value. In the polio group there was a linear increase ($r + 0.90$), but all values were significantly less than in the normal and atrophied groups. In the tabetic group neither an increase nor a decrease could be observed (fig. 3).

Discussion

If the trainability of the test muscle in the different groups is judged on the basis of the gain in strength and circumference, as it is by HETTINGER [1], the atrophied muscles may be better trained than the normal. However, an analysis of static endurance indicates that this value is always less than in normal muscles. Particularly noticeable in atrophied muscles is a curtailment of static endurance at the expense of a rapid and marked development of strength.

In our opinion strength endurance gives the best insight in the trainability of muscle [2]. In the groups with atrophied muscles these values are always less than normal, due to the more rapid decline of static endurance. Probably this special type of training develops strength only at the cost of static endurance. The values of static endurance and strength endurance in the tabetic group show no specific pattern. Trainability seems not only to depend on intact motor pathways but also on intact afferent nerves.

Unlike normal muscle there is a recruitment of synchronously contracting motor units in the atrophied and poliomyelitic groups during training. More muscle fibers per unit of time are thus stressed by the stimulus of training than in normal muscle. This may explain the more rapid increase of thigh circumference in atrophied and poliomyelitic muscle.

With exception of the tabetic, the electrical activity nearly paralleled the average change in circumference of the thigh. The difference in normal muscle in the first week of training may be due to the fact that the electrical activity also contains the impulse frequency of a certain number of motor units. Therefore, it is valid to conclude that from the beginning of decrease in electrical activity muscle contraction becomes more economical.

The tabetic muscle shows no gain in strength during training, no regular pattern in the other values. The interruption of the afferent part of the reflex arch results in a complete cessation of afferent impulses. This diminishes the level of excitation in the motor cells. The activity of the central motor pathways seems to be insufficient to keep the cells in a normotrophic state or to permit the training of the muscle to a higher level of strength. The afferent impulses thus appear to have a certain trophotrophic function on the motor cell of the spinal cord.

References

1   HETTINGER, T.: Isometrisches Muskeltraining (Thieme, Stuttgart 1968).
2   STOBOY, H. and FRIEDEBOLD, G.: Changes in muscle function in atrophied muscles due to isometric training. Bull. N.Y. Acad. Med. *44:* 553 (1968).
3   STOBOY, H.; NÜSSGEN, W. und FRIEDEBOLD, G.: Das Verhalten der motorischen Einheiten unter den Bedingungen eines isometrischen Trainings. Int. Z. angew. Physiol. *17:* 391 (1959).

Author's address: Prof. H. STOBOY, Orthopädische Klinik und Poliklinik der Freien Universität, Berlin, Oskar-Helene Heim, *D-1 Berlin 33* (FRG)

Medicine and Sport, vol. 8: Biomechanics III, pp. 497–500 (Karger, Basel 1973)

Analysis of Warning Sounds Preceding Bone Rupture

Granulometric Distribution of the Elements Originating from the Fragmentation of the Heads of Femurs

A. Leduc and G. Panou

I. Introduction

Since the last century, the femur has been the subject of many studies. In the beginning these were orientated towards measurements of deformations provoked by various influences.

The many studies devoted to the femur have yielded a better understanding of the mechanical behaviour of this bone. They have also shown the need to make use of new means of investigation calculated to supply information in fields not explored by techniques already in use. From this angle it seemed advisable, to us, to register and analyse the sounds emitted by bones when loads causing rupture are imposed on them.

In fragmentation, granulometric analysis makes it possible in certain cases to draw conclusions concerning the rupture process. We thought there would be profit in using this method, voluntarily limiting ourselves, in the initial stage, to the crushing of the head of the femur, a relatively compact region.

II. Experimental Results

A. Experimental Technique

The femurs used were taken from adult cadavers, then fixed (age 55–80 years). We took no account of the age or sex of the donors. These femurs were subjected to compression under a press. During compression, the variation of the load applied in relation to the time was carefully noted, this made it possible to determine with precision the force engendering a specific sound.

After rupture the femur heads were detached and compressed until they were crushed. The fragments obtained were passed through a sieve after washing and drying.

B. Analysis of the Recording of the Sounds

We tested 19 femurs. In order to localise certain sounds with greater precision, we amputated the distal one-third of several bones (femurs No. 10 to 19). The femurs No. 1–9 were tested in the complete state.

The registrations (fig. 1) show the repetition of several peaks linked with a particular manner of rupture. We were able to identify the following peaks.

1. Sound situated between 325 and 330 kg ($\triangledown$): this announces the rupture of the distal one-third.

2. Sounds situated between 420 and 430 or 530 and 535 kg ($+$): it is clear that these sounds announce a rupture in the middle portion of the diaphysis.

3. Sounds situated between 620 and 625, 655 and 660 and 680 and 685 kg ($+$): these peeks appear in various cases. The manner of rupture of these bones shows that we here have sounds emitted by a centre situated at the upper curve, in the regions of the head and the neck.

4. Sounds situated between 610 and 615, 640 and 645 and 700 and 705 kg ($\blacktriangledown$): these sounds are found in recordings 4, 6, 14 and 19. It is, therefore, a question of an emission localised at the greater trochanter.

5. Sound situated between 810 and 815 kg ($\bullet$): this sound is emitted when the first signs of rupture begin to manifest themselves in the lesser trochanter.

6. Sound situated between 745 and 750 kg ($\diamond$): we suppose that this sound is emitted from the head of the femur, as bones 2 and 6 have suffered lesions not apparent in this region.

7. Sound situated between 945 and 950 kg ($\times$): this sound appears every time that the load applied attains or exceeds the level indicated, this independently of the manner of rupture.

C. Granulometric Study

Among the numerous formulae proposed to represent the granulometric curves most often met with, we will refer later to the formula known in fragmentation under the distribution name of Gaudin and Schuhmann

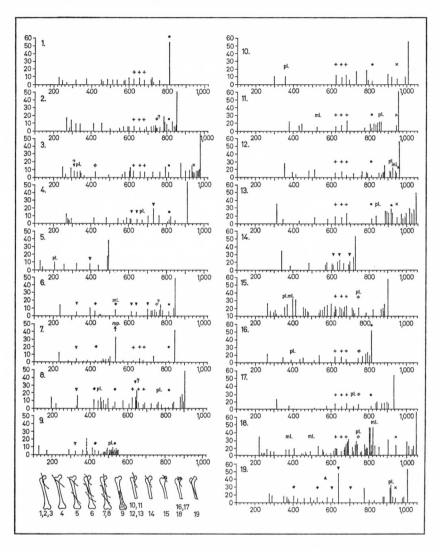

Fig. 1. Description see text.

(GS): $P(D) = (D/D_0)^n$, where D = dimension and D_0, n = constants. It was recently shown that fragmentation with the well-known Adreasen's process yields endoclastic grains which strictly obey the GS law.

Curves C, D and E of figure 2 represent the granulometric distributions of the fragments created by crushing under the press 9, 10 and 6 femur heads,

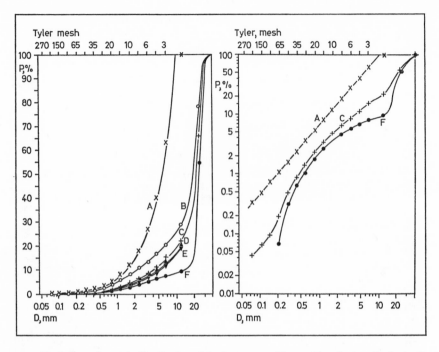

Fig. 2. Granulometric distribution of the fragments of femoral heads. A = Endoclastic fragments (internally broken). B = Fragments of heads and necks. C, D and E = fragments produced by the rupture of 9, 10 and 6 femoral heads, respectively. F = Exoclastic fragments (surface broken).

respectively. These curves are very close to each other and it may, therefore, be admitted that the granulometric distribution of the femur head fragments remains practically constant.

By referring to figure 2b it will be noted that the distribution of the endoclastic grains conforms with the GS law; the parameters of this distribution are $D_0 = 1$ cm, $n = 1.17$. The verification of this law by the endoclastic grains, despite the very special texture of the femoral head, allows the conclusion that it is very probable that this process is respected.

If, in addition to the head, we keep a part of the femur neck, the granulometric distribution is different (curve B).

Author's address: Prof. A. LEDUC, Vrije Universiteit Brussel, Kinesitherapie, Hegerlaan 28, *B-1050 Brussels* (Belgium)

Medicine and Sport, vol. 8: Biomechanics III, pp. 501–505 (Karger, Basel 1973)

Muscular Performance Test Results and Back Injuries

A Preliminary Report[1]

J. Spencer Felton and E. R. O'Connell

University of Southern California School of Medicine and Department of Physical Education, California State College, Los Angeles, Calif.

Among the largest of public employers, the County of Los Angeles has 72,000 workers scattered at many work sites in an area of over 4,000 mi². There are approximately 2,200 separate job classifications and, of considerable concern medically, is the physical status of nearly 8,200 safety personnel, a generic designation for Law Enforcement Officers, Firefighters, Deputy Marshals, District Attorney's Investigators and Beach Lifeguards.

Because of a specific statute in the State of California, this group of workers, when presenting a heart disorder, tuberculosis, pneumonia or a hernia, is considered to have sustained the pathologic change in the course of employment. This legal variance provides a different setting for the diagnosis, treatment, and retirement of such workers, which is extraordinarily costly to the employer. As the employer in such cases is in the public sector, the costs are borne eventually by the taxpayer.

In an effort to offer preventive services to the thousands of workers of this county, an Occupational Health Service was established in late 1968, in which a full program of occupational medicine was formulated, including such elements as physical examinations, environmental health capability, health education, occupational health nursing services, emergency medical care, immunizations and multiphasic testing. Most recently added to the program has been an activity in physical fitness to minimize the costs inherent in back injuries and heart disease sustained by safety members.

---

1   Presented at the 3rd International Seminar on Biomechanics, October 1, 1971, Rome, Italy.

Physical Fitness Program

Nearly a year ago, a special cardiopulmonary laboratory was established to provide exercise testing of safety personnel. The protocol consists of a complete medical examination, including a resting ECG, a battery of blood chemistry studies, urinalysis, audiometry, vision testing and X-ray examination of the chest.

On receipt of the results, the subject is examined in the cardiopulmonary laboratory, where an estimate of body composition is conducted through mensuration of skin folds at various sites. The height and weight are recorded, and special tests of spinal mobility and muscular strength are conducted, to be described shortly. In addition, vascular measurements are conducted at the time of positional change, lung function is determined, and an exercise test protocol is carried out to determine cardiovascular tolerance for work at heart rates of 100, 120, 140 and 160 bpm, each heart rate increment being maintained for 5 min. For the latter procedure, the examinee rides a heart rate controlled bicycle ergometer, with the subject pedalling at the rate of 60–60 revolutions/min against resistance automatically applied by the bicycle pedal apparatus. The heart rates mentioned previously are preset, and programmed into the equipment, so that the test is carried out smoothly, unless an untoward event makes stopping the test advisable.

Of particular concern, however, in this presentation, is a description of the modification of established tests, and the design of new tests to measure spine mobility and to provide some generalizations regarding data on the mobility of the spine, which have been gathered from over 200 safety employees.

Methodology

To reduce the prevalence of back strains and sprains occurring among safety employees, a functional evaluation of the basic movements of the spine was designed to measure the range of joint motion in the 4 basic movement patterns. The excursion of body movements was measured in centimeters, and the 4 basic movements involved are lateral flexion, forward flexion, rotation and extension.

Lateral Flexion

Lateral flexion is a measure of the ability of the individual to bend to the side. Since the spine is multijointed, it would be difficult to measure the angular movement of each joint. A simple and highly functional testing device was constructed to measure the distance that the hand travels as the result of sideward bending from a standardized standing position.

Forward Flexion

True trunk flexion is a difficult measure to obtain because of the many intervertebral joints, and the problem is stabilizing the hips to prevent hip flexion. A more functional measure, which includes that of trunk flexion as well as of hip and shoulder movement, is referred to as the sit-and-reach test.

Trunk Rotation

Reference to the measurement of trunk rotation is almost nonexistent in literature. Since rotation is one of the basic movements of the trunk, considerable emphasis was placed on the design of an instrument which would give objective and reliable measures.

Trunk Extension

The measurement of true trunk extension is equally as difficult as that of true trunk flexion. A functional testing procedure described in the literature[1], which also includes the mensuration of a degree of erector spinae muscle strength, was modified by using a more standardized position and means of stabilization to provide as objective a measure as possible.

Muscular Strength

Since the law enforcement officer and the firefighter are required to deal with resistance or opposing forces, be it a criminal or a fire hose, the use of muscular strength becomes an important factor in the performance of their duties. Grasping objects or individuals firmly can aid in controlling the behavior of a suspect or a 'live' hose. Lifting with the hip extensor muscles in the standing position is necessary at times in rescuing an injured person from a crashed vehicle, following an accident. In addition, the stabilization of one's body position in holding a fire hose or subduing a suspect in an alteration requires strong knee extensor muscles.

Grip Strength

Right and left hand grip strength scores are obtained by having the subject stand with the arm pendant and with the hand holding the dynamometer.

Hip Extensor Strength

The strength of the hip extensor muscles is measured in the standing position, the subject standing with the hips flexed to 156°, the arms pendant and the trunk and legs extended.

Knee Extensor Strength

The strength of the thigh muscles is obtained by attaching a belt around the hips of the subject with the knees flexed to 120°.

Discussion

It would appear that a more functional measure of spine mobility would be useful in field evaluations. X-rays do not always reveal the cause of back problems. Range of movement appears to decrease with age up to the 50s, and since there is a high prevalence of injuries in the 30- and 40-year-old age groups, perhaps individuals who score low on spine mobility measures might be considered a greater risk in potential back injuries.

At the present time, a longitudinal study has been initiated to correlate the results of the measurement of flexibility with the occurrence of back injuries among safety personnel. It is not possible to predict positive correlations between decrement in flexibility with the sustaining of a soft tissue back injury because of the variety of traumas suffered by safety personnel. Law enforcement officers involved in altercations with suspects or prisoners, and firefighters engaged in a large, totally consuming, watershed-area fire, frequently are subjected to energy expenditure which is totally unplanned for physically, emotionally or positionally. As data are developed from workmen's compensation records of back injuries, from disability retirement records of the residua of back injuries, the data will be juxtaposed to substantiate the worth of the studies of flexibility currently conducted as part of the physical fitness program.

Summary and Conclusions

1. A physical fitness program has been initiated as part of an occupational health program for law enforcement officers, firefighters, district attorneys' investigators, beach lifeguards and deputy marshals.

2. The program consists of a complete medical examination with adjuvant tests and a sophisticated cardiopulmonary evaluation, added to which is a study of muscular strength and joint flexibility.

3. The back injuries sustained by personnel in these categories are of great significance because of the constant risk of the group to such hazards and because of the high costs attendant upon the care, of disability payments for and possible retirement because of such injuries.

4. Scores from the measurements of spine mobility appear to be similar for movements to the right and to the left.

5. Approximately a 10-percent decrement in spine mobility scores occurs between the 20s and 30s, as well as between the 30- and 40-year-old age groups. The decrement is reduced more than half (4%) between the 40- and 50-year-old age groups.

6. Back strains and sprains appear to be more frequent in the 30s (40%) and in the 40s (40%), with the lower percentage (7) occurring in the 50-year-old age group.

7. Back strains and sprains occur more frequently with a decrement in spine mobility scores, but detailed documentation of such a possible correlation will be presented in future reports.

8. It is recommended that the kinds of tests and the modifications developed by the Occupational Health Service of the County of Los Angeles be instituted in other preventive medicine programs in the public sector, so that comparative studies can be effected.

Acknowledgements

Gratitude is expressed to Mr. PHILLIP C. THOMAS and Mr. ROBERT A. WISWELL, Exercise Physiologists, for their participation in the study project. The activities of the Cardiopulmonary Laboratory are coordinated by HARRIETT R. VOSS, M.D.

Reference

KASCH, F. W. and BOYER, J. L.: Adult fitness Principles and practice, p. 22 (Greeley, Colorado 1968).

Author's address: Dr. JEAN SPENCER FELTON, M. D., Clinical Professor of Community Medical and Public Health, University of Southern California School of Medicine, *Los Angeles, Calif.* (USA)

Medicine and Sport, vol. 8: Biomechanics III, pp. 506–511 (Karger, Basel 1973)

Some Biomechanical Aspects of the Rachis

R. Hernández-Gómez

Rehabilitation Service, Centro Nacional de Especialidades de la Seguridad Social, and Centro Nacional de Reconocimientos e Investigaciones Médicodeportivas, Madrid

Until lately, the vertebral column has been studied by means of an analytic system, which investigated what occurred in every vertebra and in every intervertebral space, considering these as functional units.

This present work will show in brief some data which can be obtained by assuming, when studying the human rachis, that all vertebrae are working in such a way that the result represents an authentic and specific functional unity, in many aspects similar to an architectural column, i.e. when considering the rachis as a whole.

Due to the characteristic posture of the human being, the vertebral column works with a fixed lower extremity whilst the upper extremity remains free. Thus, the human rachis acts as an axis, capable of transmitting movements at a distance, which means that it possesses all the characteristics known in mechanics as 'arbor'. As any other loaded axis, the arbor that constitutes our vertebral column is submitted both to forces of torsion (rotation) and to forces of flexion, which are directly related through the formula:

$$M_{fl} = \frac{3}{8} M_t + \frac{5}{8} \sqrt{M_{fl}^2 + M_t^2},$$

where M_{fl} and M_t, respectively, represent the movements of flexion and torsion of the arbor. This signifies that the movements of flexion and torsion are directly and reciprocally influencing each other in all the postural and motoric situations of the vertebral column.

The general idea is that the vertebral column works as a solid (full) column through the successive supports represented by the vertebral bodies, which are effectively and unquestionably solid pieces. But, the existence of the

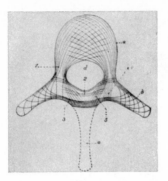

Fig. 1. The intertransverse fascicles, anterior (2) and posterior (3), which are conforming the posterior half of a empty column, being the anterior vertebral body (a). Both are united by the oblique fibers (4 and 5). [After TESTUT].

posterior arches, peduncles and laminae and, most of all, of the intervertebral posterior articulations, must make us think that not all the vertebral support is realized at the exclusive expense of the vertebral bodies, admitting thereby that in fact the human rachis works as an empty column, an authentic empty cylinder, as demonstrated in figure 1. This also indicates the reality of the posterior support, through the existence of the anterior and posterior intertransverse fascicles. Based upon this hypothesis, let us see what may occur when the human vertebral column is considered to be an empty column (fig. 2, 3).

The work coefficient of a column may be expressed by the Love formula:

$$K_w = \frac{1250}{1.45 + 0.00337 \cdot (L/D)^2},$$

where K_w represents the work coefficient, L the length and D the diameter of the column. When an empty and a solid column have equal diameters, the weight of the first one will be less, which will represent an advantage for a possible mobile capacity. On the other hand, the solid column will have a detriment from the motoric point of view. In the case of equality of its full sections, i. e. equality of material weight, between a full and an empty column, D, in the Love formula, will reach its greatest values in the empty column, which means that the value of the L/D ratio will be less and consequently the absolute value of K_w will be higher. It can be then deduced that in the case of equal weights, the empty column is more resistant than the full column, con-

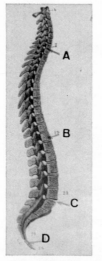 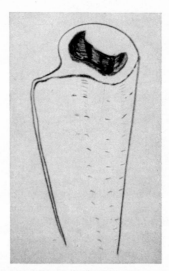

2,3

Fig. 2. Sagittal cut view of the rachis showing it as a empty cylinder, as is in our opinion. The numbers express the vertebrae, the letters the rachidian levels. [After TESTUT].

Fig. 3. Our own interpretation of the rachis as an empty column.

sidering both have the same length. This mechanical fact, which is not frequently studied, demonstrates undoubtedly that the human rachis has a mechanical advantage in working as an empty column. An advantage which may be added to the fact that this type of work allows more lightness and – as the column may be considered as a mobile stalk – more capacity of displacement.

The rachis in the standing position possesses two supporting extremities or endings, a lower one, which is fixed, and an upper one, which remains free. Let us study what might happen at the level of these two endings. During the standing and sitting positions and during a great number of postural and motor alterations and realizations, the lower part of the rachis acts mechanically as a pivot, considering that the loads to which it is subjected have a normal direction to the section of the supported arbor. That means, in principle, that in the human rachis, contrary to what happens in the rachis of the four-footed animals, we will have a clear predominance of the pressure forces. And, what is more, based on the fact that the vertebral column works as an empty arbor we will have to admit that its pivot is also an empty pivot, supplied with internal and external diameters. These diameters are reciprocally related among themselves and also related with the pressure to which the pivot is subjected, according to the following formula which is used in the calculation of material resistance:

$$D = \sqrt{D'^2 + \frac{4P}{\pi \cdot k_c}},$$

where D represents the external diameter, D' the internal diameter and P the pressure; $\pi = 3.1416$ and $k_c =$ the proper constant of each material characteristic. Considering that P, D and D' are in direct relation, we can easily deduce that the pressure which an empty pivot is capable of resisting is proportional to the values of the two diameters, external and internal. Augmenting both simultaneously will be dangerous, thus reducing the span bone thickness, which is only possible in the superior vertebrale on the cervical level. It will also be dangerous to decrease the internal diameter, considering that the spinal medulla needs protection. But it is perfectly possible to augment the rachis capacity of resistance to pressure by exclusively increasing the external diameter. Figure 4, showing the diagram of a vertebra in transverse section emphasizes the presence of the vertebral bodies as a way to augment, with their masses, the external diameter of the whole rachis arbor, without compression of the spinal cord.

In animals, where the direction of the load to be supported by the vertebral column is parallel to the rachis section (the ending supports in these cases are known, in mechanics, as gudgeons), the necessity of augmenting the rachis diameter disappears and, for that reason, the vertebral bodies are much more reduced than in human beings. In fact, the module of resistance of an empty gudgeon is inversely proportional to the external diameter, as expressed by the following formula:

$$R = 0.1 \cdot \frac{D^4 - D'^4}{D}.$$

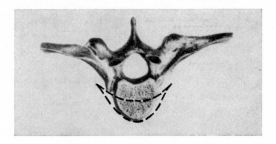

Fig. 4. We think that the vertebral body enlarges the external diameter of the rachidian cylinder. The ribs prolong the action of each vertebra as a beam. [After FICK].

Fig.5. When a column is fixed at its inferior point, being free upward, it has a deformity so that is a curve equivalent to 2/3 of the total length. A, upper point. F, lower point. P, pressure on the free point. [From STEINDLER, after BOYD and FOLK].

Because of this animals do not need and do not have large vertebral bodies.

We previously said that the human rachis works fundamentally under pressure. Normally, the architectonic columns maintain a harmonious proportion between their length and diameter (4 height diameters in Doric columns and either 9½ or 10 in the Ionic and Corinthian columns). When a disproportion exists between the total length of the column and its diameter, that is in columns excessively tall, a phenomenon occurs called 'bulge' in which the trunk of the column is obliged to curve. After Euler, a column with a fixed lower part and a free upper one, produces a curve by bulging which represents two-thirds of the total length of the trunk at rest. Figure 5 [from BOYD and FOLK] clearly demonstrates this. So, the fact that the human rachis is predestined to bulge (65–70 cm in length with an average external diameter of some 7 cm) makes us think that it really needs to bulge. Indeed, as is well-known, the human rachis resistance increases in relation to the number of curves, according to the formula $n^2 + 1$. The architectonic disproportion motivating the bulge represents another factor in the formation of these curves, as do also the different loads and postural necessities of the head, thorax, abdomen and sacrum base obliquity. We think that the scoliotic curves also yield an increase of the rachis resistance in a frontal plane similar to the one that is provided by the cipholordotic curves in a sagittal plane, always in accordance with the formula of the number of curves. What occurs is that, once a determined threshold is exceeded or, on the other hand, if this threshold is not reached, the pressure forces not only do not produce the stimulation of the bone tissue growth, but determine its proper atrophy; a law to which we have given the name of the great researcher, JORES, and which unifies the seeming antagonism existing between the well-known laws of Wolff and of Volkmann-Delpech. Thus, the typical rachis deformities of scoliosis appears and only disappear if we succed in modifying appropriately, during the rachis stages of growth, the values and directions of these pressure forces.

A last aspect to be commented on is the behavior of the vertebral units inside the rachis complex. To consider the column as a whole, an authentic unity, does not imply a rejection of the analysis of the comportment of its constitutive elements. In our opinion, every vertebral unit acts mechanically as a beam, that is, as a transversal bar destinated to support and secure an architectonic structure. The first of these beams, starting with the lower end, should be the one represented by the sacrum, which leans upon two sacroiliac joints. Or otherwise, the pelvis, resting on the hip joints, can be considered to possess this priority. We cannot give a more detailed mechanical analysis of this problem. But what, indeed, interests us is to emphasize the fact that the ribs, at the dorsal level, mean the architectonic continuation of every vertebral beam, increasing tremendously its arm lever (fig. 4). This situation, beneficial at the beginning, leads with time to some additional deformities, as occur in scoliosis, which are most difficult to correct. Up to a certain degree, we can obtain some correcting actions of the muscles acting upon the vertebral beams; but these techniques, obviously kinesitherapeutics *id est* kinesiologic with therapeutic bearing after our own concept of kinesitherapy, yield the danger of being confused with some apparently similar techniques, as gymnastics or physical exercises. Concretely it is easy to find, at least in my country, people pretending to correct vertebral column deformities by swimming. But if we review the real mechanical situation of the human rachis, as is intended in this brief study, it would be better not to make this, or any other, possible mistake.

Author's address: Dr. RICARDO HERNÁNDEZ GÓMEZ, Avda. General Perón 13, *Madrid* (Spain)